Redox Systems Under Nano-Space Control

Toshikazu Hirao
(Editor)

Redox Systems Under Nano-Space Control

With 133 Figures, 97 Schemes, 3 Structures and 19 Tables

 Springer

Toshikazu Hirao
Department of Applied Chemistry
Graduate School of Engineering
Osaka University
Yamada-oka, Suita, Osaka 565-0871
Japan
e-mail: hirao@chem.eng.osaka-u.ac.jp

Chemistry Library

Library of Congress Control Number: 2005937596

ISBN-10 3-540-29579-8 Springer Berlin Heidelberg New York
ISBN-13 978-3-540-29579-2 Springer Berlin Heidelberg New York
DOI 10.1007/b96698

Springer is a part of Springer Science+Business Media
springer.com
© Springer-Verlag Berlin Heidelberg 2006
Printed in Germany

The use of general descriptive names, registered names, trademarks, etc. in this publication does not imply, even in the absence of a specific statement, that such names are exempt from the relevant protective laws and regulations and therefore free for general use.

Product liability: The publishers cannot guarantee the accuracy of any information about dosage and application contained in this book. In every individual case the user must check such information by consulting the relevant literature.

Cover design: *design & production* GmbH, Heidelberg
Typesetting and production: LE-TEX Jelonek, Schmidt & Vöckler GbR, Leipzig
Printed on acid-free paper   2/3141/YL - 5 4 3 2 1 0

Redox Systems Under Nano-Space Control

Supramolecular chemistry has permitted a variety of conceptually novel artificial systems in various fields. Quite recently, the dynamic behavior is required by using the properties of these systems. From these points of view, the construction of functionalized redox systems under nano-space control appears to be essential for efficient electron and/or hole transfer in pioneering organic synthesis and nanostructured materials. For this purpose, coordination-induced, metal-assembled, self-assembled, and molecular chain-induced highly regulated spaces in a nano level play an important role through fusion of techniques in modern chemistry including supramolecular and bio-inspired chemistry. This book consist of three parts: redox systems via d,π-conjugation, coordination control, and molecular chain control, mainly dealing with the precise synthesis of π-conjugated and d,π-conjugated systems, metallohosts, metal clusters, self-assembled monolayers and antibody systems, and their application. Furthermore, the recent progress of rotaxanes and catenanes, dendrimers, and star-shaped polymers is also described as important part of this book. These systems are expected to achieve the dynamic redox functions for highly selective and versatile electron-transfer reactions and functionalized nano-device materials. For future investigation, their functions should be more beautiful. In this sense, I hope that this book will deepen the readers' background and widen the scope of nanoscience.

Contents

List of Contributors

Shigehisa Akine
Department of Chemistry,
University of Tsukuba,
Tsukuba, Ibaraki 305-8571, Japan
Tel: +81-29-853-4507
Fax: +81-29-853-4507

Chun-Hsien Chen
Department of Chemistry,
National Tsing Hua University,
Hsin Chu, Taiwan
Tel: +886 3 573 7009
Fax: +886 3 571 1082
chhchen@mx.nthu.edu.tw

Pierre H. Dixneuf
Institut de Chimie de Rennes,
UMR 6509
CNRS-Université de Rennes 1:
Organométalliques et Catalyse,
Campus de Beaulieu,
35042 Rennes, France
Pierre.Dixneuf@univ-rennes1.fr

Amar H. Flood
California NanoSystems Institute
and Department of Chemistry
and Biochemistry,
University of California,
Los Angeles, 405 Hilgard Avenue,
Los Angeles, CA 90095, USA
Tel: (+1) 310-206-7078
Fax: (+1) 310-206-1843
amarf@chem.ucla.edu

Shigeki Habaue
Department of Chemistry
and Chemical Engineering,
Faculty of Engineering,
Yamagata University,
Yonezawa 992-8510, Japan
Fax: +81-238-26-3116
habaue@yz.yamagata-u.ac.jp

Masa-aki Haga
Department of Applied Chemistry,
Faculty of Science and Engineering,
Chuo University, 1-13-27 Kasuga,
Bunkyo-ku, Tokyo 112-8551, Japan
Tel: +81-3-3817-1908
Fax: +81-3-3817-1908
mhaga@chem.chuo-u.ac.jp

Tomohiko Hamaguchi
Department of Chemistry,
Faculty of Science,
Fukuoka University,
Nanakuma 8-19-1, Jonan-ku,
Fukuoka 814-0180, Japan

Akira Harada
Department of Macromolecular
Science, Graduate School of Science,
Osaka University, Toyonaka,
Osaka 560-0043, Japan
Tel: +81-6-6850-5445
Fax: +81-6-6850-5445
harada@chem.sci.osaka-u.ac.jp

Bunpei Hatano
Department of Chemistry
and Chemical Engineering,
Faculty of Engineering,
Yamagata University,
Yonezawa 992-8510, Japan
Fax: +81-238-26-3116

Toshikazu Hirao
Department of Applied Chemistry,
Graduate School of Engineering,
Osaka University
Yamada-oka, Suita,
Osaka 565-0871 (Japan)
Tel: +81-6-6879-7413
Fax: +81-6-6879-7415
hirao@chem.eng.osaka-u.ac.jp

Takane Imaoka
Keio University,
Faculty of Science and Technology,
Department of Chemistry,
3-14-1 Hiyoshi, Kohoku-ku,
Yokohama 223-8522, Japan
Tel: +81-45-566-1718
Fax: +81-45-566-1718
imaoka@chem.keio.ac.jp

Tasuku Ito
Department of Chemistry,
Graduate School of Science,
Tohoku University,
Sendai 980-8578, Japan
ito@agnus.chem.tohoku.ac.jp

Masami Kamigaito
Department of Applied Chemistry,
Graduate School of Engineering,
Nagoya University,
Nagoya 464-8603, Japan
Tel: +81-52-789-5400
Fax: +81-52-789-5112
kamigait@apchem.nagoya-u.ac.jp

Nobuhiro Kihara
Department of Applied Chemistry,
Graduate School of Engineering,
Osaka Prefecture University,
Sakai, Osaka 599-8531, Japan
Tel: +81-72-254-9295
Fax: +81-72-254-9910
kihara@chem.osakafu-u.ac.jp

Shunsaku Kimura
Department of Material Chemistry,
Graduate School of Engineering,
Kyoto University,
Kyoto-Daigaku-Katsura,
Nishikyo-ku, Kyoto 615-8510, Japan

Kazuya Kitagawa
Department of Material Chemistry,
Graduate School of Engineering,
Kyoto University,
Kyoto-Daigaku-Katsura,
Nishikyo-ku, Kyoto 615-8510, Japan

Yi Liu
California NanoSystems Institute
and Department of Chemistry
and Biochemistry,
University of California,
Los Angeles, 405 Hilgard Avenue,
Los Angeles, CA 90095, USA
Tel: (+1) 310-206-7078
Fax: (+1) 310-206-1843
yliu@chem.ucla.edu

Tomoyuki Morita
Department of Material Chemistry,
Graduate School of Engineering,
Kyoto University,
Kyoto-Daigaku-Katsura,
Nishikyo-ku, Kyoto 615-8510, Japan

Toshiyuki Moriuchi
Department of Applied Chemistry,
Graduate School of Engineering,
Osaka University
Yamada-oka, Suita,
Osaka 565-0871 (Japan)
Tel: +81-6-6879-7413
Fax: +81-6-6879-7415
moriuchi@chem.eng.osaka-u.ac.jp

Tatsuya Nabeshima
Department of Chemistry,
University of Tsukuba,
Tsukuba, Ibaraki 305-8571, Japan
Tel: +81-29-853-4507
Fax: +81-29-853-4507
nabesima@chem.tsukuba.ac.jp

Shie-Ming Peng
Department of Chemistry,
National Taiwan University,
Taipei, Taiwan
smpeng@ntu.edu.tw
Tel: +886 2 2363 8305
Fax: +886 2 2363 6359

Stéphane Rigaut
Institut de Chimie de Rennes,
UMR 6509
CNRS-Université de Rennes 1:
Organométalliques et Catalyse,
Campus de Beaulieu,
35042 Rennes, France
Stephane.Rigaut@univ-rennes1.fr

Mitsuhiko Shionoya
Department of Chemistry,
Graduate School of Science,
The University of Tokyo, Hongo,
Bunkyo-ku, Tokyo 113-0033, Japan
Tel: +81-3-5841-8061
Fax: +81-3-5841-8061
shionoya@chem.s.u-tokyo.ac.jp

J. Fraser Stoddart
California NanoSystems Institute
Department of Chemistry
and Biochemistry,
University of California,
Los Angeles, 405 Hilgard Avenue,
Los Angeles, CA 90095-1569 (USA)
Tel: +(310) 206-7078
Fax: (+1) 310-206-1843
stoddart@chem.ucla.edu

Timothy M. Swager
Department of Chemistry,
Massachusetts Institute
of Technology, Cambridge,
Massachusetts 01239, USA
tswager@mit.edu

Toshikazu Takata
Department of Organic
and Polymeric Materials,
Tokyo Institute of Technology,
Ookayama, Meguro,
Tokyo 152-8552, Japan
Tel: +81-3-5734-2898
Fax: +81-3-5734-2888
ttakata@polymer.titech.ac.jp

Kentaro Tanaka
Department of Chemistry,
Graduate School of Science,
The University of Tokyo, Hongo,
Bunkyo-ku, Tokyo 113-0033, Japan

Daniel Touchard
Institut de Chimie de Rennes,
UMR 6509
CNRS-Université de Rennes 1:
Organométalliques et Catalyse,
Campus de Beaulieu,
35042 Rennes, France
Daniel. touchard@univ-rennes1.fr

Chih-Chieh Wang
Department of Chemistry,
Soochow University,
Taipei, Taiwan
Tel: +886 2 2881 9471
Fax: +886 2 2881 1053
ccwang@mail.scu.edu.tw

Michael O. Wolf
Department of Chemistry,
University of British Columbia,
Vancouver, BC, V6T 1Z1, Canada
Tel: 604-822-1702
Fax: 604-822-2847
mwolf@chem.ubc.ca

Hiroyasu Yamaguchi
Department of Macromolecular
Science, Graduate School of Science,
Osaka University, Toyonaka,
Osaka 560-0043, Japan
Tel: +81-6-6850-5447
Fax: +81-6-6850-5446
hiroyasu@chem.sci.osaka-u.ac.jp

Tadashi Yamaguchi
Department of Chemistry,
Waseda University,
Okubo, Shinjuku-ku,
Tokyo 169-8555, Japan

Kimihisa Yamamoto
Keio University,
Faculty of Science and Technology,
Department of Chemistry,
3-14-1 Hiyoshi, Kohoku-ku,
Yokohama 223-8522, Japan
Tel: +81-45-566-1718
Fax: +81-45-566-1718
yamamoto@chem.keio.ac.jp

Kazuyuki Yanagisawa
Department of Material Chemistry,
Graduate School of Engineering,
Kyoto University,
Kyoto-Daigaku-Katsura,
Nishikyo-ku, Kyoto 615-8510, Japan

Chen-Yu Yeh
Department of Chemistry,
National Chung Hsing University,
Taichung, Taiwan
Tel: +886 4 2285 2264
Fax: +886 4 2286 2547
cyyeh@dragon.nchu.edu.tw

Part I

Redox Systems via d,π-Conjugation

Chapter 1

Conjugated Complexes with Quinonediimine Derivatives

Toshiyuki Moriuchi · Toshikazu Hirao

Department of Applied Chemistry, Graduate School of Engineering, Osaka University, Yamada-oka, Suita, Osaka 565-0871, Japan

Summary An architecturally controlled formation of conjugated complexes with redox-active π-conjugated quinonediimine (qd) ligands by regulation of the coordination mode is described in this chapter. The qd moiety also serves as a redox-active π-conjugated bridging spacer for the construction of redox-active conjugated complexes. Incorporated metals play an important role as a metallic dopant to form a multiredox and multimetallic system. Conjugated complexes provide redox-switching systems based on redox properties of the qd spacer.

Abbreviations

bpy	2,2′-Bipyridine
CD	Circular dichroism
Cp	Cyclopentadienyl
DABCO	1,4-Diazabicyclo[2.2.2]octane
DMF	N,N-Dimethylformamide
DMSO	Dimethyl sulfoxide
en	Ethylenediamine
EPR	Electron paramagnetic resonance
ESI-MS	Electrospray ionization mass spectrometry
ET	Electron transfer
Et	Ethyl
Fc	Ferrocene
FT-IR	Fourier transform infrared spectrometry
ICD	Induced circular dichroism
L	Liter(s)
Me	Methyl
mol	Mole(s)
NMR	Nuclear magnetic resonance
NOE	Nuclear Overhauser effect
OAc	Acetate
pd	1,4-Phenylenediamine
pda	Phenylenediamide dianion
Ph	Phenyl
POT	poly(o-toluidine)

qd   Quinonediimine
sq   Semiquinonediimine radical anion
THF   Tetrahydrofuran
tpy   2,2′:6′,2″-Terpyridine
UV-vis   Ultraviolet-visible

1.1
Introduction

π-Conjugated polymers have attracted much attention in their application to electrical materials depending on their electrical properties [1]. The function of π-conjugated polymers is expected to be modified dramatically by the incorporation of metallic centers into the polymers [2, 3]. The conjugated polymeric complexes, in which transition metals are incorporated in the main chain, have been focused as one approach [3]. The interest in these conjugated polymeric complexes has been concerned mainly with the nature of interactions between the metal centers through a π-conjugated chain. Polyanilines are a promising electrically conducting π-conjugated polymer with chemical stability. Polyanilines exist in three different discrete redox forms, which include the fully reduced leucoemeraldine base form, the semioxidized emeraldine one, and the fully oxidized pernigraniline one [1b]. Two nitrogen atoms of the quinonediimine (qd) moiety of the emeraldine base form have been revealed to be capable of participating in the complexation to afford novel conjugated complexes [2c–i]. Multicoordination permits the construction of multiredox systems with electronic communication between metals through the π-conjugated spacer. The polymer complex can effectively serve as an oxidation catalyst [2c–e, 2g], in which the qd moiety is considered to contribute to a reversible redox process in the catalytic cycle of a transition metal. A d,π-conjugated cluster-type catalytic system is considered to be constructed for the first time. The qd moieties are reduced to semiquinonediimine radical anions (sq) and phenylenediamide dianions (pda), both of which possess binding capability to metals. The combination of this redox behavior and complexation with transition metals is expected to provide an efficient redox system. 1,2-Benzoquinonediimines have received extensive interest as a redox-active ligand in this context [4]. However, transition-metal complexes with 1,4-benzoquinonediimine have been investigated in only a few cases, and the coordination behavior of 1,4-benzoquinonediimine has hitherto remained unexplored [5]. 1,4-Benzoquinonediimines are present as a mixture of *anti* and *syn* isomers. The regulation of the coordination mode of the qd moiety is a promising approach to an architecturally controlled formation of redox-active conjugated polymeric or macrocyclic complexes with 1,4-benzoquinonediimines.

This chapter sketches an outline of conjugated complexes with redox-active π-conjugated qd derivatives, including redox properties and structures of the conjugated complexes.

1.2
Architecturally Controlled Formation of Conjugated Complexes with 1,4-Benzoquinonediimines

The formation of conjugated complexes is demonstrated by the use of a redox-active π-conjugated molecule, N,N'-bis(4'-dimethylaminophenyl)-1,4-benz-oquinonediimine (L^1) [6], which is a model molecule of polyaniline. The complexation with a palladium complex bearing a tridentate ligand, which has one open coordination site, is considered to give a bimetallic conjugated complex. In this context, the palladium(II) complex [$(L^2)Pd(MeCN)$] (1) bearing one interchangeable coordination site is prepared by treatment of Pd(OAc)$_2$ with the N-heterocyclic tridentate podand ligand, N,N'-bis(2-phenylethyl)-2,6-pyridinedicarboxamide (L^2H_2) [7]. The X-ray crystal structure of 1 [8] indicates that the open coordination site is occupied by an ancillary acetonitrile (Fig. 1.1) [9]. The macrocyclic tetramer [$\{(L^2)Pd\}_4$] (2) is obtained quantitatively by removal of a labile acetonitrile ligand (Scheme 1.1) [10]. The coordination of the amide oxygen atom to the palladium center is observed in the X-ray crystal structure (Fig. 1.2). Interestingly, treatment of 2 with acetonitrile leads to the reversible formation of 1.

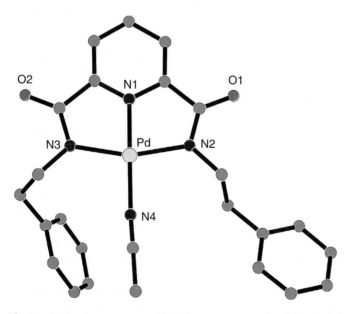

Fig. 1.1. Molecular structure of 1 (hydrogen atoms omitted for clarity)

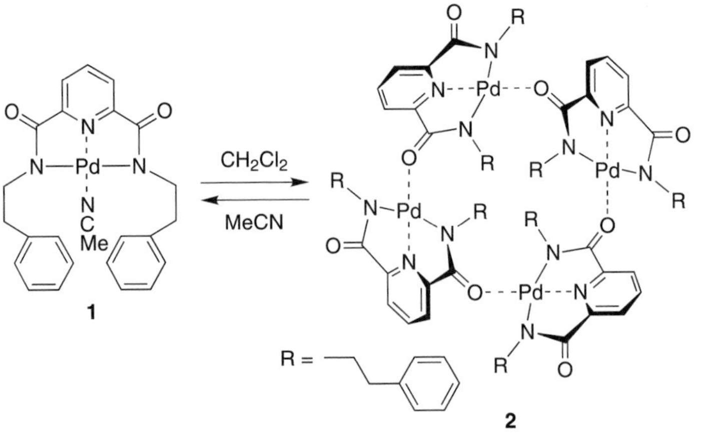

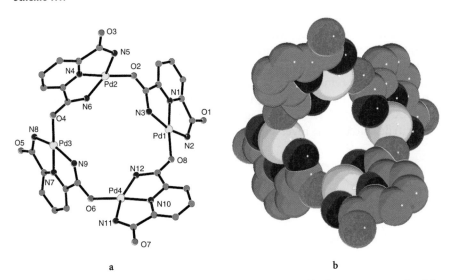

Scheme 1.1.

Fig. 1.2. (a) Molecular structure of **2** (phenylethyl moieties and hydrogen atoms omitted for clarity). (b) Space-filling representation of molecular structure of **2** (phenylethyl moieties and hydrogen atoms omitted for clarity)

The reaction of L^1 with two equimolar amounts of the palladium(II) complex **1** leads to the formation of the 1:2 conjugated homobimetallic palladium(II) complex $[(L^2)Pd(L^1)Pd(L^2)]$ (**3**, Scheme 1.2) [11]. The X-ray crystal structure of the C_2-symmetrical 1:2 complex **3anti** reveals that the two $[(L^2)Pd]$ units are bridged by the qd moiety of L^1 with a Pd–Pd separation of 8.17 Å, as depicted in Fig. 1.3. The steric interaction between the hydrogen atoms at C(32) and at C(38) causes the phenylene ring of L^1 to rotate away from the orientation parallel to the qd moiety.

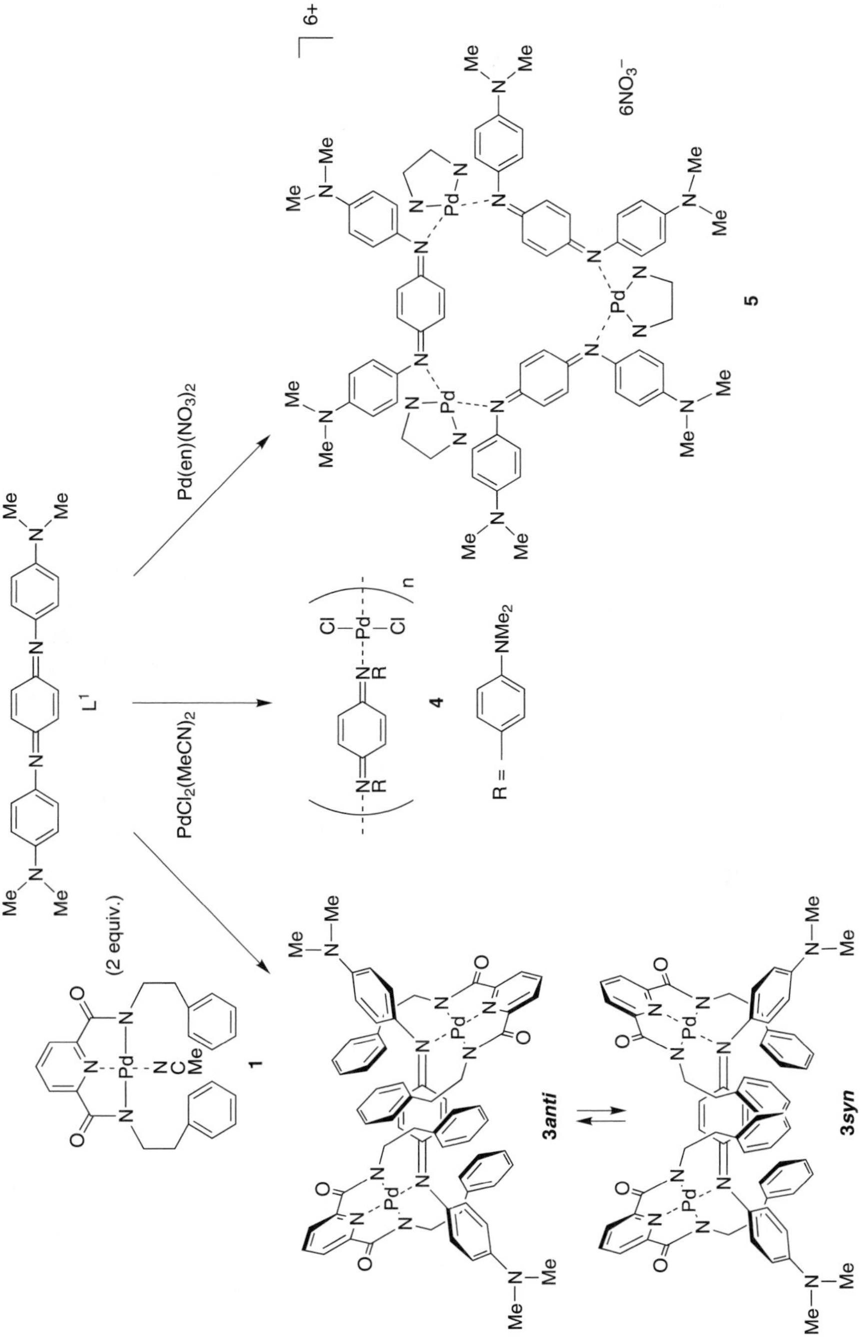

Scheme 1.2.

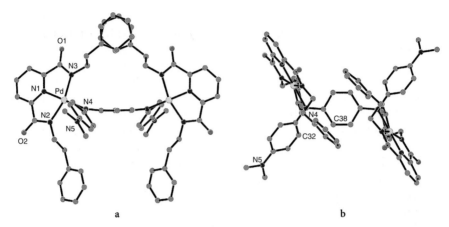

Fig. 1.3. (a) *Top view* and (b) *side view* of the molecular structure of **3anti** (hydrogen atoms are omitted for clarity)

Variable-temperature ^{1}H NMR studies of the conjugated complex **3** indicate interesting molecular dynamics in solution. As the temperature is lowered, the peaks of the *syn* isomer **3syn** appear and increase gradually. The conjugated complex **3** prefers an *anti* configuration at temperatures above 220 K and, conversely, the *syn* configuration at below this temperature. The conjugated complex **3** in dichloromethane shows three separate redox waves ($E_{1/2}$ = −1.49, −0.85, and 0.20 V vs. Fc/Fc$^+$) (Fig. 1.4). The waves at −1.49 and −0.85 V are assigned to the successive one-electron reduction of the qd moiety to give the corresponding reduced species. This result is in sharp contrast to the redox behavior of L^1 in dichloromethane, in which an irreversible reduction wave is observed at −1.67 V. Generally, the generated radical anion appears to be unstable, although this depends on the availability of a proton source [12]. In

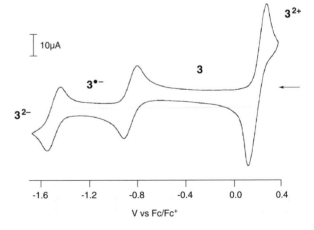

Fig. 1.4. Cyclic voltammogram of **3** (1.0×10^{-3} M) in dichloromethane (0.1 M nBu$_4$NClO$_4$) at a glassy carbon-working electrode with scan rate = 100 mV/s under Ar

the case of reduction of the complex **3**, the added electrons are considered to be delocalized over the Pd^{II}-qd d-π^* system. Compared with the uncomplexed one, the complexed qd becomes stabilized as an electron sink. Accordingly, the redox properties of the qd moiety are modulated by complexation with the palladium complex **1**, affording a multiredox system. The most positive anodic peak with twice the height ($E_{1/2} = 0.20$ V) is attributable to the one-electron oxidation process of the two terminal dimethylamino groups. A substantial positive shift of this oxidation wave compared with the free qd L^1 ($E_{1/2} = -0.08$ V) is consistent with the coordination of L^1 to palladium. Redox behavior of the conjugated complex **3** is depicted schematically in Scheme 1.3. The ESR measurement indicates that the unpaired electron is located mostly on the qd moiety, although some delocalization onto the metal is revealed by the weak satellite lines due to ^{105}Pd coupling.

On the other hand, treatment of L^1 with $[PdCl_2(MeCN)_2]$ having two coordination sites in acetonitrile leads to the formation of the conjugated polymeric complex **4**, in which the palladium centers are incorporated in the main chain (Scheme 1.2) [13]. In **4**, both *syn* and *anti* isomers of the qd moieties are likely to be present.

To regulate the coordination mode of the qd moiety, a metal-directed strategy for the construction of metallomacrocycles is embarked upon using $[Pd(NO_3)_2(en)]$, which has *cis* binding sites as a "metal clip". The conjugated trimetallic macrocycle $[\{Pd(en)(L^1)\}_3](NO_3)_6$ (**5**) is obtained quantitatively by treating L^1 with an equimolar amount of $[Pd(NO_3)_2(en)]$ (Scheme 1.2) [13]. The X-ray crystal structure of **5** confirms a trimetallic macrocyclic skeleton and the coordination of both qd nitrogen atoms to the palladium centers in the *syn* configuration with a Pd–Pd distance of 7.68 Å (Fig. 1.5). The noteworthy structural feature is that the qd planes are inclined about 28° to the plane defined by the six nitrogen atoms of the qd moieties (Fig. 1.6). An open cavity possessing different faces is formed with the cone conformation. This inclination of the qd planes is probably due to the coordination of the nitrogen atoms to the palladium centers. Another remarkable feature of the structure is the orientation of the phenylene rings of L^1 in a face-to-face arrangement at a distance of about 3.5 Å at each corner of the triangle, which suggests a π–π stacking interaction. Macrocycle **5** accommodates two methanol molecules at the top and bottom of the cavity, which permits the encapsulation of 1,2-dimethoxybenzene as a guest molecule.

In recent years there has been increased interest in chiral induction of polyanilines because of their potential use in diverse areas such as surface-modified electrodes, molecular recognition, and chiral separation [14]. Chiral polyaniline is formed by a chiral acid dopant [15]. The use of chiral complexes induces chirality to the π-conjugated backbone of polyaniline, giving the corresponding chiral d,π-conjugated complexes. Complexation of L^1 with two equimolar amounts of chiral palladium(II) complexes, $[((S,S)-L^3)Pd(MeCN)]$ (**(S,S)-6**) and $[((R,R)-L^3)Pd(MeCN)]$ (**(R,R)-6**), yields the chiral 1:2 con-

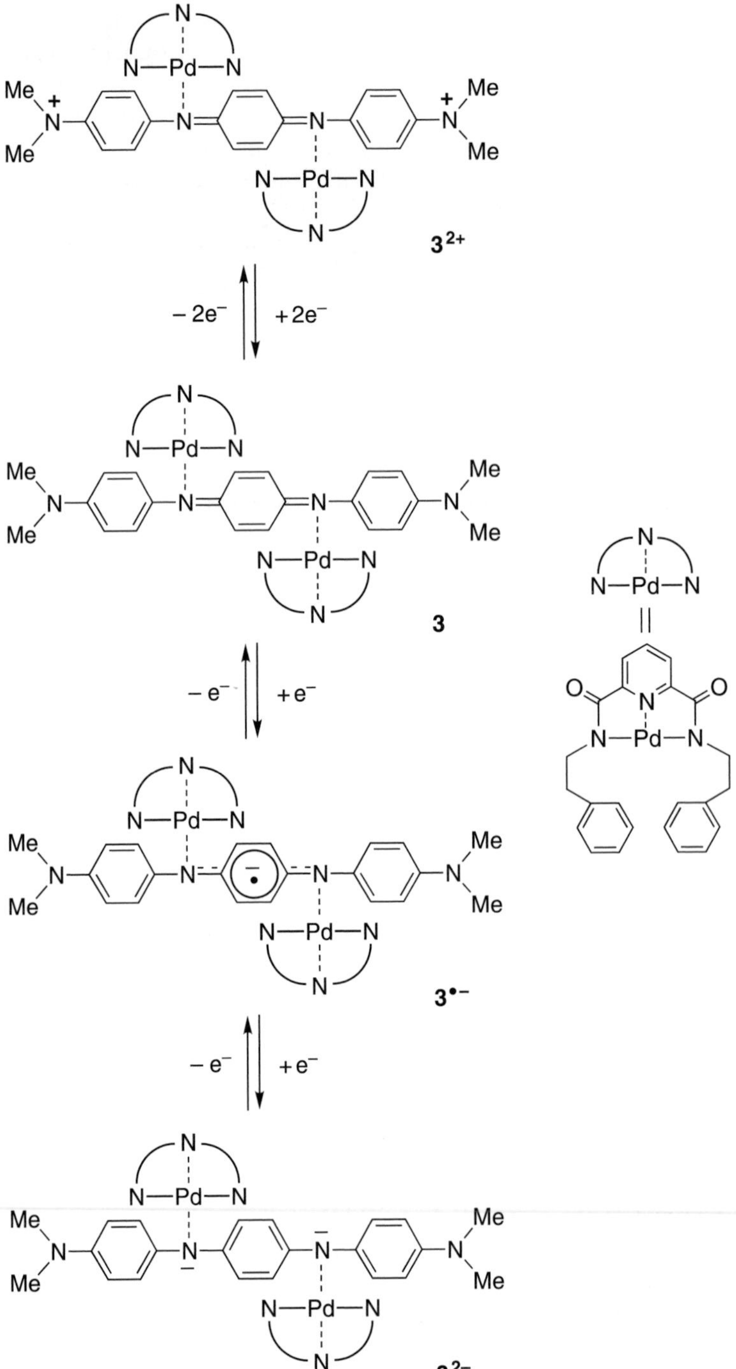

Scheme 1.3.

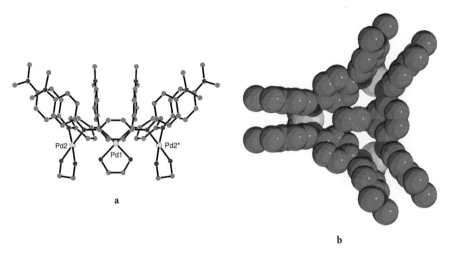

Fig. 1.5. (a) Molecular structure of **5** (hydrogen atoms and NO_3^- ions omitted for clarity). (b) Space-filling representation of molecular structure of **5** (hydrogen atoms and NO_3^- ions omitted for clarity). Two methanol molecules are located at the *top* and *bottom* of the cavity

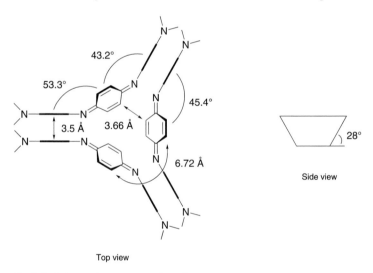

Fig. 1.6. Schematic representation of **5**

jugated palladium(II) complex $[((S,S)\text{-}L^3)Pd(L^1)Pd((S,S)\text{-}L^3)]$ (**(S,S)-7**) or $[((R,R)\text{-}L^3)Pd(L^1)Pd((R,R)\text{-}L^3)]$ (**(R,R)-7**), respectively (Scheme 1.4) [16]. Although variable-temperature ^1H NMR studies of the conjugated complex **(S,S)-7** indicate the existence of *syn* and *anti* isomers in solution, the mirror image relationship of the CD signals around CT band (600 to 900 nm) of the qd moiety is observed between **(S,S)-7** and **(R,R)-7** in dichloromethane, as shown in Fig. 1.7. This induced CD (ICD) is not observed in the case of **6**. These

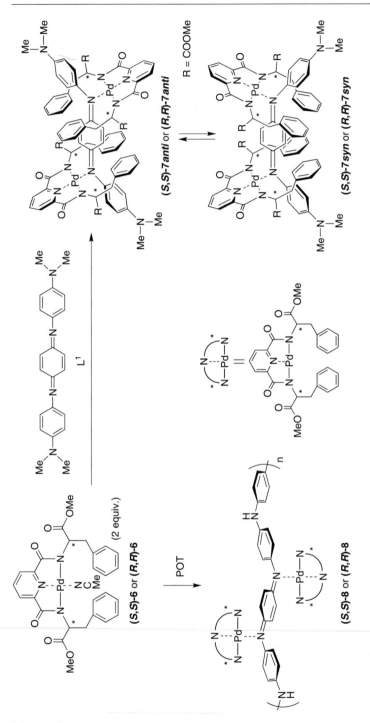

Scheme 1.4.

Fig. 1.7. CD spectra of **6** (1.0×10^{-4} M) and **7** (0.5×10^{-4} M) in dichloromethane

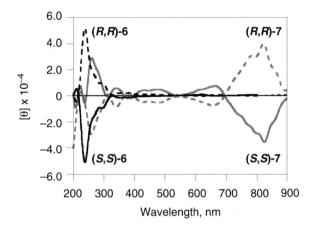

results suggest that chirality of the qd moiety appears to be induced through chiral complexation.

The X-ray crystal structure of the C_2-symmetrical 1:2 complex (**R,R**)-**7syn** reveals that the two [(L³)Pd] units are bridged by the qd moiety of L¹ with a Pd–Pd separation 7.59 Å, as depicted in Fig. 1.8. Each phenylene ring of L¹ has an opposite dihedral angle of 47.3° with respect to the qd plane, causing a propeller twist of 75.6° between the planes of the two phenylene rings. The chirality of the podand moieties of [(L³)Pd] is considered to induce a propeller twist of the π-conjugated molecular chain.

The chiral conjugated polymer complex [POT–((**S,S**)-L³Pd)] ((**S,S**)-**8**) is obtained by the treatment of the emeraldine base of poly(*o*-toluidine) (POT) in THF with (**S,S**)-**6** (Scheme 1.4) [16]. The UV-vis spectrum of (**S,S**)-**8** in THF shows a broad absorption around 500 to 900 nm, which is probably due

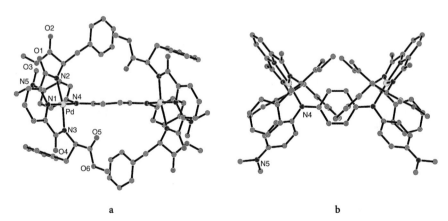

a   b

Fig. 1.8. a *Top* and **b** *side views* of molecular structure of (**R,R**)-**7syn** (hydrogen atoms omitted for clarity)

Fig. 1.9. CD spectra (*top*) of **8** and UV-vis spectra (*bottom*) of (*S,S*)-**8** and POT in THF (1.3×10^{-3} M)

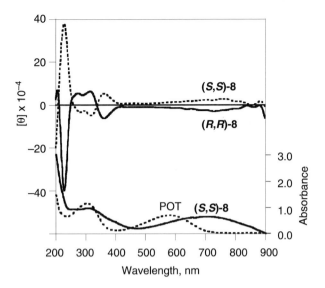

to a low-energy charge-transfer transition with significant contribution from palladium (Fig. 1.9). Furthermore, the mirror image relationship of the ICD signal around 500 to 850 nm is observed with (*R,R*)-**8** (Fig. 1.9), supporting the chirality induction in the case of POT. The random twist conformation of POT might be transformed into a helical conformation with a predominant screw sense through complexation.

The redox properties of pd and coordination ability of the corresponding oxidized qd permit the in situ oxidative complexation of pd with the palladium(II) complex **1** to form the conjugated bimetallic complex [(L²)Pd(qd)Pd(L²)] (**9**) (Scheme 1.5) [10]. In the ¹H NMR spectrum of the conjugated complex **9**, two sets of peaks based on *syn* and *anti* qd isomers are observed in DMSO-d₆. The cyclic voltammogram of the conjugated complex **9** in DMSO exhibits two separate redox waves at $E_{1/2} = -1.52$ V and -0.78 V vs. Fc/Fc⁺ assignable to the successive one-electron reduction of the qd moiety. The X-ray crystal structure of **9anti** revealed that the two [(L¹)Pd] units are bridged by the qd spacer (Fig. 1.10). Each benzene ring of the podand moiety of [(L¹)Pd] is oriented in a near face-to-face arrangement at a distance of ca. 3.8 Å with the benzene ring of another [(L¹)Pd] to form a pseudomacrocycle.

A conjugated complex containing an [M^{n+}(qd)M^{n+}] unit may, in principle, be converted to two other valence isomers, [M^{(n+1)+}(sq)M^{n+}] and [M^{(n+1)+}(pda)M^{(n+1)+}], that differ only in the electron distribution between the qd moiety and metals. This valence isomerization is considered to depend on the redox properties of both components. Vanadium compounds can exist in a variety of oxidation states and generally convert between the states via a one-electron redox process [17]. Vanadium compounds in low oxidation states can serve as one-electron reductants, as exemplified by the redox

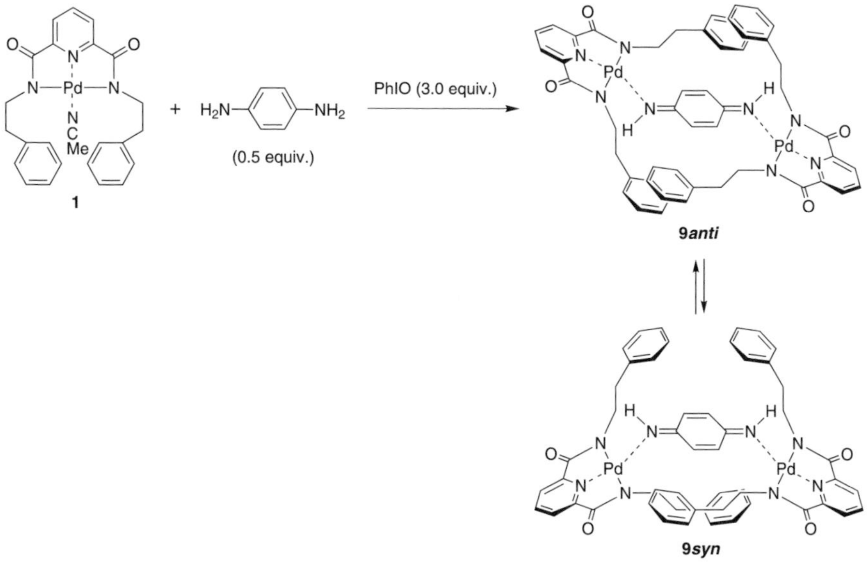

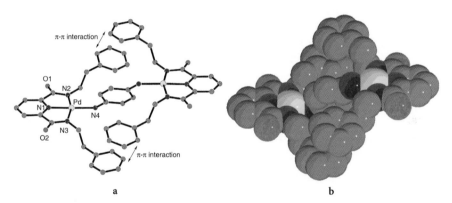

Scheme 1.5.

Fig. 1.10. (a) Molecular structure of **9anti** (hydrogen atoms omitted for clarity). (b) Space-filling representation of molecular structure of **9anti** (hydrogen atoms omitted for clarity)

process of V(III) to V(IV). The π-conjugated molecule L^1 and POT undergo complexation with VCl$_3$ together with redox reaction, affording the conjugated complexes **10** and **11**, respectively. The complexation proceeds via reduction of the qd moiety with oxidation of V(III) to V(IV), in which the vanadium species is considered to play an important role in both complexation and reduction processes (Scheme 1.6) [18]. The cyclic voltammogram of L^1 exhibits only one redox couple ($E_{1/2} = -0.08$ V) corresponding to the two one-electron-transfer processes of the two terminal dimethylamino groups in DMF. The addition

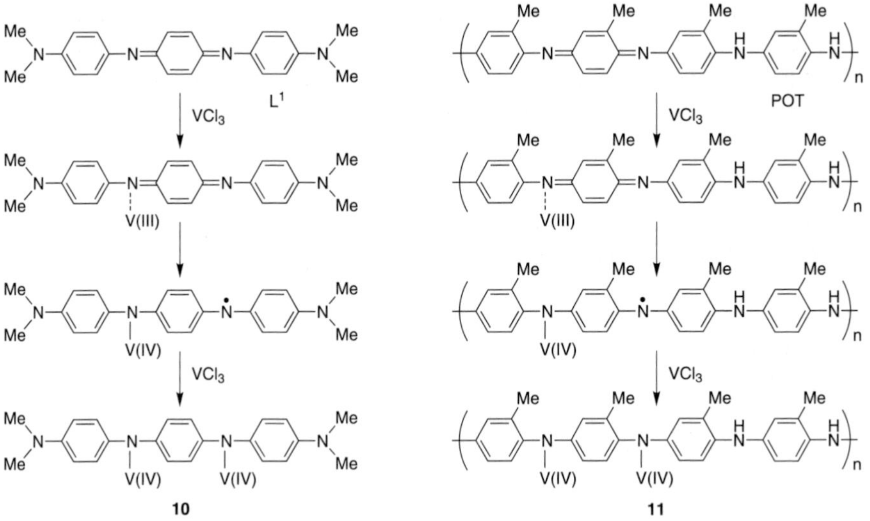

Scheme 1.6.

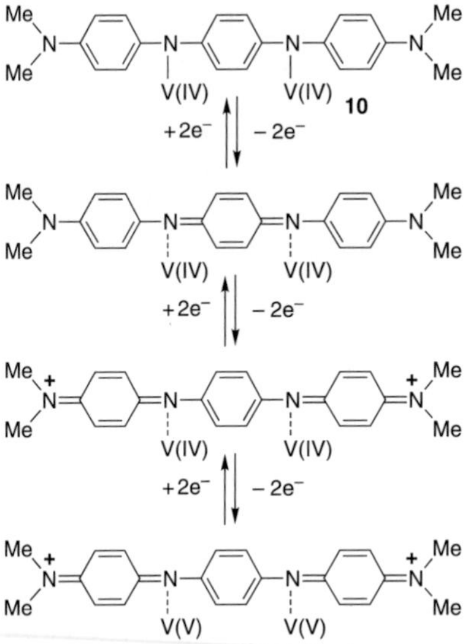

Scheme 1.7.

of a 2 molar equiv of VCl_3 to the solution of L^1 leads to the appearance of new redox couples ($E_{1/2}$ = −0.19 and 0.01 V). The redox couple at −0.19 V is considered to be attributable to the vanadium-bound reduced qd moiety. The plausible redox processes are shown in Scheme 1.7.

1.3
Redox-Switching Properties of Conjugated Complexes with 1,4-Benzoquinonediimines

Bimetallic complexes composed of π-conjugated bridging spacers and terminal redox-active transition metals have received much attention as functional materials with focus on the electronic communication through a π-conjugated spacer [19]. However, redox-switching systems for such transition metals have been investigated in only a few cases [20]. The introduction of terminal redox-active ferrocenyl units into a 1,4-phenylenediamine (pd) bridging spacer provides a novel redox-switching system for electronic communication through the π-conjugated bridging spacer [21]. The pd derivative L^4_{red} bearing terminal ferrocenyl units is synthesized by the $Pd_2(dba)_3$/(±)-BINAP-catalyzed amination of 4-bromophenylferrocene with pd (Scheme 1.8). The X-ray crystal structure of L^4_{red} reveals a twist conformation of the pd moiety as shown in Fig. 1.11. The pd derivative L^4_{red} is readily oxidized with PhIO to give the qd derivative L^4_{ox} as *syn* and *anti* (1:1 ratio) qd isomers (Scheme 1.8).

The pd derivative L^4_{red} exhibits four one-electron redox waves ($E_{1/2}$ = −0.13, −0.03, 0.16, and 0.49 V vs. Fc/Fc$^+$) in dichloromethane. The waves at −0.13 and −0.03 V are assigned to the successive one-electron oxidation of the ferrocenyl moieties. It should be noted that electronic communication between the ferrocenyl moieties is observed through the pd bridging spacer. The extent of the ferrocene–ferrocene interaction is estimated from the wave splitting, $\Delta E_{1/2}$ = 0.10 V. The corresponding equilibrium constant (K_c) for the comproportionation reaction ([Fc–Fc] + [Fc$^+$–Fc$^+$] = 2[Fc$^+$–Fc]) is 49. The waves at 0.16 and 0.49 V are assigned to one-electron oxidation processes of the pd moiety. This result is in sharp contrast to the redox behavior of L^4_{ox} in dichloromethane, in which the redox of the qd moiety and ferrocenyl ones is observed as an irreversible reduction wave at −1.66 V and simultaneous one-electron redox wave at 0.005 V, respectively. In the case of the oxidized form L^4_{ox}, electronic communication between the terminal ferrocenyl moieties is suppressed. These results indicate that a redox switching of the electronic communication through the redox-active pd bridging spacer is achieved in this system. Such communication is not observed with the π-conjugated qd molecule L^5 [22].

Complexation of L^4_{ox} with the palladium(II) complex 1 leads to the formation of the 1:2 conjugated palladium(II) complex [(L^2)Pd(L^4_{ox})Pd(L^2)] (12)

Scheme 1.8.

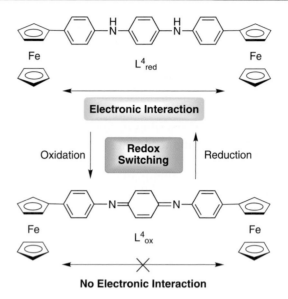

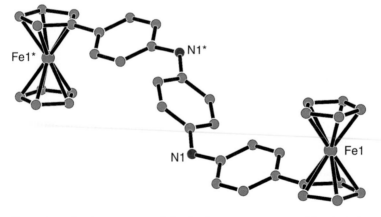

Fig. 1.11. Molecular structure of L^4_{red} (hydrogen atoms omitted for clarity)

(Scheme 1.9) [21]. The X-ray crystal structure of **12** reveals that the two [(L^1)Pd] units are bridged by the qd spacer to form the 1:2 complex **12anti** in an *anti* configuration as shown in Fig. 1.12. In the case of **12**, no electronic communication is observed between the terminal ferrocenyl moieties.

The ruthenium(II) complex **13$_{red}$** bearing *N,N'*-bis(4-aminophenyl)-1,4-phenylenediamine moieties can be chemically oxidized to the corresponding qd derivative **13$_{ox}$** (Scheme 1.10) [23], while the oxidized form **13$_{ox}$** is reduced to the reduced form **13$_{red}$** with NH$_2$NH$_2$ · H$_2$O. In the emission spectrum of **13$_{red}$** excited at 477 nm, almost complete quenching is observed in acetonitrile. An efficient photoinduced electron transfer is likely to operate in complex **13$_{red}$**, where the reduced form of the π-conjugated pendant groups serves as an electron donor. Use of the oxidized form **13$_{ox}$** also results in a quenched spectrum upon excitation at 477 nm. Taking the reported electron-transfer mechanism of complexes bearing a viologen or benzoquinone moiety into

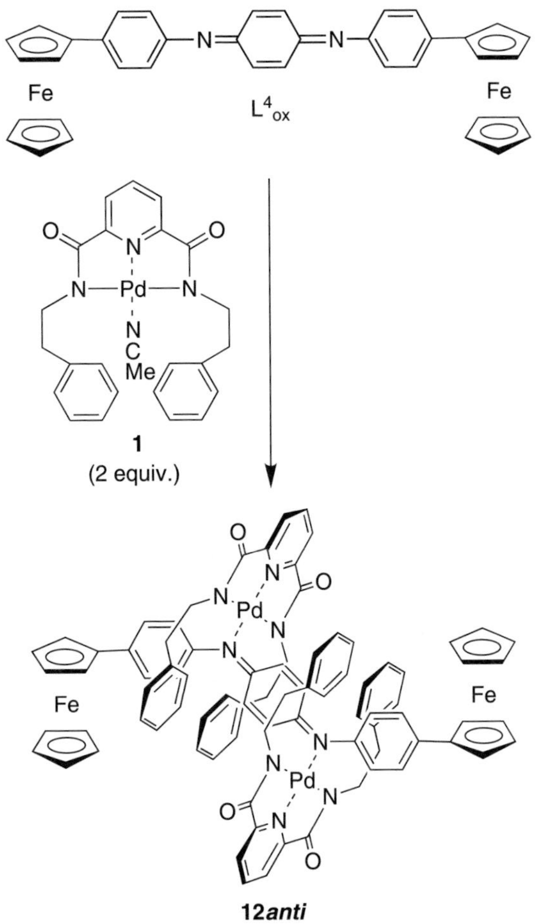

Scheme 1.9.

12anti

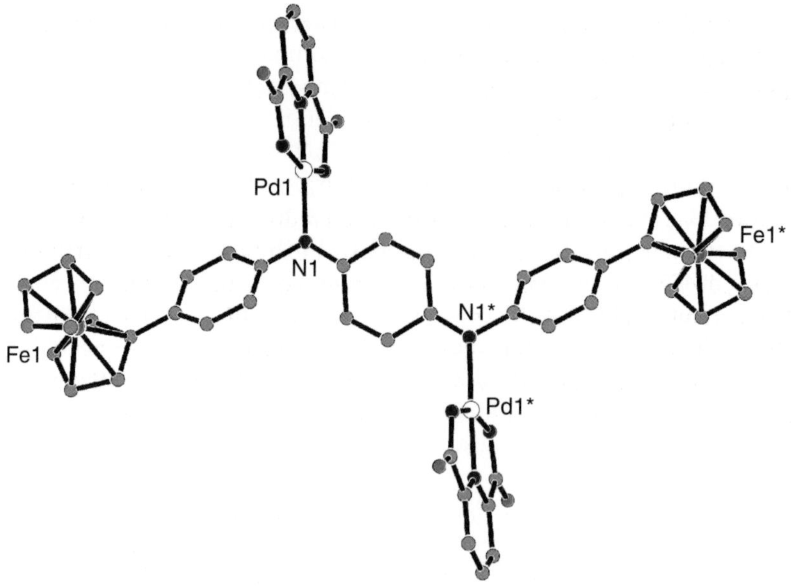

Fig. 1.12. Molecular structure of **12***anti* (phenylethyl moieties and hydrogen atoms omitted for clarity)

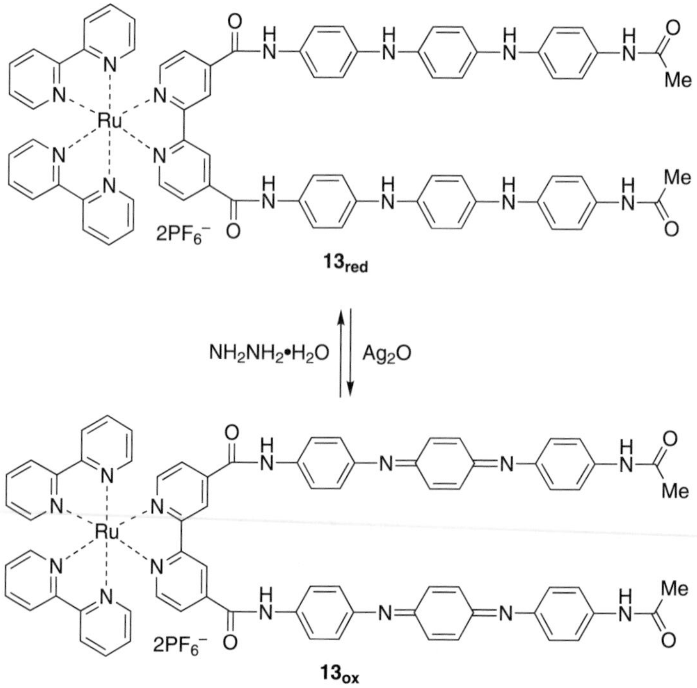

Scheme 1.10.

account [24], this result might be explained by electron transfer in a direction opposite to that of 13_{red} or energy transfer.

Although polynuclear bipyridyl ruthenium(II) complexes linked by a bridging spacer have been investigated electrochemically and photophysically to provide electronic and photoactive devices [19a, 25], only a few cases have focused on redox-active bridging spacers [26]. In combination with a redox-active pd function, such complexes afford a novel redox-active donor–acceptor system. The dinuclear ruthenium(II) complex 14 exhibits a redox-switchable photoinduced electron-transfer system as observed in the ruthenium complex 13 (Scheme 1.11) [27]. In the emission spectrum of 14_{red} excited at 450 nm in dichloromethane, almost complete quenching is observed. The quenching of 14_{red} is probably attributed to the photoinduced electron transfer from the pd moiety to the bpy-Ru moieties, wherein the pd moiety serves as an electron donor. The emission is also quenched in the case of the oxidized form 14_{ox}.

Photoirradiation of (acetonitrile)(2,2′-bipyridine)(2,2′:6′,2″-terpyridine) ruthenium(II)hexafluorophosphate [Ru(tpy)(bpy)(CH$_3$CN)](PF$_6$)$_2$ in the presence of pd leads to the formation of the conjugated ruthenium(II) complex [(tpy)(bpy)Ru(pd)Ru(tpy)(bpy)](PF$_6$)$_4$ (15_{red}) in a one-pot reaction (Scheme 1.12) [28]. The X-ray crystal structure of 15_{red} reveals that the two

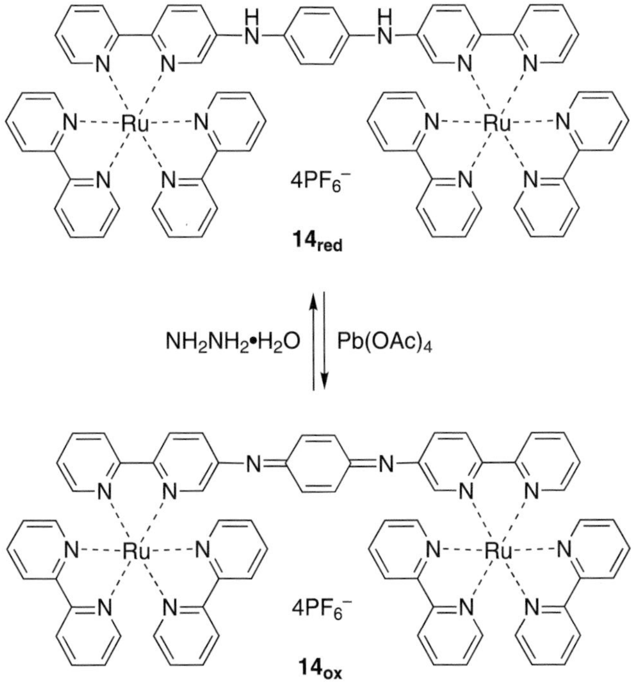

Scheme 1.11.

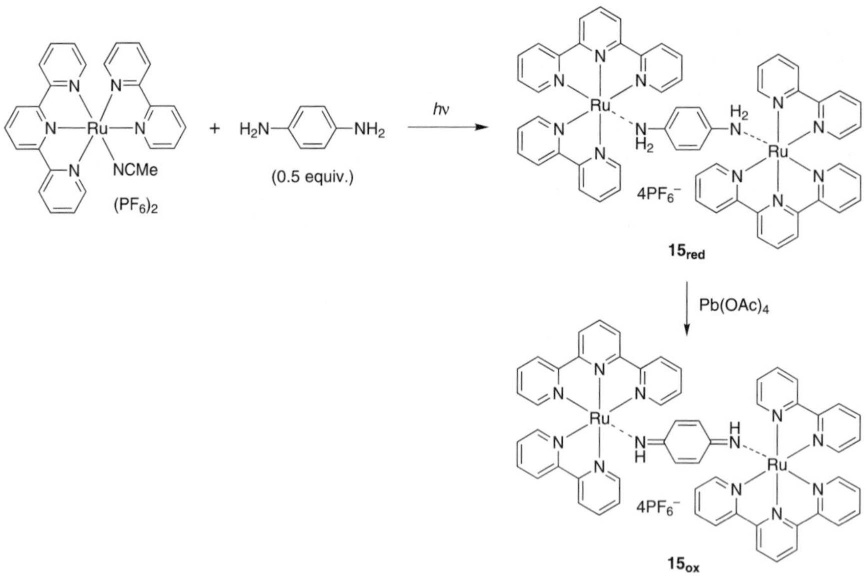

Scheme 1.12.

[(tpy)(bpy)Ru] units are bridged by the pd spacer to form the C_2-symmetrical 1:2 complex **15$_{red}$** in an *anti* configuration, in which the qd bridging spacer is sandwiched between two tpy moieties of each [(tpy)(bpy)Ru] unit through π–π interaction (Fig. 1.13). The redox switching of the emission properties of **15** is also possible. The reduced form **15$_{red}$** shows the emission at 605 nm in acetonitrile. In contrast, almost complete quenching is observed in the emission spectrum of the oxidized form **15$_{ox}$**.

Photoinduced electron transfer is observed with the porphyrin bearing four dimensionally orientated redox-active π-conjugated phenylenediamine

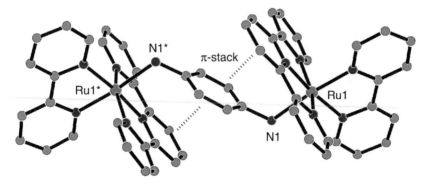

Fig. 1.13. Molecular structure of **15$_{red}$** (hydrogen atoms omitted for clarity)

strands. Electron transfer in an opposite direction might be possible by oxidation of the phenylenediamine moieties. Treatment of the zinc porphyrin **16** with 0.5 molar equiv of a bidentate ligand, 1,4-diazabicyclo[2.2.2]octane (DABCO), affords the sandwich dimer complex **17**, in which the zinc porphyrin moieties are surrounded by π-conjugated pendant groups (Scheme 1.13) [29]. The connection of the zinc porphyrin **18** with a π-conjugated bridging ligand, 4,4′-bipyridine, gives the sandwich dimer complex **19** (Scheme 1.14) [30].

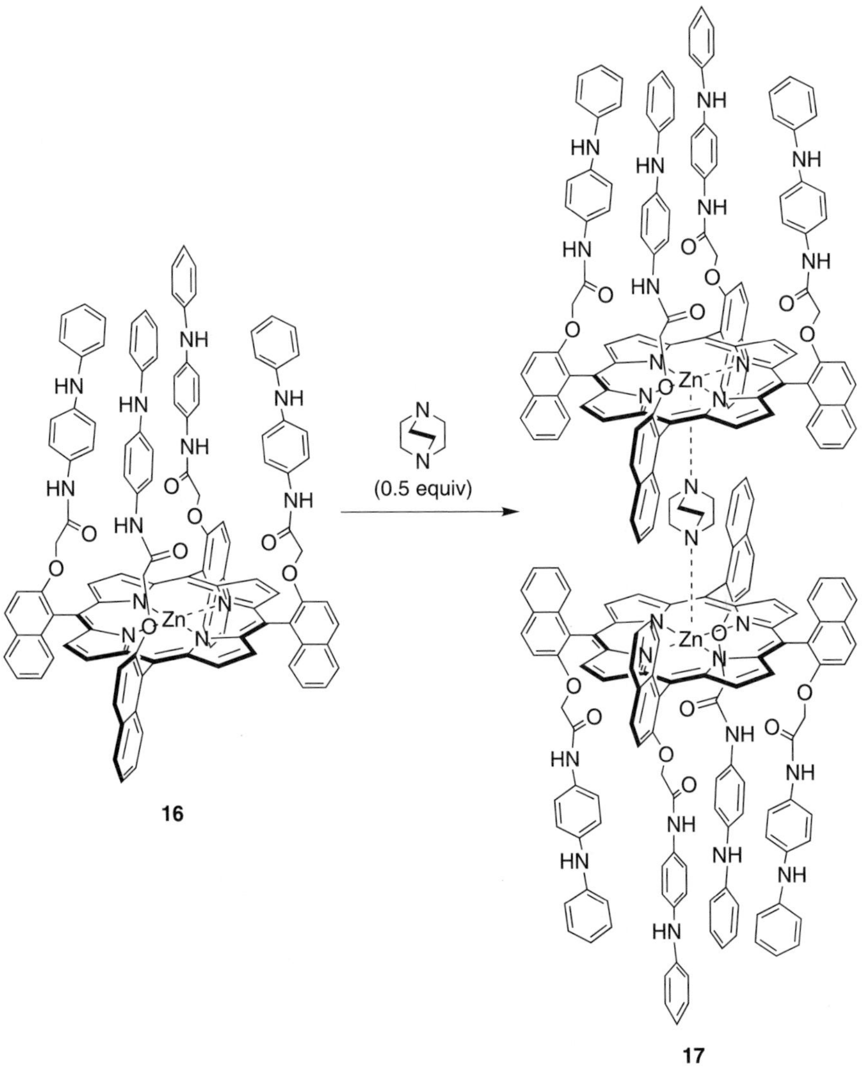

Scheme 1.13.

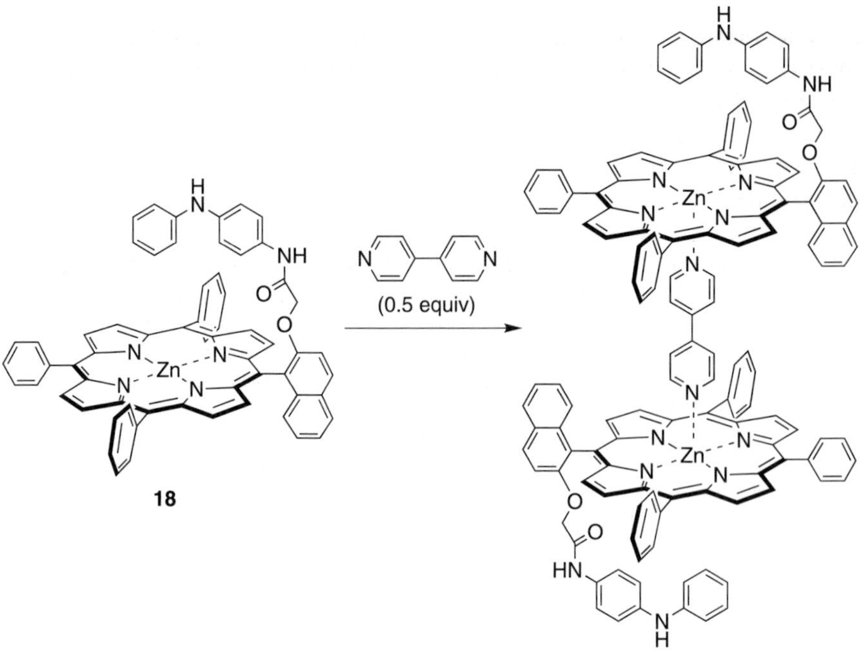

Scheme 1.14.

1.4
Conclusion

π-Conjugated qd derivatives are recognized as a π-conjugated bridging lig-
and based on the coordination ability of the qd nitrogen atoms. Another
advantage depends on their redox properties. The qd moieties are reduced
to semiquinonediimine radical anions (sq) and phenylenediamide dianions
(pda), both of which also possess the ability to bind to metals. The combination of this redox behavior and complexation with transition metals provides
an efficient redox system. The formation of conjugated complexes with redox-
active π-conjugated qd derivatives is controlled by the coordination mode of
the qd moiety. Incorporated metals play an important role as metallic dopants,
and the complexed qd becomes stabilized as an electron sink. Furthermore, chirality induction to the π-conjugated backbone of polyaniline is performed by
the complexation with chiral complexes, affording the chiral d,π-conjugated
complexes. Another noteworthy feature of the conjugated complexes is the
redox-switching properties based on the redox control of the qd spacer. The
conjugated complexes composed of the redox-active conjugated qd ligands are
envisioned to provide not only functional electronic materials but also redox
catalysts.

Acknowledgement This work was financially supported in part by a Grant-in-Aid for Scientific Research on Priority Areas from the Ministry of Education, Culture, Sports, Science and Technology, Japan.

1.5
References

1. a) MacDiarmid AG, Yang LS, Huang WS, Humphrey BD (1987) Synth Met 18:393. b) Epstein AJ, Ginder JM, Richter AF, MacDiarmid AG (1987) Conducting Polymers, Alcacer L (ed), Reidel, Holland, p 121. c) Salaneck WR, Clark DT, Samuelsen EJ (1990) Science and Application of Conductive Polymers. Adams Hilger, New York. d) Ofer D, Crooks RM, Wrighton MS (1990) J Am Chem Soc 112:7869. e) Gustafsson G, Cao Y, Treacy GM, Klavetter F, Colaneri N, Heeger AJ (1992) Nature 357:477. f) Bradley DDC (1993) Synth Met 54:401. g) Miller JS (1993) Adv Mater 5:587. h) Miller JS (1993) Adv Mater 5:671. i) Bloor D (1995) Chem Br 31:385. j) Jestin I, Frère P, Blanchard P, Roncali J (1998) Angew Chem Int Ed 37:942, and references therein. See also the Nobel lectures of Heeger AJ, MacDiamid AG, and Shirakawa H (2001) Angew Chem Int Ed 40:issue 14.
2. a) Ramaraj R, Natarajan P (1991) J Polym Sci A 29:1339. b) Yamamoto T, Yoneda Y, Maruyama T (1992) J Chem Soc Chem Commun 1652. c) Hirao T, Higuchi M, Ikeda I, Ohshiro Y (1993) J Chem Soc Chem Commun 194. d) Hirao T, Higuchi M, Hatano B, Ikeda I (1995) Tetrahedron Lett 36:5925. e) Higuchi M, Yamaguchi S, Hirao T (1996) Synlett 1213. f) Higuchi M, Imoda D, Hirao T (1996) Macromolecules 29:8277. g) Higuchi M, Ikeda I, Hirao T (1997) J Org Chem 62:1072. h) Hirao T, Yamaguchi S, Fukuhara S (1999) Tetrahedron Lett 40:3009. i) Hirao T, Yamaguchi S, Fukuhara S (1999) Synth Met 106:67. j) Murray RW (1984) Electroanalytical Chemistry, Bard AJ (ed), Marcel Dekker, New York, Vol 13, p 191. k) Abruna HD (1988) Electroresponsive Molecular and Polymer System, Skotheim TA (ed), Marcel Dekker, New York, Vol 1, p 97.
3. a) Sonogashira S, Takahashi S, Hagihara N (1977) Macromolecules 10:879. b) Hmyene M, Yassar A, Escorne M, Percheron-Guegan A, Garnier F (1994) Adv Mater 6:564. c) Altmann M, Bunz UHF (1995) Angew Chem Int Ed Engl 34:569. d) Lavastre O, Even M, Dixneuf PH, Pacreau A, Vairon J-P (1996) Organometallics 15:1530. e) Sheridan JB, Lough AJ, Manners I (1996) Organometallics 15:2195. f) Hirao T, Kurashina M, Aramaki K, Nishihara H (1996) J Chem Soc Dalton Trans 2929. g) Nguyen P, Gómez-Elipe P, Manners I (1999) Chem Rev 99:1515 and references therein. h) Richard P, Kingsborough RP, Swager TM (1999) J Am Chem Soc 121:8825. i) Würthner F, Sautter A, Thalacker C (2000) Angew Chem Int Ed 39:1243.
4. a) Balch AL, Holm RH (1966) J Am Chem Soc 88:5201. b) Christoph GG, Goedken VL (1973) J Am Chem Soc 95:3869. c) Warren LF (1977) Inorg Chem 16:2814. d) Vogler A, Kunkely H (1980) Angew Chem Int Ed Engl 19:221. e) Belser P, von Zelewsky A, Zehnder M (1981) Inorg Chem 20:3098. f) Gross ME, Ibers JA, Trogler WC (1982), Organometallics 1:530. g) Miller EJ, Brill TB (1983) Inorg Chem 22:2392. h) Pyle AM, Barton JK (1987) Inorg Chem 26:3820. i) Masui H, Lever ABP, Auburn PR (1991) Inorg Chem 30:2402. j) Masui H, Lever ABP, Dodsworth ES (1993) Inorg Chem 32:258. k) Mitra KN, Goswami S (1997) Chem Commun 49. l) Metcalfe RA, Lever ABP (1997) Inorg Chem 36:4762. m) Jüstel T, Bendix J, Metzler-Nolte N, Weyhermüller T, Nuber B, Wieghardt K (1998) Inorg Chem 37:35. n) Mitra KN, Peng S-M,

Goswami S (1998) Chem Commun 1685. o) Rall J, Stange AF, Hübler K, Kaim W (1998) Angew Chem Int Ed 37:2681. p) Dollberg CL, Turro C (2001) Inorg Chem 40:2484. q) Herebian D, Bothe E, Neese F, Weyhermüller T, Wieghardt K (2003) J Am Chem Soc 125:9116.

5. a) Herington EFG (1959) J Chem Soc 3633. b) Rieder K, Hauser U, Siegenthaler H, Schmidt E, Ludi A (1975) Inorg Chem 14:1902. c) Cheng H-Y, Lee G-H, Peng S-M (1992) Inorg Chim Acta 191:25. d) Conner D, Jayaprakash KN, Gunnoe TB, Boyle PD (2002) Organometallics 21:5265.

6. Wei Y, Yang C, Ding T (1996) Tetrahedron Lett 37:731.

7. The series of N-heterocyclic multidentate podand ligands have been used as complex catalysts in oxygenation with molecular oxygen: a) Hirao T, Moriuchi T, Mikami S, Ikeda I, Ohshiro Y (1993) Tetrahedron Lett 34:1031. b) Hirao T, Moriuchi T, Ishikawa T, Nishimura K, Mikami S, Ohshiro Y, Ikeda I (1996) J Mol Catal A: Chemical 113:117.

8. Moriuchi M, Bandoh S, Miyaji Y, Hirao T (2000) J Organomet Chem 599:135.

9. a) Huck WTS, Hulst R, Timmerman P, van Veggel FCJM, Reinhoudt DN (1997) Angew Chem Int Ed Engl 36:1006. b) Kickham JE, Loeb SJ, Murphy SL (1997) Chem Eur J 3:1203. c) Cameron BR, Loeb SJ, Yap GPA (1997) Inorg Chem 36:5498.

10. Moriuchi T, Kamikawa M, Bandoh S, Hirao T (2002) Chem Commun 1476.

11. Moriuchi T, Bandoh S, Miyaishi M, Hirao T (2001) Eur J Inorg Chem 651.

12. Wnek GE (1986) Synth Met 15:213.

13. Moriuchi T, Miyaishi M, Hirao T (2001) Angew Chem Int Ed 40:3042.

14. a) Moutet JC, Saint-Aman E, Tranvan F, Angibeaud P, Utille JP (1992) Adv Mater 4:511. b) Guo H, Knobler CM, Kaner RB (1999) Synth Met 101:44. c) Egan V, Bernstein R, Hohmann L, Tran T, Kaner RB (2001) Chem Commun 801.

15. a) Majidi MR, Kane-Maguire LAP, Wallace GG (1995) Polymer 36:3597. b) Majidi MR, Kane-Maguire LAP, Wallace GG (1996) Polymer 37:359. c) Reece DA, Kane-Maguire LAP, Wallace GG (2001) Synth Met 119:101. d) MaCarthy PA, Huang J, Yang S-C, Wang H-L (2002) Langmuir 18:259. e) Yang Y, Wan M (2002) J Mater Chem 12:897. f) Li W, Wang H-L (2004) J Am Chem Soc 126:2278.

16. Shen X, Moriuchi T, Hirao T (2004) Tetrahedron Lett 45:4733.

17. Hirao T (1997) Chem Rev 97:2707.

18. Hirao T, Fukuhara S, Otomaru Y, Moriuchi T (2001) Synth Met 123:373.

19. For reviews on this subject, see: a) Balzani V, Juris A, Venturi M, Campagna S, Serroni S (1996) Chem Rev 96:759. b) Harriman A, Sauvage J-P (1996) Chem Soc Rev 41. c) Paul F, Lapinte C (1998) Coord Chem Rev 178/180:431. d) Ziessel R, Hissler M, El-ghayoury A, Harriman A (1998) Coord Chem Rev 178/180:1251. e) McCleverty JA, Ward MD (1998) Acc Chem Res 31:842. f) Schwab PFH, Levin MD, Michl J (1999) Chem Rev 99:1863.

20. a) Auburn PR, Lever ABP (1990) Inorg Chem 29:2551. b) Joulié LF, Schatz E, Ward MD, Weber F, Yellowlees LJ (1994) J Chem Soc Dalton Trans 799. c) Keyes TE, Forster RJ, Jayaweera PM, Coates CG, McGarvey JJ, Vos JG (1998) Inorg Chem 37:5925.

21. Moriuchi T, Takagai Y, Hirao T, to be submitted.

22. Moriuchi T, Shen X, Saito K, Bandoh S, Hirao T (2003) Bull Chem Soc Jpn 76:595.

23. Hirao T, Iida K (2001) Chem Commun 431.

24. a) Goulle V, Harriman A, Lehn J-M (1993) J Chem Soc Chem Commun 1034. b) Yonemoto EH, Riley RL, Kim YI, Atherton SJ, Schmehl RH, Mallouk TE (1992) J Am Chem Soc 114:8081.

25. For reviews on this subject, see: a) De Cola L, Belser P (1998) Coord Chem Rev 17:301. b) Belser P, Bernhard S, Blum C, Beyeler A, De Cola L, Balzani V (1999) Coord Chem Rev 190–192:155. c) Brigelletti F, Flamigni L (2000) Chem Soc Rev 29:1.

26. a) Auburn PR, Lever ABP (1990) Inorg Chem 29:2551. b) Hartl F, Snoeck TL, Stufkens DJ, Lever ABP (1995) Inorg Chem 34:3887. c) Keyes TE, Forster RJ, Jayaweera PM, Coates CG, McGarvey JJ, Vos JG (1998) Inorg Chem 37:5925. d) Staffilani M, Belser P, De Cola L, Hartl F (2002) Eur J Inorg Chem 335.
27. Shen X, Moriuchi T, Hirao T (2003) Tetrahedron Lett 44:7711.
28. Moriuchi T, Shiori J, Hirao T, to be submitted.
29. Hirao T, Saito K (2002) Synlett 415.
30. Hirao T, Naka S, Saito K, unpublished results.

Chapter 2

Realizing the Ultimate Amplification in Conducting Polymer Sensors: Isolated Nanoscopic Pathways

Timothy M. Swager

Department of Chemistry, Massachusetts Institute of Technology, Cambridge, MA 01239, USA

Summary The ability of electronic polymers (molecular wires) to amplify sensory signals has been well documented in both optical and electrical transduction schemes. For optical (fluorescent) sensors a continous film with facile three-dimensional transport is optimal. In resistivity-based sensors the ultimate gain is realized when the carriers are constrained to nanoscopic pathways. In this architecture the carriers are forced to traverse barriers that are erected as a result of molecular recognition events. Without restricted nanoscopic pathways the carriers can simply go around the barrier, and although the resistivity will be increased, the full potential of the gain is not realized. In this chapter, I briefly describe systems that fulfill the desired restrictions in carrier transport and present analysis that supports the assertion that chemoresitive materials with isolated pathways offer great potential for the creation of ultrasensitive sensors.

Abbrevations

CP   Conducting polymer
PPE   Poly(phenylene ethynylene)
AFM   Atomic force microscope
Ppy   Poly(pyrrole)
C-Py Canopied poly(pyrrole)

2.1
Dimensionality in Molecular-Wire Sensors

Nanomaterials have been at the heart of most recent sensor innovations. The enhancements offered by nanomaterials are multifaceted and can be as simple as producing a larger surface-to-volume ratio and as complex as creating molecular electronic circuitry. Some years ago my group demonstrated sensory-response amplification using molecular wires [1]. The concepts are universal and applied to sensors that functioned in either resistive or emissive modes. This chapter will focus largely on resistive systems wherein the molecular wire is a conducting polymer (CP). As illustrated in Fig. 2.1, the conceptual inspiration is to create a device wherein charge passes through a molecular wire comprised of repeating units that are effectively variable resistors. Recognition elements attached to these repeating units are designed to give an increase in resistance upon binding of the target analyte. If all of the repeating units of

Fig. 2.1. Conceptual illustration of a molecular-wire sensor. In an analyte-free enviornment the wire's receptors are vacant and the wire has a very low resistance. Upon binding a target analyte, a local activation barrier is erected and transport of the carriers (in this case radical cations) through the point of analyte binding becomes a highly activated (resistive) process. Ideally the wire's resistance increases to a point where it is rendered insulating by a single binding event

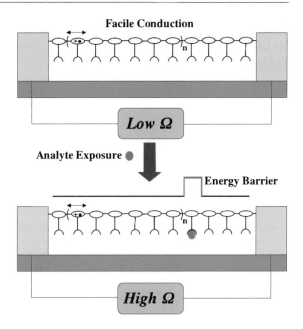

the wire are competent analyte binding sites, then we expect to see a response that reflects the additivity of the association constants of all of the receptors in the wire. In other words, if we have 10^9 receptors wired in series each having a binding constant of 10^6, we would expect to see a sensory response with an effective association constant of 10^{15}. Antibodies can provide this level of association constant; however, in this case the association is effectively irreversible and the response is a function of the exposure time to the analyte. Therefore a system working in a nonequilibrium (irreversible mode) is properly called a dosimetric sensor. To use such a system for continuous monitoring of an analyte's concentration, the system must be subjected to other periodic reset events (i.e., antibody denaturing).

A difficulty with demonstrating molecular-wire sensors, which is often a point of contention in molecular electronic-device claims, is that it is extremely difficult to make verifiable contacts to single molecules. Even if one achieves a stable contact with the electrode, the prospects for mass production of reliable devices is a daunting challenge. Furthermore, whenever a single molecule or CP is examined, it will necessarily display stocastic behavior because its individual environment is dynamic and randomly fluctuating [2]. As a result of these inherent problems, we chose to initially demonstrate the intrinsic amplifying ability of molecular wires by making use of the ability of semiconductive electronic polymers to transport optically generated excitations (excitons). In this way we could determine amplification within a dilute solution of noninteracting molecular wires. Given that we measure the response of a large number of individual molecules, we will not observe stocastic effects. Furthermore the input and output signals are photons, and therefore the above mentioned difficulties of electrode contacts are mitigated.

There are many differences between an optical molecular-wire demonstration and the electrical device conceputalized in Fig. 2.1. One obvious difference is that the excitons in the optical system can undergo spontaneous emission, which leads to a finite lifetime of the carriers. A second, less appreciated, aspect is that excitons in a structurally homogenous electronic polymer do not display a directional predispostion and wander aimlessly in a random-walk process. In other words, the exciton does not systematically sample each receptor in turn, and the detection of the analyte requires a chance encounter. In solution, the governing physics of this process yields limited amplification, and we determined that excitations can visit approximately 140 polymer repeating units in poly(phenylene ethynylenes) (PPEs) [3]. This significant, but limited, amplification is due to the fact that undirected one-dimensional diffusion (electronic polymer in solution) is extremely inefficient and to sample 140 repeating units in a one-dimensional random walk, the system has to make 140^2 energy jumps [4]. Hence in an isolated photonic molecular wire the excitons revisit the same units many times. Fortunately, the situation is greatly improved in thin films (Fig. 2.2), wherein the random-walk statistics of diffusion allow for the excitons to visit far more receptors. Exciton diffusion is also facilitated in thin films by facile dipolar energy transfer between polymer segments. We have extablished the role of dimensionality in precisely constructed thin films and observed an increase in the exciton energy transfer efficiency upon transitioning from two dimensions to three dimensions [5]. Although the precise numbers are difficult to determine, the amplification in thin films likely exceeds 10^4 and has enabled unique sensors to be constructed. In particular, these designs have been used to create the most sensitive explosives sensors ever produced, which rival the detection limits displayed by trained canines [6]. To summarize, molecular-wire amplification in a fluorescent mode requires the maximal freedom for ex-

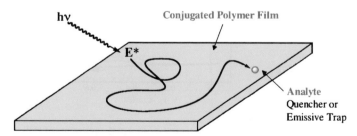

Increases in Diffusion Path Length Enhance Sensitivity

Fig. 2.2. Schematic representation showing exciton diffusion in a thin film. The diffusion path length determines the number of analyte binding sites an exciton can visit. Longer path lengths are enhanced in 3D films

citon diffusion that is achieved in a three-dimensional continuous material (Fig. 2.2). However, in the molecular-wire-resistivity schemes an external voltage provides a vectorial (directional) movement of the charge in one direction (Fig. 2.1).

2.2
Analyte-Triggered Barrier Creation in Conducting Polymers

The resistive sensory scheme shown in Fig. 2.1 requires that an analyte binding event create an impediment to charge transport. An example of such a designed system and its sensory response to sodium ions is given in Fig. 2.3. The molecular wire in this case is based upon polythiophene, and the electrochemically injected carriers are cationic [7]. The polymer is designed to be highly specific to sodium ions, and the binding thereof creates the desired energy barrier shown in Fig. 2.1 by a combination of three effects. In the bound form the lone pairs of the phenolic crown-ether oxygens are coordinated to the sodium ions, which reduces their ability to stabilize proximate cationic carriers (either a cation radical or dication). In addition to lone-pair effects (delocalization),

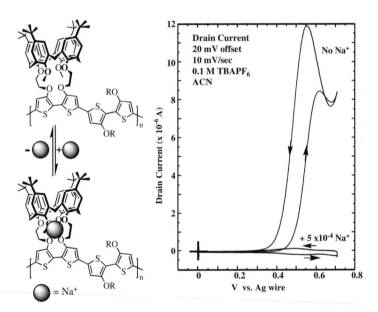

Fig. 2.3. The polymer shown on the *left* ceases to conduct in the presence of sodium ions due to the formation of large barriers to electrical transport (see text). The *in situ* conductance (which is proportional to the drain current) shows the polymer to be a good conductor when doped by applying a potential above 0.4 V in the absence of sodium ions. The polymer effectively becomes an insulator in the presence of a solution that is half millimolar in sodium ions

the positive charge of the sodium ion will have repulsive electrostatic interactions with the cationic carriers. To further block carrier transport, the binding of the specific alkali metal ions to polymers of this type introduces a large dihedral angle between the planes of the two thiophene rings that are part of the crown-ether macrocycle [8]. The combination of these three effects is profound, and the electrical chemosensor response is shown in Fig. 2.3 [7]. The data shown in Fig. 2.3 are obtained from a microelectrochemical transistor device that effectively measures the conductance (Ω^{-1}) in situ as a function of doping level (number of carriers). The doping level is a function of applied electrochemical potential (vs. Ag/Ag$^+$) applied to two individually addressable contact electrodes interconnected by the polymer to form a circuit. At low applied potentials there are no carriers and the polymer is an insulator (no drain current), and at higher potentials carriers are oxidatively injected and the conductance increases. The conductance hysteresis in the absence of sodium ions is due to structural relaxations in the polymer that occur with doping. Upon the addition of sodium ions the conductance drops dramatically and the resultant current is that of the polymer's cyclic voltammogram. This Faradaic current is always present as an artifact of the measurement but is typically obscured by the large conductance. The fact that the polymer is virtualy an insulator in the presence of sodium ions allows for this contribution to the electrical behavior to be readily observed.

In spite of the large resistivity increase (conductance decrease) of the CP in Fig. 2.3, as well as systems from our laboratory [9–11], the responses represent only a fraction of the potential sensitivity. The reason for this limited response is that the CPs are deposited in a continuous film over the electrodes and virtually allow for an infinite number of carrier paths between the electrodes. As shown in Fig. 2.4 the analyte-induced barriers can only effectively impede the carrier transport by forcing them into less direct alternate routes between the cathode

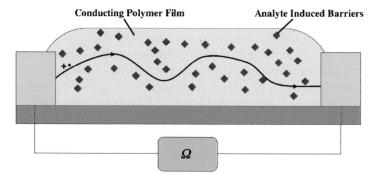

Fig. 2.4. Schematic representation of analyte-induced impediments to carrier (in this case cation radicals or polarons) transport through a continiuos film of a CP. An increase in the resistance (Ω) is observed due to the longer path necessary for the carrier to travel from one electrode to the other

and anode. This elongated carrier path length will increase the resistance (resistance scales linearly with distance), but only with a large number of bound analytes will the material cease to conduct as in Fig. 2.3.

2.3
Isolated Nanoscopic Pathways

Given that a continuous film of a CP is not the optimal morphology for creating the ulimate resistive molecular-wire sensors and stringing single molecular wires between electrodes is presently impractial, my group's efforts have been focused on creating structures that offer an intermediate situation. Our scheme is to create networks of nanoscopically structured molecular wires wherein the carriers are restricted from essentially an infinite number of pathways in a continuous film to a finite number of nanoscopic pathways. Differential resistance measurements can be made with sub-ppm resolution with inexpensive electronics, and in principle we could achieve facile single molecule dection with systems having millions of parallel independent (insulated) pathways. In the remainder of this chapter, I will detail promising methods that can be used to create architectures capable of producing sensory devices having insulated nanoscopic electrical-conduction pathways.

2.4
Langmuir–Blodgett Approaches to Nanofibrils

One method we have used to create the required architectures involves a combination of self-assembly and mechanical methods to organize electronic polymers into aligned nanofibril assemblies. This approach builds on Wegner's pioneering work on the assembly of polymers into aligned films using the Langmuir–Blodgett technique [12]. In those studies it was demonstrated that polymers dispersed at the air–water interface could be transfered to give aligned polymer films on solid substrates by the Langmuir–Blodgett method. The polymer requirements were that they have a high density of sidechains that cylindrically enshroud the polymer's backone and an anisotropic shape persistent structure. The high density of side chains prevents strong intermolecular interactions between the rigid unsaturated polymer backbones, which allows the anisotropic macromolecules to dynamically reorientaate in the presence of a flow field. Wegner termed polymers having these properties hairy rods, and the monolayer at the air–water interface is a two-dimensional nematic liquid crystal provided that the polymers have a sufficiently high aspect ratio. The prototypical hairy-rod systems were shish-kebab polymers based upon octa-sidechain phthalocyanine repeating units and polyphenylenes with two side chains per unit [12]. A key feature of both of these hydrophobic systems is that they do not have a planar boardlike structure (hence the rod designation), which often creates strong interpolymer cofacial π-electron interactions.[1] Such strong associations create rigid films at the air–water interface and chain alignment cannot be achieved by flow fields. This design element is very restrictive and effectively rules out highly conductive CPs with planar conformations and extended delocalization.

To expand the scope to more delocalized electronic polymers that can be assembled into two-dimensional liquid crystals at the air–water interface and to gain further control over the film structures, we began experiments on PPEs with surfactant characteristics [13–15]. Depending upon the regiochemical disposition of the hydrophobic and hydrophilic groups, two different organizations are possible (Fig. 2.5). The edge-on structure promotes the strong cofacial π-aggregates discussed above, and, in analogy with the parameters originally given by Wegner, this system does not align in flow fields. To obtain a phase wherein the polymers are ordered with their aromatic planes parallel to the air–water interface, monomers having symmetric para substituted rings were used and alternating copolymers of hydrophobic and hydrophilic side-chain monomers were synthesized. We refer to this as the face-on organization due to the orientation of the aromatic rings to the interface (Fig. 2.5). In the face-on phase the interpolymer interactions are very weak and are only due to edge-to-edge associations. Face-on structures therefore produce two-dimensional liquid crystalline monolayers.

A large number of polymers have been investigated and demonstrate the generality of these designs [14]. We find that with mechanical annealing (repeated compression and expansion of the monolayer) the polymers at the air-water interface display alignment with their long axes perpendicular to the direction of compression (Fig. 2.6). With this prealignment higher anisotropy is obtained after transfer to a glass slide by the Langmuir–Blodgett method.

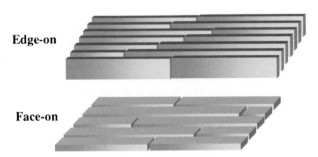

Edge-on

Face-on

Fig. 2.5. Two different organizations of conjugated polymers at the air–water interface. The edge-on structure gives strong interpolymer interactions and surface crystallites that lack the ability to be oriented by mechanical forces. The face-on structures are true two-dimensional liquid crystals that may be readily aligned by flow fields

---

[1] Note that polyphenylenes containing substituents are extremely nonplanar polymers with very limited conjugation.

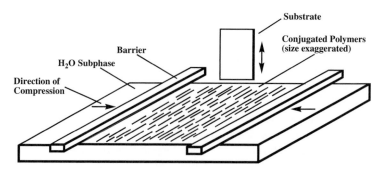

Fig. 2.6. Two-dimensional liquid crystalline films of surfactant CPs can be oriented by repeated mechanical annealing (compresssion and expansion) prior to dip-coating an aligned film on a glass slide. The dipping creates the bulk of the alignment due to the flow field associated with the transfer

Tranfer of aligned CP films to glass slides or silicon wafers with a natural oxide coating by the Langmuir–Blodget method produces thin films with high optical anisotropy. If the silicon wafer is untreated thereby presenting polar Si–OH groups, then the hydrophilic side chains (generally ethylene oxide groups) bind tightly to the substrate and the films faithfully give a nanoscopically smooth conformal monolayer on the starting substrate. However in the event that the wafer is treated with $((i\text{-}Pr)_3Si)_2NH$ to render it hydrophobic then the films undergo a structural reorgnization to minimize their surface energy [15]. The driving force for this transformation is the high surface energy resulting from the hydrophobic chains interacting weakly with the nonpolar surface and the polar ethylene oxide chains being forced to interface with air. Removal of the film from bulk water creates a high-energy state with uncompensated dipoles. The polymers then roll up into nanofibrils and thereby place the polar ethylene oxide moieties in their core. An example of this type of process is shown for a poly(phenylene ethynylene) in Fig. 2.7. The nanofibrils have their hydrophobic chains organized outward and the surface has a hydrophobic (high contact angle) nature. As can be determined by atomic force microscopy these fibers are globally aligned over the entire macroscopic substrate, and optical measuremtents reveal that the polymer chains retain their alignment after the surface reconstrution. In Fig. 2.7 we show AFM data for a PPE containing a crown-ether macrocycle as its polar enity. The macrocycles serve as a buttress to prevent strong interpolymer interactions that often give rise to reduced fluorescence quantum yields. These macrocycles are, however, not a requirement for the structural reoganization [14].

This Langmuir–Blodgett method affords an excellent route for producing large area arrays of nanofibers and effectively limits the number of pathways that charge carriers can use to traverse a film. However, there are multiple polymers in each nanofibril, which can still allow for limited carrier transport around a barrier erected by analyte binding.

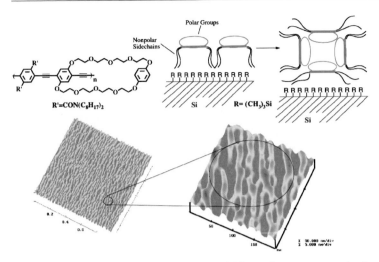

Fig. 2.7. (*Top*) PPE structure and general model for surface reconstruction by viewing along a projection perpendicular to the aligned polymer chain. The weak anchoring between the hydrophobic surface (R groups) allows for the polymers to fold into nanofibrils to allow for the assembly of the polar groups. (*Bottom left*) AFM image large area and smaller area (tapping mode, scan size 200 nm, set-point 1.265 V, scan rate 2.001 Hz) of aligned nanofibrils of the PPE shown. Note from the larger area AFM that the nanofibrils are homogenously aligned in the direction of the dipping

2.5
Molecular Scaffolds for the Isolation of Molecular Wires

Conceptually the ultimate mechanism for creating restricted pathways in CPs is to isolate individual wires through molecularly designed insulation. We have approached the challenge of making insulated molecular wires by using three-dimensional molecular scaffolds assembled about a CP's backbone.

One appoach involves applying the assembly methods pioneered by Sauvage for the formation of metal-assembled rotaxanes [16]. The metallorotaxane structural motif that can be used to create a three-dimensionally defined structure that can isolate CPs and also introduce electronically active (redox) metal centers in precise spatial geometries. Our approach to these types of isolated CPs involved an interplay of directed template-assembly and sequential electrochemical polymerizations as shown in Scheme 2.1 [17]. An electropolymerizable rotaxane structure is quantitatively assembled using a metal template. This structure is specially designed such that the blue thread is twice as long as the red monomer and the tetrahedral coordination geometry about the metal centers ensures that the blue and red units are aligned in a parallel fashion. The red portion of the rotaxane structure is very electron-rich with ethylene dioxythiophene groups and a central dialkoxy phenyl group. These

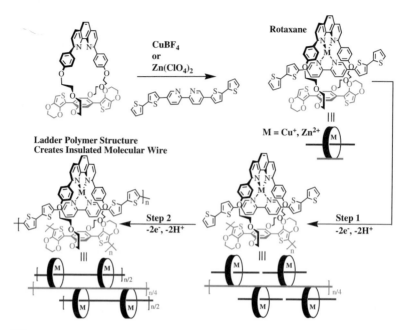

Scheme 2.1.

types of monomers are known to polymerize at low potentials to give highly conductive polymers ($\sigma \approx 10$–100 S/cm). The blue threading polymer has un-subsituted thiophenes and an electron-withdrawing bipyridyl core, and hence this molecule only oxidatively polymerizes at higher potentials. The resulting blue CP is less interesting and only becomes a very modest conductor ($\sigma \approx 0.001$ S/cm) when doped at high potentials.

Due to the different oxidation characteristics of the red and blue moieties of the rotaxane shown in Scheme 2.1, it is possible to squentially oxidatively polymerize the different thiophene monomers. In Step 1 the potential is cycled between -0.2 V and 0.8 V vs. Fc/Fc$^+$, the red polymer, which is selectively polymerized. The irreversible charge passed is associated with removal of two electrons followed by C–C bond formation and the release of two protons for each thiophene–thiophene linkage. Integration of the irreversible current gives a faithful accounting of the amount of polymer deposited as an insoluble film on the electrode. This result can be further validated by comparisons of the reversible electroactivity of deposited polymers containing copper ions relative to those containing zinc ions. The charge associated with the reversible Cu^{+1}/Cu^{+2} redox wave in the deposited films can also be used to quantify the immobilized CP. As expected, the amount of additional electroactivity is in agreement with the amount of polymer calculated from the irreversible charge associated with polymerization.

Once the prepolymer has been deposited, the electrode is transferred to a fresh monomer-free electrolyte and the blue polymer is then polymerized in step 2 by cycling the potential between −0.2 V and 1.3 V vs. Fc/Fc$^+$. The amount of irreversible charge passed in step 2 is the same as that passed for the red monomer in step 1, thereby indicating that the polymerization has reached at least 95% completion and that the ladder structure as drawn in Scheme 2.1 is a proper representation of the resultant polymer structure. A computationally optimized three-demensional model of an oligomer is shown in Fig. 2.8.

The resultant CP has a very novel rotaxane ladder structure where at lower potentials the central red polymer can be doped to a highly conductive state, while the blue polymers are completely insulating. The CP's structure prevents direct interactions between neighboring red polymers, and we estimate that they are separated by at least 10 Å. For charge to be effectively transported between conductive red polymers at low potentials, while the blue polymer is insulating, an additional transport mechanism is needed as revealed in Fig. 2.9. At intermediate potentials, where only the red polymer is doped, the zinc complex displays low conductivity, whereas the copper analog is highly conductive. As shown schematically, the copper's electroactivity creates a pathway for charge transport between the highly conducting red polymers, and, as expected, the conductivity peak at $\approx$ 0.2 V vs. Fc/Fc$^+$ is coincident with the Cu^{+1}/Cu^{+2} couple. At high potentials (> 0.4 V vs. Fc/Fc$^+$) the blue polymers of both the zinc- and copper-based materials are oxidized (doped) and thereby provide a conductive pathway between the red polymers. In effect the insulation becomes conducting and the molecular wires are interconnected. It is noteworthy that the bulk conductivity of the ladder CP at an applied potential of 0.8 V (both blue and red polymers oxidized and conducting) exceeds the intrinsic conductivity of the blue polymer by a factor of 10^4. This reflects the fact that the majority of the carriers' conduction path is provided by the more conductive red polymer and carriers need only pass through the less conductive blue polymer occasionally and for short distances to get between red polymers.

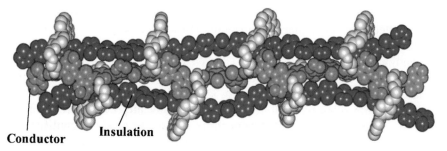

Conductor   **Insulation**

Fig. 2.8. Space-filling representations of the rotaxane ladder polymer metallated with copper. The structure shown is a computationally optimized structure of an undoped octamer

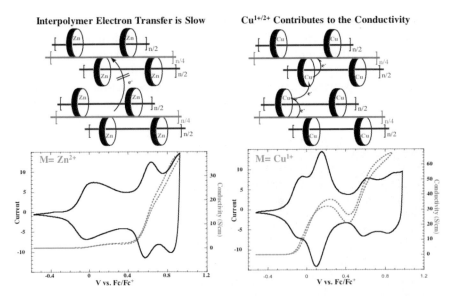

Fig. 2.9. Cyclic voltammograms and conductivity profiles as a function of applied potential of ladder polymers metallated with zinc and copper. The zinc material only has high conductivity when the blue polymer is oxidized over 0.5 V, whereas the copper-containing polymer shows conductivity mediated by the Cu^{+1}/Cu^{+2} redox couple

An advantage of the architecture of the ladder polymer is that if a limited number of copper ions is inserted, then the system can be placed at the percolation threshold and a molecular-wire response will be realized. The percolation threshold refers to a case where there are just enough copper ions to create a continuous facilitated conduction path between two electrodes. Below that threshold there are no pathways and above it there are many. Hence analytes that bind copper ions and prevent their self-exchange redox conduction effectively remove conduction pathways and give the amplification promised by Fig. 2.1. Studies in our laboratory have shown that these systems have an extreme sensitivity to anions, such as fluoride, which apparently bind to the copper ions and disrupt their reversible electrochemical activity. This sensitivity reflects the molecular-wire principles described here.

Fused polycyclic ring systems can also be used to control the distance between the conjugated backbones of CPs. We intially introduced this concept for the design of fluorescence sensory polymers and demonstrated that fusion of triptycene residues containing [2.2.2] bicyclic ring systems provided responses with size-exclusion behavior. Specifically, molecules over a specific size were found to be excluded from the material and give a limited sensory response [18, 19]. In an effort to extend these design concepts to produce insulated molecular wires with similar abilities to discriminate based upon size, we have designed and synthesized canopied-polypyrrole (C-PPy) [20]. The

C-PPy monomer (Fig. 2.10) is designed to have orient pendant phenyl rings directly over the polymer's backbone. The [2.2.1] bicyclic ring system is structurally rigid and ensures that the phenyl rings must be positioned as shown in Fig. 2.10.

The C-PPy displays a number of features that reveal the conjugated backbones to be relatively isolated (insulated) from each other. C-PPy's optimal

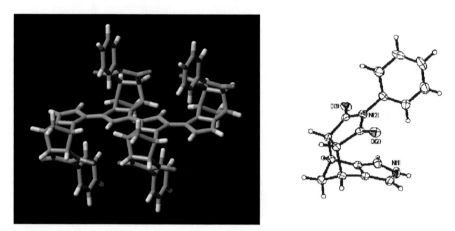

Fig. 2.10. X-ray crystal structure (*right*) of the C-PPy monomer and a computationally minimized structure of a tetramer (*left*). Notice that phenyl rings are placed above the polypyrroles conjugated backbone

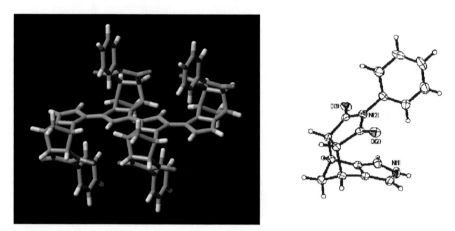

Scheme 2.2.

conductivity of 40 S/cm is lower than polypyrrole's (PPy), which can be as high as 10^3 S/cm. In contrast to PPy it displays a well-defined maximum conductivity at partial oxidation, and a combination of optical and electron spin resonance studies indicate that the conduction is afforded by polarons (radical cations) rather than bipolarons (dications) that are responsible for carrying the majority of the current in conducting PPy. Polarons are often unstable beyond low levels of doping in CPs due to condensation to form bipolarons. It has been suggested that these species are generally interchain species and can be considered π-dimers of radical cations [21]. We have shown that steric bulk can favor the formation of polarons over bipolarons [22, 23], and hence the behavior of the C-PPy is consistent with a polymer that cannot achieve strong interchain electronic interactions.

The interstitial space and lack of interpolymer interactions endow our C-PPy with excellent sensory properties. To examine this behavior, we have examined the response of this material to added base relative to the response of PPy. It is known that in its oxidized form PPy can be made to be an insulating

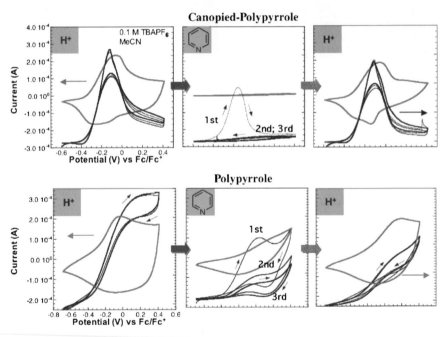

Fig. 2.11. Electrochemical response (*red*) and conductivity response (*blue*) of C-PPy (*top*) and PPy (*bottom*) upon exposure to pyridine and trifluoroacetic acid. As indicated in Scheme 2.2 with oxidation, the N–H groups in both polymers become acidic, and for C-PPy the electroactivity and conductivity are completely absent after exposure to pyridine. Note that C-PPy responses are completely restored with acid treatment and the conductivity trace even has the same shape. PPy, on the other hand, has a limited response, and the conductivity and electroactivity never recover even with continued cycling in acidic media

material due to deprotonation of the nitrogens as shown in Scheme 2.2. PPy never responds fully and is highly irreversible, whereas the C-PPy displays a complete and reversible response both in conductivity and resistivity, as shown in Fig. 2.11.

The lack of reversibility of PPy to acid and base cycling is likely due to complex conformational changes that are induced by the combination of electrochemical cycling and the acid base reactivity. Polymer films are nonequilibrium systems, and it appears that even gentle deprotonation/protonation events give significant adverse effects to the behavior of PPy. We speculate that strong interchain interactions can produce physical crosslinks that prevent the polymers from returning to their synthesized conformations. In the case of C-PPy there are no prospects for the strong interchain interactions associated with cofacial π–π bonding between neighboring polymer chains. As a result, C-PPy is able to respond in a completely reversible fashion. The excellent behavior of this system is further revealed in that it also displays a reversible pH response in aqueous solutions [23]. The response in this case clearly reflects the molecular-wire nature that has been determined by statistical mechanics [24].

2.6
Summary and Future Prospects

We are at the early stages in the development of molecular-wire-based sensors. The potential is great, but so are the challenges for the design of robust and selective sensory responses. It is my hope that this chapter has illuminated for the reader the role of dimensionality and structure in the design of these materials. Many new ways of realizing the amazing sensitivity gains possible with molecular wires await the creative inventor. The future is very exciting with important applications that can improve humankind's security, health, and environment.

2.7
References

1. Swager TM (1998) Acc Chem Res 31:201
2. Vanden Bout DA, Yip WT, Hu D, Fu DK, Swager TM, Barbara PF (1997) Science 277:1074
3. Zhou Q, Swager TM (1995) J Am Chem Soc 117:7017
4. Montroll EW (1969) J Phys Soc Jpn 26(Suppl):6
5. Levitsky IA, Kim J, Swager TM (1999) J Am Chem Soc 121:1466
6. Cumming JC, Aker C, Fisher M, Fox M, la Grone MJ, Reust D, Rockley MG, Swager TM, Towers E, Williams V (2001) IEEE Trans Geosci Remote Sens 39:1119
7. Marsella MJ, Newland RJ, Carroll PJ, Swager TM (1995) J Am Chem Soc 117:9842
8. Marsella MJ, Swager TM (1993) J Am Chem Soc 115:12214
9. Marsella MJ, Carroll PJ, Swager TM (1994) J Am Chem Soc 116:9347
10. Takeuchi M, Shioya T, Swager TM (2001) Angew Chem Int Ed 40:3372

11. Vigalok A, Swager TM (2002) Adv Mater 14:368
12. Wegner G (1992) Thin Solid Films 216:105
13. Kim J, Swager TM (2001) Nature 411:1030
14. Kim J, Levitsky IA, McQuade DT, Swager TM (2002) J Am Chem Soc 124:7710
15. Kim J, McHugh S, Swager TM (1999) Macromolecules 32:1500
16. Dietrichbuchecker CO, Sauvage JP (1987) Chem Rev 87:795
17. Buey J, Swager TM (2000) Angew Chem Int Ed 39:608
18. Yang JS, Swager TM (1998) J Am Chem Soc 120:5321
19. Yang YS, Swager TM (1998) J Am Chem Soc 120:11864
20. Lee D, Swager TM (2003) J Am Chem Soc 125:6870
21. Miller LL, Mann KR (1996) Acc Chem Res 29:417
22. Kingsborough RP, Swager TM (1999) J Am Chem Soc 121:8825
23. Lee D, Swager TM ((2005) Chem Mater 17:4622
24. Sung J, Silbey RJ (2005) Anal Chem 77:6169

Chapter 3

Metal-Containing π-Conjugated Materials

MICHAEL O. WOLF

Department of Chemistry, University of British Columbia, Vancouver, BC, V6T 1Z1, Canada

Summary π-conjugated polymers and oligomers, materials with an extended π-system along the backbone, may be coupled to metal complexes to give hybrid materials with new properties. Phosphines are used as ligands for the attachment of transition metal centers to conjugated backbones, with a variety of coordination modes observed, depending on the available coordination sites at the metal and the position of the phosphine substituent. Several examples are discussed, including a system that may be chemically switched between coordination modes. In some cases, supramolecular interactions such as π-stacking may lead to control of charge mobilities and other properties in metal-substituted oligomers. Chemical and electrochemical routes to metal nanoparticle-containing conjugated materials are summarized, along with the conductivity and optical properties of these hybrid materials. Applications of these classes of materials in chemical sensors, polymer light-emitting devices, electrocatalysts, cathode materials, and as photoconductors are discussed.

Abbreviations

CP	Conjugated polymer
ε	Extinction coefficient
$E_{1/2}$	Half-wave oxidation or reduction potential
FET	Field-effect transistor
$\lambda_{\max}$	Maximum absorption or emission wavelength
NP	Nanoparticle
OT	Oligothiophene
P,C	Phosphine, carbon coordination
P,S	Phosphine, sulfur coordination
PLED	Polymer light-emitting device
PTh	Polythiophene
PT_3	5-Diphenylphosphino-2,2′:5′,2″-terthiophene
PT_3P	5,5″-Bis(diphenylphosphino)-2,2′:5′,2″-terthiophene
sh	shoulder
UV-vis	Ultraviolet-visible

3.1
Introduction

3.1.1
π-Conjugated Materials

π-conjugated polymers (CPs) and oligomers are materials with an extended π-system along the backbone. The materials possess many remarkable properties, including high charge carrier mobilities, electrical conductivities (doped), electrochromism, and electroluminescence [1]. These properties have been taken advantage of in exploration of potential applications including in chemical sensors, light-emitting devices, and field-effect transistors. Many efforts have been devoted to synthesizing new conjugated polymers and oligomers in an effort to increase their processibility, optimize the desirable properties, and explore new properties. In Fig. 3.1 are shown examples of some of the CPs that have been prepared and studied.

Coupling π-conjugated materials to metal complexes gives hybrid materials in which the properties of the metal complex may be coupled to those of the conjugated backbone [2]. For example, these materials could be used in energy-harvesting devices such as solar cells or polymer-based light-emitting devices, where high charge carrier mobilities of the conjugated material may be combined with either the light-absorbing or emitting metal groups, giving improved device performance [3, 4]. In addition to an electronic role, metal complexes may also be used to geometrically orient π-conjugated materials in specific three-dimensional arrangments in the solid state. Careful consideration of the electronic interactions and excited states is necessary for the design of functional materials of this type.

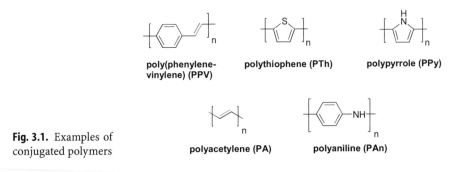

Fig. 3.1. Examples of conjugated polymers

poly(phenylene-vinylene) (PPV)

polythiophene (PTh)

polypyrrole (PPy)

polyacetylene (PA)

polyaniline (PAn)

3.1.2
Nanomaterials

Nanomaterials and their applications are of intense current interest. The new optoelectronic, chemical, and mechanical properties that occur in nanoscale materials lead to many possible uses in fields such as electronics, nonlinear optics, coatings, and sensors [5]. Metallic nanoparticles (NPs) show properties differing from the bulk materials as a consequence of quantum effects [6,7]. Synthetic methods have been devised to prepare these NPs in stable form, usually using a surface-stabilizing group to give soluble, relatively stable materials. Numerous studies have focused on studying the properties of these nanomaterials and effectively exploiting these properties. Hybrid nanocomposites in which metallic nanoparticles are embedded in a π-conjugated material are of interest because the electronic, optical, and chemical behavior of the two materials may be modified by the presence of the other component.

3.2
Metal-Complex-Containing Conjugated Materials

3.2.1
Preparation

Coupling metal complexes to π-conjugated backbones may be accomplished using a variety of approaches. Metal-containing monomers may be synthesized

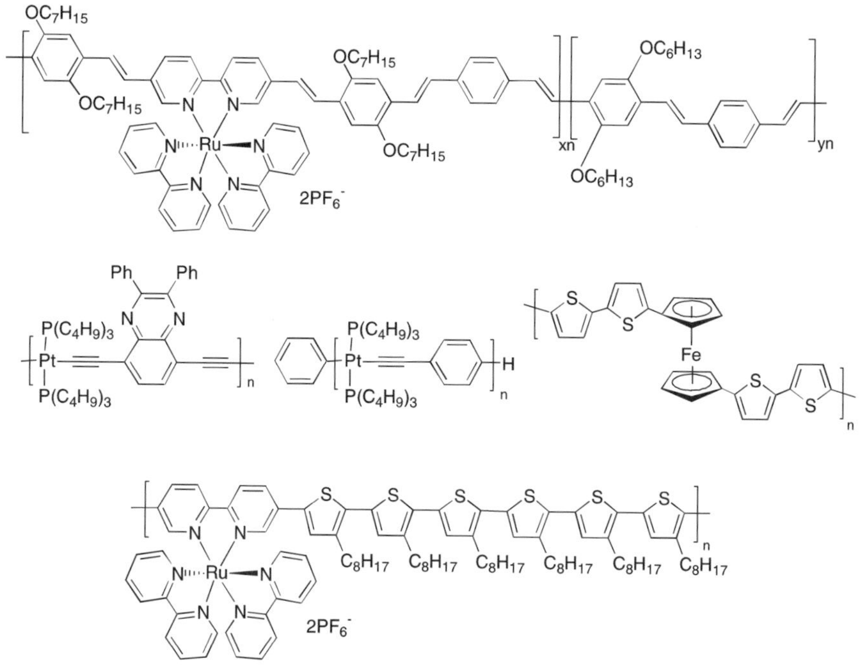

Fig. 3.2. Some metal-containing conjugated polymers

and subsequently polymerized, or, alternatively, metals may be coupled to an already polymerized π-conjugated backbone. The first method requires the metal complex to be stable under the polymerization conditions and has been used more commonly, while the latter method requires an efficient metallation reaction to give good yields of the metallized polymer. Figure 3.2 shows some examples of polymers prepared using these techniques [8–12].

Phosphines are useful as ligands for the attachment of transition metal centers to conjugated backbones. Our group has devoted efforts toward using

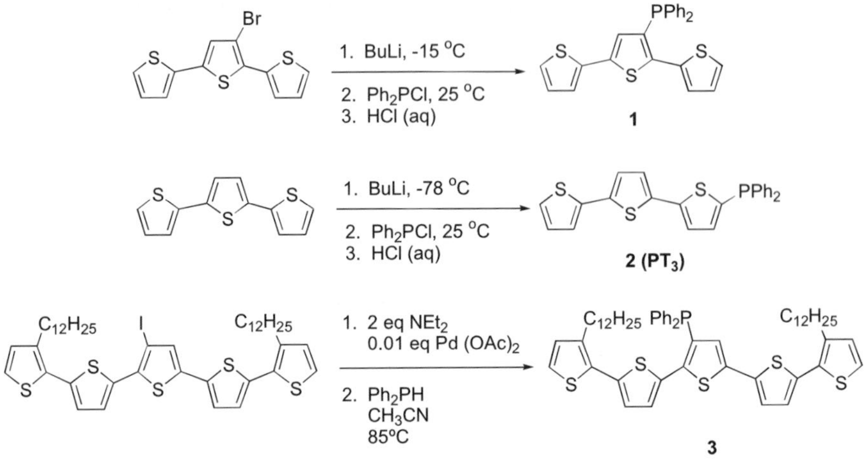

Fig. 3.3. Synthesis of diphenylphosphino-substituted terthiophenes

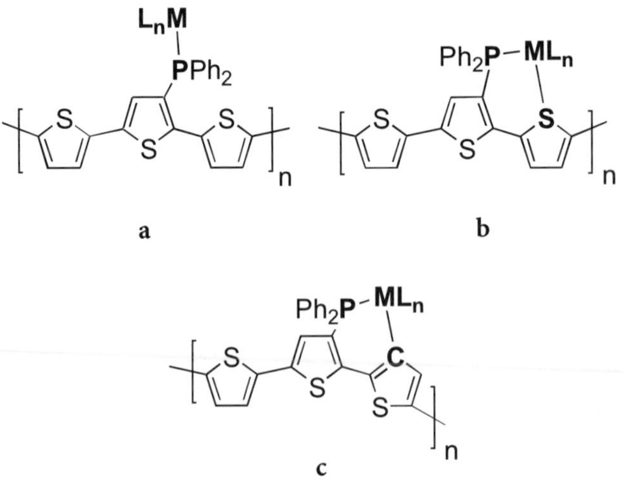

Fig. 3.4. Examples of possible coordination modes for metal complexes tethered to PTh: **a** monodentate (*P*), **b** bidentate (*P, S*), and **c** bidentate (*P, C*)

phosphine groups to attach metal complexes to oligothiophenes, and these groups have been selected as they may be readily synthesized and polymerized chemically or electrochemically. Diphenylphosphino groups can be attached via reaction of an oligothienyl anion with PPh_2Cl or Pd-catalyzed coupling of a halo(oligothiophene) to PR_2H (Fig. 3.3) [13, 14].

Transition metals may complex to phosphine-substituted oligothiophenes in a variety of coordination modes, depending both on the available coordination sites at the metal and the position of the phosphine substituent (Fig. 3.4). These include monodentate phosphine coordination (*P*), bidentate phosphine-thiophene sulfur (*P, S*), and bidentate phosphine-thiophene carbon (*P, C*) modes. Examples of these are shown in Fig. 3.4. Coordination in an η^5 fashion to an oligothiophene has also been observed [15].

3.2.2
Properties

The electronic properties of hybrid metal-complex-containing conjugated materials may be assessed using a number of tools. Commonly, electrochemical and spectroscopic (absorption, time-resolved, and steady-state emission) methods are used in combination to develop an understanding of the electronic structure. The oligomers 4 and 5 shown in Scheme 3.1 can be reversibly switched from the *P, S* to *P, C* modes and vice versa via reaction with acid or base [14, 16]. The effect of these two coordination modes on the oligomers is interesting, as shown by the data in Table 3.1. The thienyl rings are more highly conjugated due to increased coplanarity of adjacent rings in the more electron-rich *P, C* coordination mode. This results in large decreases in the oxidation and reduction potentials of the thienyl chain, as well as a red shift in the thienyl $\pi \to \pi^*$ absorption.

The three-dimensional organization of conjugated oligomers in the solid state is interesting as features such as π–π stacking are relevant to achieving the high charge carrier mobilities required for applications such as FETs [17].

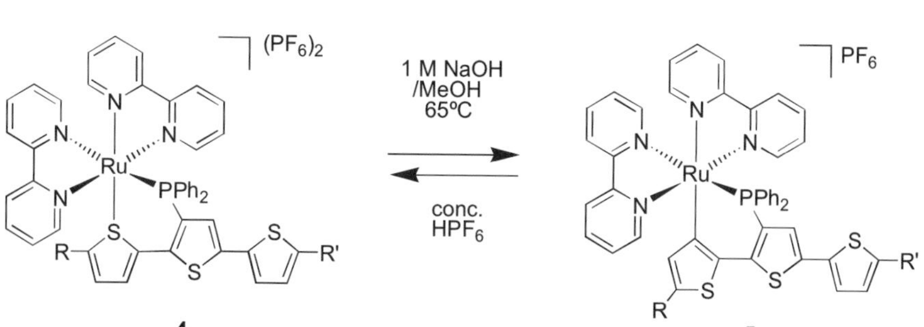

Scheme 3.1. Reversible switching of coordination mode in Ru complexes 4 and 5

Table 3.1. Electrochemical, UV-vis, and luminescence spectral data for **4** and **5**

Compound	$E_{1/2,ox} \pm 0.01$ V	$E_{1/2,red} \pm 0.01$ V	UV-vis λ_{max}, nm $[(\varepsilon \pm 0.01 \times 10^4)\,M^{-1}cm^{-1}]$	Luminescence λ_{max}, nm
4	+1.48 (Ru$^{II/III}$)	−1.28 (bpy$^{0/-}$)	280 (3.79 × 10^4) 320 (sh) (1.83 × 10^4) 393 (1.80 × 10^4)	–
5	+0.57 (Ru$^{II/III}$)	−1.53 (bpy$^{0/-}$)	295 (4.63 × 10^4) 347 (2.00 × 10^4)	754
	+1.11 (PT$_3^{0/+}$)	−1.78 (bpy$^{-/2-}$)	456 (1.84 × 10^4) 617 (sh) (2.25 × 10^3)	

The effect of alkyl groups on the solid-state structure and charge mobilities has been examined [18]; however, the use of metal complexes to organize oligothiophenes in the solid state is still new. Many of the principles relevant to two- and three-dimensional organization of conjugated oligomers may be found in the work of Stang and Fujita [19, 20]. The complex PdCl$_2$(PT$_3$) exhibits intermolecular π–π interactions (3.65-Å interplanar distance between inner thiophene rings) in the solid state (Fig. 3.5a), [21], while (PT$_3$P)Au$_2$Cl$_2$ forms dimers resulting from two aurophilic interactions between monomers (Fig. 3.5b) [22]. This area of research is too young to determine whether these supramolecular interactions will allow control of charge mobilities and other properties; however, more structural investigations are clearly warranted.

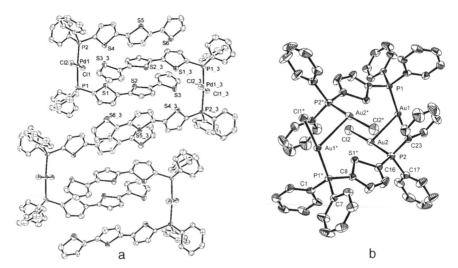

Fig. 3.5. a Solid-state packing diagram of PdCl$_2$(PT$_3$)$_2$. **b** Structure of (PT$_3$P)Au$_2$Cl$_2$. [Adapted with permission from Inorg. Chem. 44 (2005) 620]

3.3
Metal-Nanoparticle-Containing Conjugated Materials

3.3.1
Preparation

Chemical and electrochemical routes to metal-nanoparticle-containing conjugated materials have been pursued. Chemical approaches generally follow one of two approaches: a one-pot synthesis where the monomer or polymer acts as reductant, or chemical polymerization around prepared NPs. Advincula's group has demonstrated a one-pot synthesis using functionalized terthiophenes as a reductant for HAuCl$_4$ solutions [23,24]. This approach gives 6- to 100-nm-diameter NPs. Related approaches have also been used to give Pd and Pt NPs [25].

Electrochemical routes have been used to impregnate conducting polymers with metal NPs via reduction of metal salts at electrodes coated with polymer films [26,27]. Au NPs have also been tethered to a monomer containing an electropolymerizable functional group and copolymerized to give a NP-containing film [28]. Oligothiophene-capped Au NPs have also been electropolymerized to give NP-containing films (Fig. 3.6) [29].

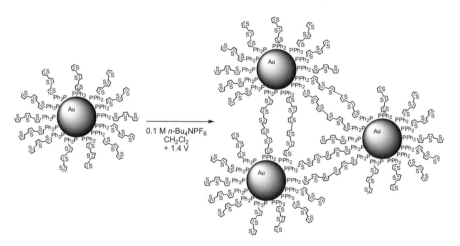

Fig. 3.6. Preparation of Au NP-containing conducting films. [Adapted with permission from Chem. Mater (2004) 16:2712]

3.3.2
Properties

The electronic and optical properties of conjugated polymer-metal nanoparticle composites are of interest. The π-conjugation between NPs can act as a molecular wire and possibly enhance the coupling between particles, resulting in greater conductivity and changes in the optical properties.

The conductivity of Au NPs linked by 6, shown in Fig. 3.7, was found to be three orders of magnitude greater than for unlinked particles, and the activation energy for tunneling is lower for the 6-linked particles [30]. Similarly, the oligothiophene-linked Au NPs shown in Fig. 3.6 are several orders of magnitude more conductive than the unlinked particles [29]. Comparison of particles linked by bis-thiols containing either a conjugated phenyl group (7) or a nonconjugated cyclohexyl group (8) (Fig. 3.7) showed that the phenyl-containing linked NPs were more conductive, had lower activation energies, and showed metallic behavior in plots of conductivity as a function of $1/T$ [31].

The surface plasmon absorption of metal NPs, observed for NPs smaller than the wavelength of incident light, occurs in the visible region of the spectrum (520–530 nm) for particles 5–50 nm in diameter [32]. This absorption is very environment sensitive, and the band has been shown to red-shift when NPs are passivated with conjugated polymers such as poly(3,4-ethylenedioxythiophene) [33]. This shift was attributed to overlap in the wave functions associated with the NP surface and the polymer. In the cross-linked NP film shown in Fig. 3.6 a broad absorption in the near-IR spectral region appears upon electrodeposition, possibly due to overlap of the linker molecular orbitals and the metal wave functions [29].

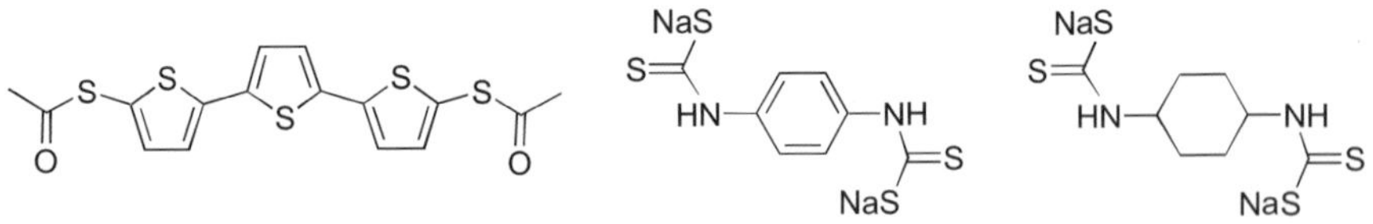

Fig. 3.7. Structures of NP linkers **6**, **7**, and **8**

3.4
Applications

Research in metal-containing conjugated polymers is still at an early stage; nonetheless, these materials are being explored for several potential applications. Chemical sensors where the metal acts as a binding site for small molecule analytes are one such application. Swager has demonstrated an example of this where NO binding to a Co-containing polymer (Fig. 3.8) gives rise to a conductivity change [34]. Other applications that have been proposed include in polymer light-emitting devices (PLEDs) and as photoconductors.

Applications for conjugated polymer–metal nanoparticle composites include in electrocatalysis [35], as cathode materials [36], and in PLEDs [37].

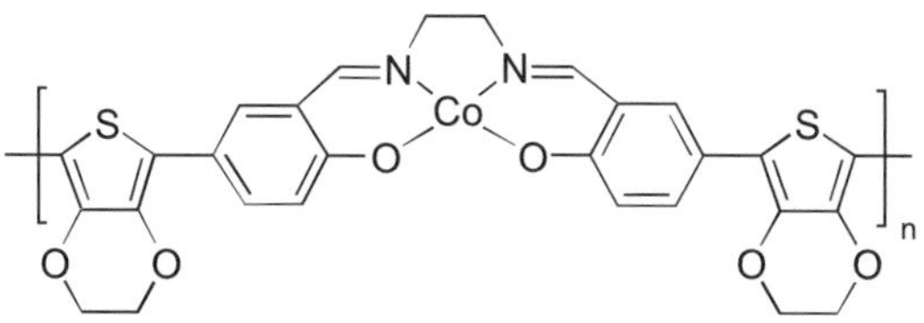

Fig. 3.8. A Co-containing polymer used in a resistivity-based NO sensor

In PLEDs containing metal nanoshells triplet excitons are preferentially quenched, resulting in reduced formation of singlet oxygen, which reacts with the polymer to give traps. This results in improved stability and higher quantum efficiencies due to improved charge injection.

3.5
Conclusions

Research in conjugated polymers containing metal complexes or nanoparticles has advanced substantially in the last few years. The study of the electronic and optical properties of these hybrid materials is still in its infancy; however, it is clear that metal groups directly bonded to the thienyl backbone may be used to alter the electronics and coplanarity of the thienyl rings. A more detailed understanding of the photophysical properties of these materials will be needed for applications of metal-containing conjugated materials in energy-harvesting and light-emitting devices. Electrodeposited metal nanoparticle films may be used in thin-film sensors, supported electrocalysis, or in batteries. Developing these and other applications is an exciting possiblity for the future of this research area.

3.6
References

1. T.A. Skotheim, R.L. Elsenbaumer, J.R. Reynolds, Handbook of Conducting Polymers, Marcel Dekker, New York, 1998
2. M.O. Wolf, Adv. Mater. 13 (2001) 545
3. T. Yamamoto, Y. Saitoh, K. Anzai, H. Fukumoto, T. Yasuda, Y. Fujiwara, B.-K. Choi, K. Kubota, T. Miyamae, Macromolecules 36 (2003) 6722
4. M. Zhang, P. Lu, X. Wang, L. He, H. Xia, W. Zhang, B. Yang, L. Liu, L. Yang, M. Yang, Y. Ma, J. Feng, D. Wang, N. Tamai, J. Phys. Chem. B 108 (2004) 13185
5. C.N.R. Rao, A.K. Cheetham, J. Mater. Chem. 11 (2001) 2887
6. M.-C. Daniel, D. Astruc, Chem. Rev. 104 (2004) 293
7. G. Schmid, M. Bäumle, M. Geerkens, I. Heim, C. Osemann, T. Sawitowski, Chem. Soc. Rev. 28 (1999) 179
8. Z. Peng, L. Yu, J. Am. Chem. Soc. 118 (1996) 3777
9. Y. Liu, S. Jiang, K. Glusac, D.H. Powell, D.F. Anderson, K.S. Schanze, J. Am. Chem. Soc. 124 (2002) 12412

10. Y. Zhu, M.O. Wolf, Chem. Mater. 11 (1999) 2995
11. J.S. Wilson, A.S. Dhoot, A.J.A.B. Seeley, M.S. Khan, A. Kohler, R.H. Friend, Nature 413 (2001) 828
12. K.A. Walters, L. Trouillet, S. Guillerez, K.S. Schanze, Inorg. Chem. 39 (2000) 5496
13. O. Clot, M.O. Wolf, B.O. Patrick, J. Am. Chem. Soc. 123 (2001) 9963
14. C. Moorlag, M.O. Wolf, C. Bohen, B.O. Patrick, J. Am. Chem. Soc. 127 (2005) 6382
15. D.D. Graf, K.R. Mann, Inorg. Chem. 36 (1997) 150
16. C. Moorlag, O. Clot, M.O. Wolf, B.O. Patrick, Chem. Commun. 24 (2002) 3028
17. C.D. Dimitrakopoulos, P.R.L. Malenfant, Adv. Mater. 14 (2002) 99
18. A. Facchetti, M. Mushrush, M.-H. Yoon, G.R. Hutchison, M.A. Ratner, T.J. Marks, J. Am. Chem. Soc. 126 (2004) 13859
19. M. Fujita, K. Umemoto, M. Yoshizawa, N. Fujita, T. Kusukawa, K. Biradha, Chem. Commun. (2001) 509
20. S. Leininger, B. Olenyuk, P.J. Stang, Chem. Rev. 100 (2000) 853
21. T.L. Stott, M.O. Wolf, A. Lam, Dalton Trans. (2005) 652
22. T.L. Stott, M.O. Wolf, Inorg. Chem. 44 (2005) 620
23. J.H. Youk, J. Locklin, C. Xia, M.-K. Park, R. Advincula, Langmuir 17 (2001) 4681
24. D. Patton, J. Locklin, M. Meredith, Y. Xin, R. Advincula, Chem. Mater. 16 (2004) 5063
25. Y. Zhou, H. Itoh, T. Uemura, K. Naka, Y. Chujo, Langmuir 18 (2002) 277
26. S. Holdcroft, B.L. Funt, J. Electroanal. Chem. Interfacial Electrochem. 240 (1988) 89
27. K.M. Kost, D.E. Bartak, B. Kazee, T. Kuwana, Anal. Chem. 60 (1988) 2379
28. Z. Peng, E. Wang, S. Dong, Electrochem. Commun. 4 (2002) 210
29. B.C. Sih, A. Teichert, M.O. Wolf, Chem. Mater. 16 (2004) 2712
30. J.-P. Bourgoin, C. Kergueris, E. Lefevre, S. Palacin, Thin Solid Films 327–329 (1998) 515
31. J.M. Wessels, H.-G. Nothofer, W.E. Ford, F. von Wrochem, F. Scholz, T. Vossmeyer, A. Schroedter, H. Weller, A. Yasuda, J. Am. Chem. Soc. 126 (2004) 3349
32. P.V. Kamat, J. Phys. Chem. B 106 (2002) 7729
33. X. Li, Y. Li, Y. Tan, C. Yang, Y. Li, J. Phys. Chem. B 108 (2002) 5192
34. T. Shioya, T.M. Swager, Chem. Commun. (2002)
35. C. Lamy, J.-M. Leger, In: Wieckowski A, Savinova ER, Vayenas CG (eds) Catalysis and Electrocatalysis at Nanoparticle Surfaces, Marcel Dekker, New York, p 907
36. J.-E. Park, S.-G. Park, A. Koukitu, O. Hatozaki, N. Oyama, Synth. Met. 140 (2004) 121
37. J.H. Park, Y.T. Lim, O.O. Park, J.K. Kim, J.-W. Yu, Y.C. Kim, Chem. Mater. 16 (2004) 688

Redox Active Architectures and Carbon-Rich Ruthenium Complexes as Models for Molecular Wires

Stéphane Rigaut · Daniel Touchard · Pierre H. Dixneuf

Institut de Chimie de Rennes, UMR 6226 CNRS – Université de Rennes 1: Sciences Chimiques de Rennes, Campus de Beaulieu, 35042 Rennes Cedex, France

Summary Recent innovations in the field of carbon-rich organometallic ruthenium systems at the frontier of materials science are presented. Redox-active ruthenium compounds are used for the synthesis of bimetallic complexes as models for conductive molecular wires including diynyl, bis-allenylidene bridges, and original odd-numbered carbon chains via innovative C–C bond coupling reactions. These systems also appear as building blocks to connect two carbon-rich chains and to mediate electron transfer processes between functionalities at the end of these chains. These models constitute new keys to reach functionalized materials composed of transition metals and carbon-rich chains and bridges in conjugated oligomers or star molecules.

Abbreviations

Bu	Butyl
Cat	Catalyst
CV	Cyclic voltametry
DBU	1,8-Diazabicyclo[5.4.0]undec-7-ene
DFT	Density functional theory
dppe	1,2-Bis(diphenylphosphino)ethane
dppm	Bis(diphenylphosphino)methane
ESR	Electron spin resonance
Me	Methyl
MM	Molecular mechanical
Nu	Nucleophile
QM	Quantum mechanical
Ph	Phenyl
i-Pr	Isopropyl
SCE	Saturated calomel electrode
Tf	Trifouromethanesulfonyl
THF	Tetrahydrofuran
V	Volt

4.1
Introduction

Nanometer-scale devices, as the possible next-generation devices of electron-ics, have recently experienced worldwide attention and rapid development [1]. In this field, coordination complexes with π-conjugated bridges have gained importance in the building of supramolecular architectures [2,3], polymers [4], wires [3,5–21], dyes [22], and material with magnetic [5b,23] or nonlinear op-tical [24] properties. Their rich redox chemistry and the unique or unusual structural and electronic features of each redox state are often the source of these applications [7,25].

Of particular interest are organometallic molecular wires with C_xH_y carbon-rich bridges connected with σ metal-carbon bonds [5,8–21] to perform the simplest function of molecular electronics, i.e., to mediate electron transfer (Scheme 4.1).

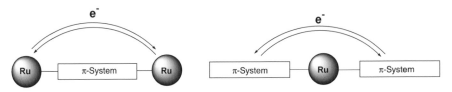

Scheme 4.1.

Several studies point out the fact that mixed valence forms of various bimetallic complexes are unstable when the bridge is too long [5a,11a]. Therefore, carbon-rich *trans*-ditopic structures [26–31] constitute an interest-ing alternative to simple metal-capped carbon chains, allowing connectivity to other components (Scheme 4.1). However, several metals in rod-shaped organometallic compounds containing two unsaturated chains disrupt the π-conjugation between two carbon-rich chains [31].

Ruthenium allenylidene and acetylide metal complexes have attracted in-terest as building units toward original carbon-rich architectures [8,14–17,32] related to materials science [24a,e,33] due to their rigidity and one-dimensional nature, and as bridges for electronic communication between a metal site and a remote functionality [22d]. In addition, metal allenylidenes are now cur-rently leading to useful applications in the field of homogeneous catalysis [34], as catalytic precursors especially for alkene metathesis [35, 36], and selec-tive propargylation [37]. Our group has been involved in the building of new carbon-rich mono- and polymetallic complexes using the stable redox-active *trans*-[ClRu(dppe)$_2$]$^+$ system [38]. Organometallic complexes generated from this fragment are attractive building blocks for several reasons: the systems display reversible redox behavior and the chloride can be easily substituted with another carbon-rich chain, a valuable property for obtaining extended systems. Thus, the ruthenium-containing terminal alkynes and diynes have the potential to build polymetallic rodlike oligomers, with stable redox systems separated by carbon-rich bridges. Rigid and flexible triynes and their activation processes offer access to star organometallics with multiple identical ruthenium units.

The purpose of this review is to present recent innovations achieved by the Rennes group in the field of carbon-rich ruthenium complexes and to show that the redox-active carbon-rich ruthenium systems with vinylidenes and allenylidene-to-acetylide linkages are valuable building blocks for access to long one-dimensional nanoscaled organometallics and to mediate original electron transfer processes. The ruthenium building blocks with their basic chemical and electrochemical properties will be discussed first, followed by a presentation of bimetallic and oligomeric models with unusual, stable topologies of type bis-allenylidene, bis-acetylide, and bis-carbyne conjugated bridges, with their properties highly dependent on the nature of the bridging ligand. The possibility of using the ruthenium center as an electron relay will be described in a bis-allenylidene with efficient electronic communication between the two carbon-rich chains. Finally, the building of tris-ruthenium complexes from triynes will show that the rigid or flexible tripodal carbon-rich bridge inhibits communication between the ruthenium units.

4.2
Ruthenium Allenylidene and Acetylide Building Blocks: Basic Properties

4.2.1
Synthetic Routes

Several routes have been opened leading to ruthenium cumulenic complexes trans-$[Cl(dppe)_2Ru(=C)_nR_1R_2]^+$ or acetylide complexes trans-$[Cl(dppe)_2Ru-(C\equiv C)_n-R]$ [16,27], following our initial work on the less stable $RuCl_2(dppm)_2$ system [38].

Among cumulene complexes $[M](=C)_nR_1R_2]$, allenylidene compounds ($n = 3$) constitute the most extensive family [39 a,c]. Their most general method of access is based on the reaction of 2-propyn-1-ol derivatives with a 16-electron metal center [40]. Several complexes starting from $RuCl_2(dppe)_2$ or from the 16-electron species $[(dppe)_2RuCl][X]$ ($X = PF_6$, BF_4, or CF_3SO_3) [27a] (Scheme 4.2) were prepared. It is worth noting that tertiary allenylidenes are more stable than secondary allenylidenes.

Ruthenium acetylide complexes were obtained either by deprotonation of vinylidene intermediates resulting from the η^2-coordination of a terminal alkyne on the 16-electron complex (Scheme 4.3) [27b] or by the classical reaction of lithium acetylide with a metal halide complex cis-$[RuCl_2(dppe)_2]$ [16]. The latter method is used mainly to obtain air-stable poly-ynyl complexes.

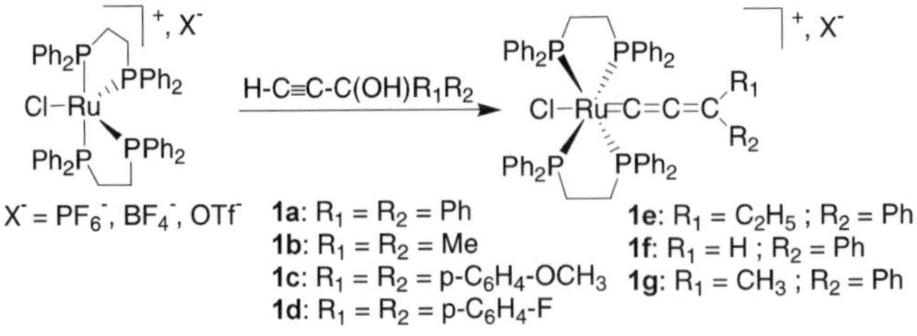

Scheme 4.2.

A major advantage of these ruthenium systems is the possible substitution of the second chlorine atom in neutral alkynyl complexes to introduce a second unsaturated carbon chain in the *trans* position [16,27a,b]. The best results to give access to mixed alkynyl allenylidene 4 and bis-alkynyl complexes 5 were obtained using the strategy depicted in Scheme 4.4, starting from the corresponding vinylidene precursor in the presence of a base, a noncoordinating salt, and a propargylic alcohol or a terminal alkyne. Symmetric acetylides can also be obtained in one step from the RuCl$_2$(dppe)$_2$ precursor as exemplified with bis-polyynes 6 and 7 (Scheme 4.4) [16].

The reactivity of metal allenylidene complexes is of special interest in a fundamental viewpoint, but also for the future building of nanoscaled architectures [39]. This reactivity is highly dependent on the electron richness of the metal moiety: electrophilic ruthenium complexes lead to highly reactive allenylidenes with electrophilic C$_\alpha$ and C$_\gamma$ centers, while the C$_\beta$ atom is a nucleophilic site. For example, they easily add alcohol to give α-, β-unsaturated alkoxy carbenes [39c,41,42]. With bulky and electron-releasing ligands such as dppe, their reactivity involves the γ-carbon atom with charged nucleophiles only [27a,43] (Scheme 4.5).

The synthesis of Fischer-type carbyne complexes has always been a laborious process [44–49]. Actually, ruthenium carbynes are very scarce [35, 46–48]

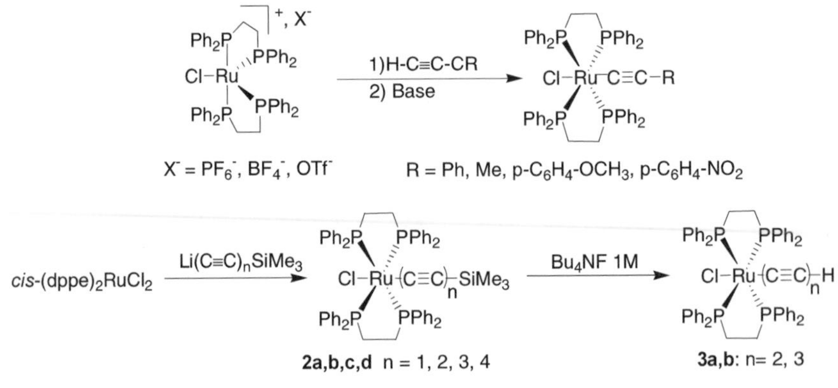

Scheme 4.3.

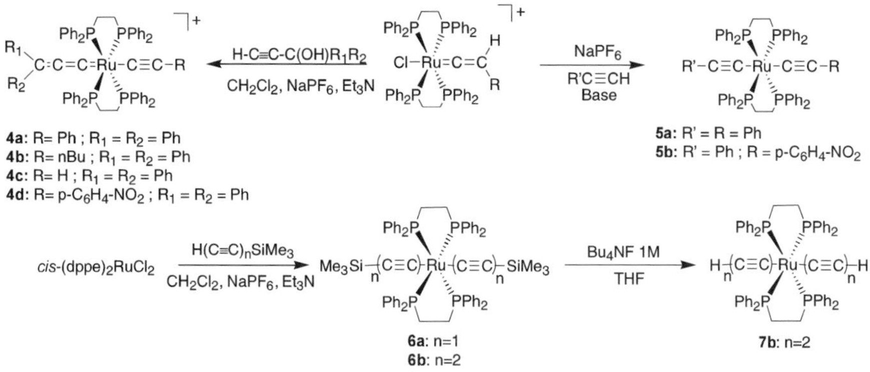
H-C≡C-C(OH)R1R2
CH2Cl2, NaPF6, Et3N
NaPF6
R'C≡CH
Base
4a: R= Ph ; R1 = R2 = Ph
4b: R= nBu ; R1 = R2 = Ph
4c: R= H ; R1 = R2 = Ph
4d: R= p-C6H4-NO2 ; R1 = R2 = Ph
5a: R' = R = Ph
5b: R' = Ph ; R = p-C6H4-NO2
cis-(dppe)2RuCl2
H(C≡C)nSiMe3
CH2Cl2, NaPF6, Et3N
Bu4NF 1M
THF
6a: n=1
6b: n=2
7b: n=2

Scheme 4.4.

unlike their osmium analogs [49], despite their interest as catalyst in olefin metathesis [47c,48b]. Synthetic routes usually involve the conversion of coordinated carbenes such as vinylidenes [47] and allenylidenes [48]. We recently reported the amphoteric properties of ruthenium allenylidenes containing a γ-methyl group to provide alkenyl carbyne ruthenium species without decoordination of any ligand (Scheme 4.6) [50]. Thus, **1b,g** were easily deprotonated on the δ carbon by action of a base to give the alkenylacetylide complexes **10b** or **10g**, a reaction that constitutes another route to ruthenium acetylide complexes. The addition of an excess of HBF_4 to **1b,g** led to the alkenylcarbyne complexes **11b,g**, and no rearrangement into an in-

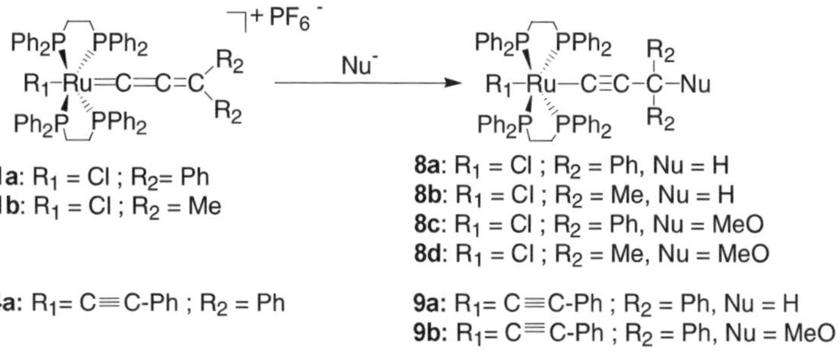
Nu-
1a: R1 = Cl ; R2= Ph
1b: R1 = Cl ; R2 = Me
4a: R1= C≡C-Ph ; R2 = Ph
8a: R1 = Cl ; R2 = Ph, Nu = H
8b: R1 = Cl ; R2 = Me, Nu = H
8c: R1 = Cl ; R2 = Ph, Nu = MeO
8d: R1 = Cl ; R2 = Me, Nu = MeO
9a: R1= C≡C-Ph ; R2 = Ph, Nu = H
9b: R1= C≡C-Ph ; R2 = Ph, Nu = MeO

Scheme 4.5.

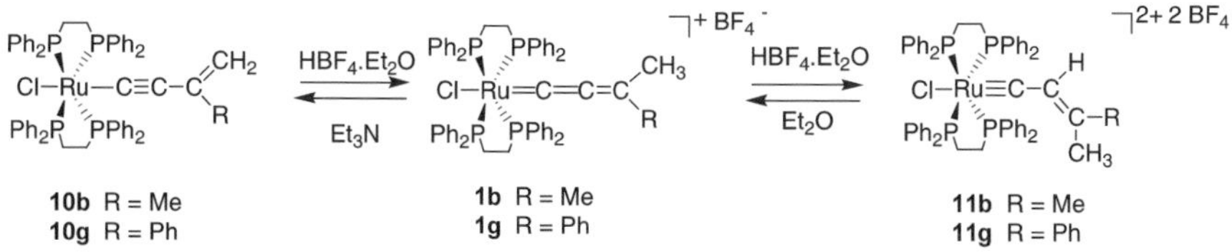
HBF4.Et2O
Et3N
HBF4.Et2O
Et2O
10b R = Me
10g R = Ph
1b R = Me
1g R = Ph
11b R = Me
11g R = Ph

Scheme 4.6.

denylidene complex was observed. In addition, UV-vis spectra show a significant 60-nm shift of the metal-to-ligand charge transfer band (MLCT) from $\lambda_{max} = 482$ nm for **1g** to $\lambda_{max} = 426$ nm for **11g** as a consequence of the conjugation change.

4.2.2
Redox Properties

4.2.2.1
Oxidation of Ruthenium Metal Acetylides: Stable Ru^{II}/Ru^{III} Systems and a New Route to Allenylidene Metal Complexes

The first oxidation of ruthenium acetylides could be viewed as essentially involving the two Ru^{II}/Ru^{III} couples, but with a rising chain character as the length increases [51–53]. For example, cyclic voltammetry (CV) shows that the *trans*-[Cl(dppe)$_2$Ru-C≡C-CHPh$_2$] (**8a**), *trans*-[Cl(dppe)$_2$Ru-C≡C-CH(CH$_3$)$_2$] (**8b**), and *trans*-[Cl(dppe)$_2$Ru-C≡C-CH$_3$] (**8c**) complexes undergo first a well-defined one-electron reversible oxidation on the CV scale (Table 4.1, Scheme 4.7) [54]. All complexes give a second irreversible oxidation at much higher potential attributed to the Ru^{III}/Ru^{IV} system.

We observed that the stability of the first cationic radical is highly influenced by the nature of R. Stable cationic radicals were formed when R = CH$_3$, H. By contrast, chemical oxidation of **8a** gave a mixture of the corresponding allenylidene **1a** and vinylidene **9** (Scheme 4.8). Oxidation of the same complex with

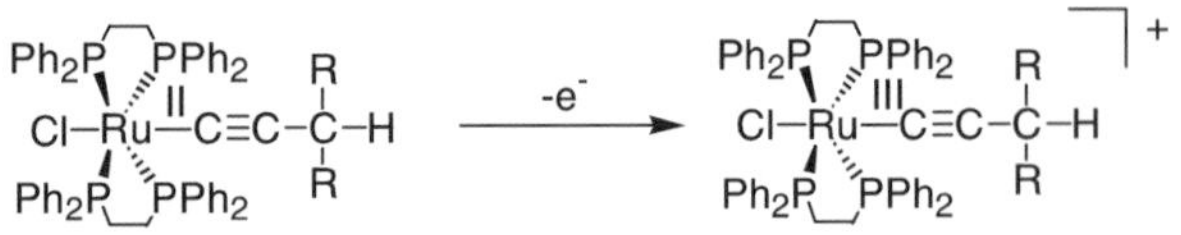

8a: R = Ph
8b: R = Me
8c: R = H

Scheme 4.7.

Table 4.1. Electrochemical data for acetylides and allenylidenes

Complex	E°_{ox1} [a]	$E_{pa\ ox2}$ [b]	E°_{RED} [a]
[Cl(dppe)$_2$Ru-C≡C-CHPh$_2$](**8a**)	0.02	1.07	
[Cl(dppe)$_2$Ru-C≡C-CH(CH$_3$)$_2$](**8b**)	−0.07	1.03	
[Cl(dppe)$_2$Ru-C≡C-CH$_3$](**8c**)	−0.07	0.99	
[Cl(dppe)$_2$Ru=C=C=CPh$_2$]PF$_6$ (**1a**)	0.99[c]		−1.03
[Cl(dppe)$_2$Ru=C=C=CMe$_2$]PF$_6$ (**1b**)			−1.27[b]

*Measures were performed in CH$_2$Cl$_2$ using Bu$_4$NPF$_6$ 0.1 M, $v = 100$ mV s^{-1}. [a] Redox processes are quasireversible $\Delta E_p \approx 60$ mV, $I_{pc}/I_{pa} \approx 1$. Redox potentials are reported in V vs. ferrocene. [b] Peak potential of irreversible processes. [c] $\Delta E_p \approx 75$ mV, $I_{pc}/I_{pa} < 1$

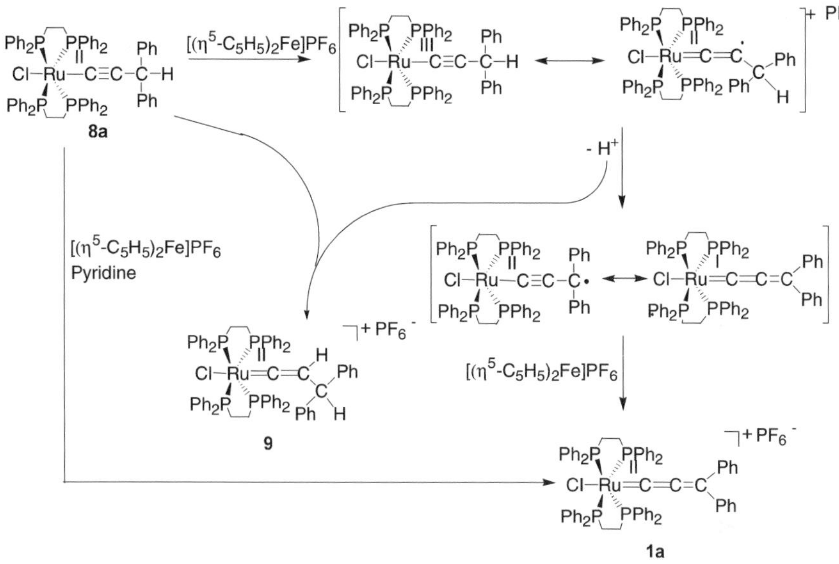

Scheme 4.8.

an excess of oxidant, in the presence of pyridine, led quantitatively to **1a** by trapping the released proton to avoid the formation of **9**. This oxidation reaction of an alkynyl-ruthenium with a hydrogen atom on a γ-carbon constitutes a novel route to allenylidene metal complexes, especially when propargylic alcohols are not available, and appeared to be valuable for the construction of a molecular-wire model (vide supra).

The neutral acetylide radical *trans*-[Cl(dppe)$_2$Ru-C≡C-C$^\bullet$Ph$_2$] resulting from the γ-proton elimination was proposed as an intermediate that is further oxidized to give the stable allenylidene **1a** on the basis of ESR spectroscopy (vide supra) [43]. With the phenyl groups, the radical is highly stabilized over the rings, and proton elimination is preferred.

4.2.2.2
Reduction of Metal Allenylidenes:
Access to Stable "Organic" Radicals and a Route to Acetylides

Stimulated by the fact that radical species have been recently proposed as intermediates in metathesis [55], and by their potential as building blocks for conductive molecular wires we studied the reduction of allenylidene complexes (Table 4.1) [43]. At 293 K, ESR experiments of the first reduced form of complex **1b** evidenced a poorly resolved quintet (g = 2.0097, a_P = 3.0 G) in a characteristic region for organic radicals, showing the hyperfine coupling with the four phosphorus nuclei (Fig. 4.1a). The reduced form of **1a** generated an intense and persistent complex signal (g = 2.0042) owing to the coupling

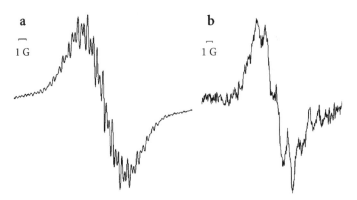

Fig. 4.1. ESR spectra resulting from reduction of *trans*-[Cl(dppe)$_2$Ru=C=C=CPh$_2$]PF$_6$ (**1a**), g = 2.0042 (**a**) and (**b**) [Cl(dppe)$_2$Ru=C=C=CMe$_2$]PF$_6$, g = 2.0097 (**1b**)

of the unpaired electron with the four phosphorus nuclei on the one hand and further coupling with phenyl hydrogen atoms of the carbon-rich chain (Fig. 4.1b) on the other.

Addition of Ph$_3$SnH to the radical led to the formation of neutral acetylide compounds **8a** and **8b**, demonstrating that radical trapping only occurs at the end of the carbon skeleton (Scheme 4.9). These data added to IR studies indicate the organic nature of the radicals and illustrate that significant radical stabilization on the cumulene chain takes place at the trisubstituted carbon atom. These results are consistent with other allenylidene systems [52a–c,56] and can be generalized to longer cumulenes as identical results were obtained with *trans*-[Cl(dppe)$_2$Ru=C=C=C=C=CPh$_2$]PF$_6$ [43].

DFT calculations have been performed on the [Cl(PH$_3$)$_4$RuC$_n$H$_2$]$^+$, Cl(PH$_3$)$_4$RuC$_n$H$_2$, and [Cl(PH$_3$)$_4$RuC$_n$H$_2$]$^-$ (n = 1–8) series and are in agreement with all our experimental data [57]. The distribution of the atomic net charges and the localization of the HOMOs and LUMOs indicate that the [Cl(PH$_3$)$_4$RuC$_n$H$_2$]$^+$ complexes are subject to nucleophilic attack at odd C atoms, except in the case of C(1). They are also subject to electrophilic attack at the even C sites, except for the CH$_2$ end of the chain. The one-electron reduction of the [Cl(PH$_3$)$_4$RuC$_n$H$_2$]$^+$ compounds leads to neutral species containing linear C$_n$H$_2$ chains, with the exception of the n = 2 com-

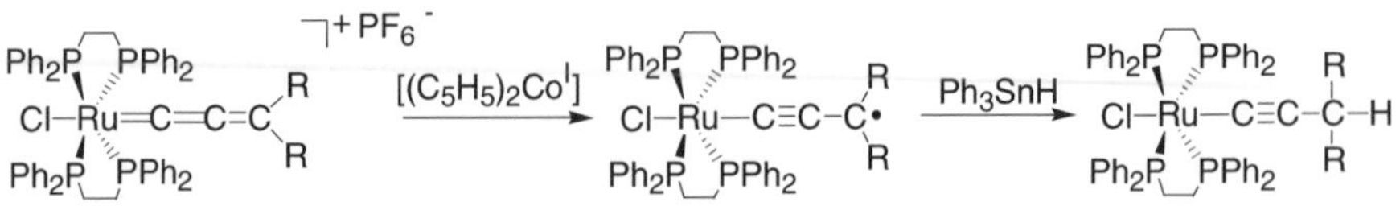

Scheme 4.9.

plex, which is bent at C(1). These $Cl(PH_3)_4RuC_nH_2$ compounds are better described as 18-electron Ru^{II} metal sites bonded to the reduced $(C_nH_2)^-$ ligand than as 19-electron Ru^I centers. These findings are also in agreement with theoretical studies on ruthenium dppm systems [56c] and on other analogous allenylidenes $[(\eta^5\text{-}C_5H_5)(CO)(PPh_3)Ru=C=C=CH_2]^+$ and $[(\eta^5\text{-}C_9H_7)(PPh_3)_2Ru=C=C=CH_2]^+$, in particular suggesting that the LUMO is strongly located on the cumulenic ligand [53].

4.3
Bimetallic Complexes from the Ru(dppe)$_2$ System

The previous properties displayed by the ruthenium allenylidene and alkynyl complexes were used for the profit of controlled building of bimetallic compounds bridged by carbon-rich bridges. Since most bimetallic systems led to oxidizable acetylides $L_nMC_xML_n$ or polyenes $L_nM(CH)_xML_n$ [21], the search for new π-conjugated bridges between two reversible, reducible metal complexes was attempted to create original binuclear allenylidene compounds [17]. These investigations led us to discover two novel methodologies of C–C bond formation giving a new class of carbon-rich homobimetallic complexes with seven conjugated carbons between remote metals and a charge highly delocalized over the extended conjugated structure [15]. A bis-acetylide was also obtained from a terminal triyne [16] in order to compare the Ru(dppe)$_2$ system with other systems [9–13].

4.3.1
A Binuclear Bis-Acetylide Ruthenium Complex

Eglinton or Hay coupling is the method of choice to obtain bimetallic complexes from terminal polyynes by coupling of the terminal carbon atoms [5]. If this method could not lead to a bimetallic complex with the diyne *trans*-[Cl(dppe)$_2$Ru-(C≡C)$_2$-H] (**3a**), the triyne *trans*-[Cl(dppe)$_2$Ru-(C≡C)$_3$-H] (**3b**) in the presence of copper(II) provided the RuC$_{12}$Ru (**11**) analog with 12 carbon atoms between the metallic centers (Scheme 4.10) [16].

Complex RuC$_{12}$Ru **11** displays characteristics very similar to those of the homonuclear complexes of comparable length (Fe, Pt, Re) [10,11a,58]. In-

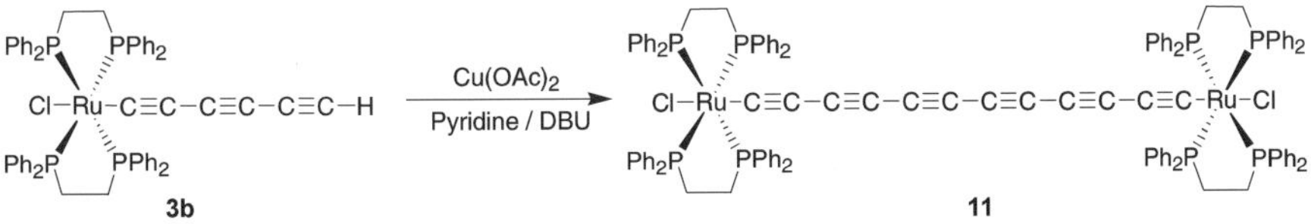

Scheme 4.10.

deed, two oxidation waves are observed at lower potential than that of *trans*-[Cl(dppe)$_2$Ru-(C≡C)$_2$-H] (**3a**) ($E°$ = 0.130 V vs. ferrocene), attributed to the two Ru^{II}/Ru^{III} oxidations in a first approximation, and thus to successive formations of the cationic species RuC$_{12}$Ru$^+$ ($E_1°$ = −0.05 V) and the dicationic species Ru$^+$C$_{12}$Ru$^+$ ($E_2°$ = 0.18 V, $\Delta E°$ = 230 mV). When two identical redox centers are isolated from each other in the same molecule, the potentials E_1 and E_2 would be expected to be not only identical or of close values but also very close to the potential of a mononuclear model. Thus the CV data evidence a substantial coupling between the two remote metal ends in **11**, i.e., the efficient conjugated path between two equal redox moieties makes the processes easier than for the mononuclear model by stabilization of the electrons along the system. Furthermore, in contrast to the reported redox data on similar or shorter complexes, these processes are fully reversible. This phenomenon is accredited to the electronic and steric protections of the chain by the dppe ligands.

4.3.2
Bis-Allenylidene Bridges Linking Two Ruthenium Complexes

Bis-propargylic alcohols separated by a conjugated linker are possible carbon-rich unsaturated bridge precursors for access to bimetallic complexes. The synthesis and electrochemical study of new stable bimetallic complexes such as *trans*-[Cl(dppe)$_2$Ru=C=C=CPh-*p*-C$_6$H$_4$-CPh=C=C=Ru(dppe)$_2$Cl](OTf)$_2$ (**14**) (Scheme 4.11) containing two identical metal moieties with a bis-allenylidene bridge were obtained via the double activation of propargylic alcohols **12a–f** with the 16 electron species [ClRu(dppe)$_2$)]OTf [17]. For stability reasons, tertiary allenylidenes were targeted, and connecting groups [T] such as oligo-thiophene, arene, or triple bond were employed. It is worth noting that, except for a very recent report [19], only two conjugated bis-allenylidene systems have been reported since our pioneering work [59], but without redox studies [60].

Electrochemical studies show that the mono-allenylidene species undergo one-electron reduction and the bis-allenylidenes two one-electron reductions (Fig. 4.2a, Table 4.2). The behavior of the conjugated bis-allenylidenes shows two interesting features: (1) the two reversible processes are easily distinguishable ($\Delta E° \geq$ 180 mV) and (2) a significant anodic shift of both reduction potentials indicates that the reductions are thermodynamically favored with respect to the mononuclear analog. No such anodic shift is observed in the nonconjugated complexes **14b** in which the allenylidene moieties are in a *meta* orientation on the phenyl linker. This indicates the efficient connection of the allenylidene moieties through the highly conjugated bridges and, more specifically, the considerable stabilization of the reduced forms by resonance.

By analogy with allenylidenes such as *trans*-[Cl(dppe)$_2$Ru=C=C=CPh$_2$]PF$_6$ (**1a**) these reductions can be ascribed to two successive reductions of the carbon-rich ligand. This was verified for **14a**. For example, the first reduced

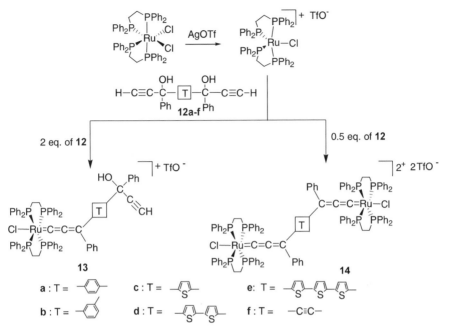

Scheme 4.11.

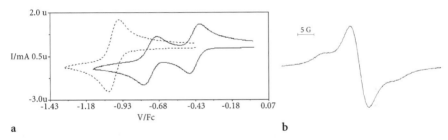

Fig. 4.2. **a** Cyclic voltammetry for monoallenylidene **13a** (...) and bis-allenylidene **14a** (−) complexes. **b** ESR spectra resulting from the monoreduction of **14a** ($g = 2,0244$)

species led to an ESR signal at room temperature ($g = 2.0244$) with shoulders attributed to the coupling of the free electron with the ^{99}Ru and ^{101}Ru isotopes of both remote ruthenium atoms (Fig. 4.2b). This result suggests that the radical is highly centered on the organic bridge and stabilized by delocalization along the carbon bridge between the ruthenium centers. The second reduction led to a diamagnetic molecule with characteristics consistent with a bis-acetylide *trans*-[Cl(dppe)$_2$Ru-C≡C-CPh-4-C$_6$H$_4$-CPh-C≡C-Ru(dppe)$_2$Cl] (**15**), which evidences a second reduction of the ligand. Hence, it is likely that all the electrons issued from reduction processes reside in an orbital with a significant carbon-rich ligand character but also with a nonnegligible metal contribution [51].

Table 4.2. Electrochemical data for **13** and **14**

	E_1° [V]	ΔE_p [mV]	E_2° [V]	ΔE_p [mV]	$E_2^{\circ} - E_1^{\circ}$ [mV]	K_c
13a	−0.910	60				
14a	−0.500	60	−0.680	60	180	1.0×10^3
13b	−0.910	60				
14b	−0.877	60	−1.056	60	180	1.0×10^3
13c	−1.000	66				
14c	−0.439	66	−0.749	90	310	1.5×10^5
13d	−1.012	60				
14d	−0.434	66	−0.746	90	320	2.2×10^5
13e	−1.027	60				
14e	−0.427	66	−0.760	69	330	3.2×10^5
13f	−0.805	60				
14f	−0.285	60	−0.645	60	360	1.0×10^6

*Sample, 1 mM; Bu$_4$NPF$_6$ (0.1 M) in CH$_2$Cl$_2$; $v = 100$ mV s^{-1}; potentials are reported in V vs. ferrocene as an internal standard. Reversible redox processes

According to common belief, the difference of electrochemical half-wave potential reflects the strengths of interactions between the allenylidene moieties. There are, however, various additional factors contributing to a stabilization of symmetrical mono- and bireduced complexes [5a]. We observed that the separation between the two processes (ΔE°) also depends on the nature of the transmitter. A ΔE° value of 180 mV ($K_c = 1.0 \times 10^3$) is observed with the nonconjugated system **14b**. Surprisingly, the conjugated p-phenyl linker (**14a**) displays the same value. Therefore, this phenomenon is principally attributed to structural rearrangement, solvatation, or electrostatic interactions. With complexes **14c–f**, ΔE° is noticeably higher, and this significant increase is ascribed to an additional contribution due to the resonance interaction between the allenylidene moieties through the linkers. Thus, in addition to the highest anodic shift, the less aromatic linkers with the shorter pathway display the higher potential differences between the reduction processes in the following order: triple bond > thiophene ≈ bithiophene ≈ terthiophene > phenyl. The above study demonstrates that these molecules constitute interesting building blocks for the construction of more complex redox-active edifices.

These carbon-rich bridged complexes show an interesting reactivity pattern. As monometallic allenylidenes, they can be either protonated or deprotonated if hydrogen atoms are present on the δ-carbon atoms. The same concept of reversible deprotonation/protonation was then applied to the binuclear complex **16** [50]. It was reversibly deprotonated into the conjugated binuclear complex **16**, and the addition of an excess of HBF$_4$ to **16** led to its complete double protonation to provide the first bis-alkenylcarbyne ruthenium complex

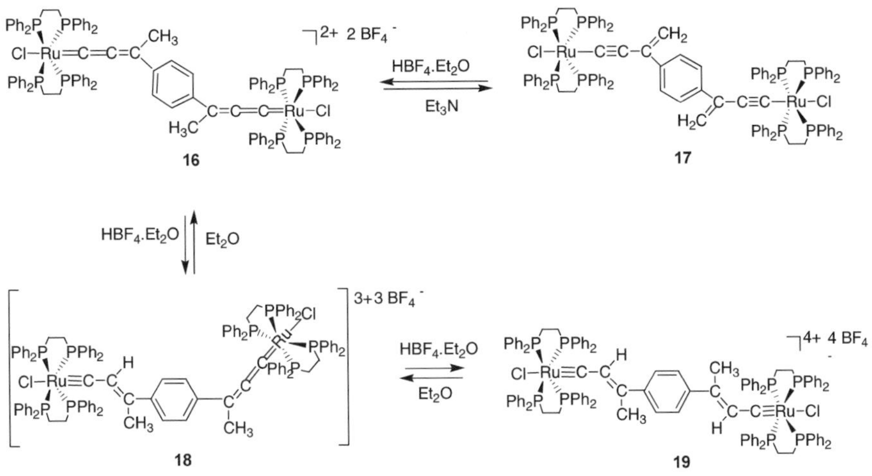

Scheme 4.12.

19 (Scheme 4.12). This double protonation of **16** (λ_{max} = 589 nm) resulted in a red shift of the MLCT band (**19**: λ_{max} = 496 nm), illustrating again the influence of the nature of the bridge on the properties of the complexes [50]. However, whereas complex **19** is stable at room temperature in the presence of an excess of a strong acid, it is very acidic and easily deprotonated by a solvent.

4.3.3
C$_7$ Bridged Binuclear Ruthenium Complexes

If a variety of bimetallic complexes allowing the exchange of electrons through an even-numbered carbon-rich bridge such as $L_nMC_xML_n$ or $L_nM(CH)_xML_n$ have been reported [3, 5–21], only a limited number of complexes with an odd-numbered linear or cyclic carbon bridge linking the two metal moieties was described [61–63]. The most general process affords rigid four-membered cyclic bridges with a delocalized C$_3$ path between metals via the regioselective [2 + 2] cycloaddition between the C=C bond of a vinylidene [M]=C=CHR and the C$_\alpha$≡C$_\beta$ bond of a metal acetylide [M]-C≡C-R (Scheme 4.13) [61]. The C$_\alpha$-C$_\beta$ bond was generally the most activated, and when the reaction was applied to an allenylidene [M=C=C=CR$_1$R$_2$] complex [61b] or a diyne [M-C≡C-C≡C-R] complex [61c], a similar C$_3$ conjugated path was formed with an exocyclic double or triple bond, inhibiting access to higher odd-numbered carbon-rich bridges.

During our investigation on bimetallic systems, we obtained novel classes of carbon-rich homobimetallic complexes via two new C–C bond-formation methodologies. The first one provides a carbon-rich homobimetallic complex

Scheme 4.13.

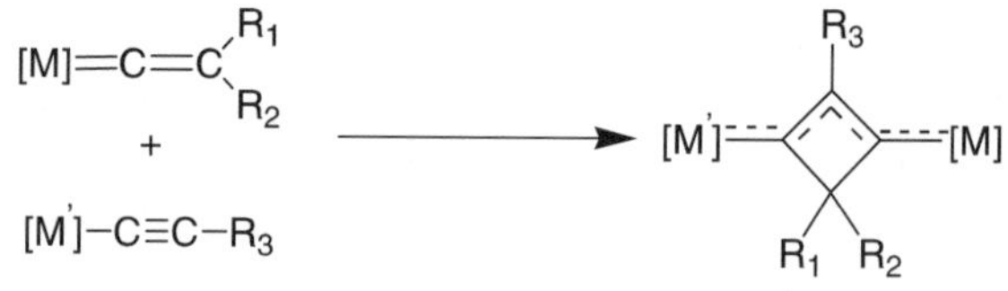

with seven conjugated carbons, including an annelated C_8H_3 bridge [15a], while the second leads to W-shaped complexes [15b].

This first method of access consists in a radical promoted coupling reaction occurring on the $C_\gamma\equiv C_\delta$ bond of a 1,3-diynylmetal derivative. More specifically, chemical oxidation of *trans*-[Cl(dppe)$_2$Ru-(C$\equiv$C)$_2$-H] (**3a**) or *trans*-[Cl(dppe)$_2$Ru-(C$\equiv$C)$_2$-SiMe$_3$] (**2b**) with ferricinium hexafluorophosphate leads to the annelated bimetallic complex **20** via a cycloaddition of the terminal $C_\gamma\equiv C_\delta$ triple bonds (Scheme 4.14). The spectroscopic and RX data evidence a structure intermediate between those of the two mesomeric forms sketched in Scheme 4.14 [15a].

The unprecedented regioselectivity observed for the cyclic addition is explained by the steric hindrance of the $C_\alpha\equiv C_\beta$ bond in the diyne complexes. It is anticipated that despite the unfavorable potential, the reaction is initiated by an electron transfer between the ferricinium cation and **3a** or **2b**, generating an electrophilic metallacumulene radical intermediate. This oxidized form should correspond to Ru$^\bullet$-C$\equiv$C-C$\equiv$CH $\leftrightarrow$ Ru$^+$=C=C=C=C$^\bullet$H with a significant contribution of the cumulenic form that can react with another molecule of acetylide, before incorporation of hydrogen atoms from the medium (Scheme 4.14). In the case of **2b**, desilylation occurs during the process.

As highly reactive metallacumulene [Ru]=C=C=C=CR$_2$ intermediates [64] are probably involved in the formation of **20**, we attempted to react the metal diyne systems with other preformed cumulenic species, such as ruthenium allenylidenes. This led us to the second carbon–carbon coupling reaction. The

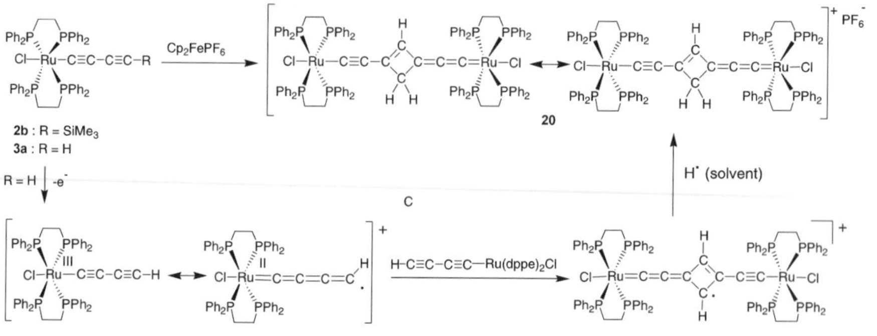

Scheme 4.14.

cationic allenylidene complexes **21a–c** were allowed to react with the diynyl compound **3a** in a metal-assisted C–C forming reaction to generate **22a–c** [15b]. These complexes were also obtained when the reaction was performed with the protected diynyl **2b**. The X-ray structure of **22a** established the delocalized nature of the products (Scheme 4.15).

A probable mechanism for this reaction is based on the fact that ruthenium allenylidenes with a –CH$_2$R$_1$ group on C$_\gamma$ are easily deprotonated into stable ruthenium acetylides. This mechanism is depicted on Scheme 4.15 for **22b**. The first step consists in a proton transfer from **21b** to the nucleophilic carbon C$_\delta$ of **3a** to form an unstable butatrienylidene complex [A] and the acetylide **23b**. A further fast addition of the nucleophilic C$_\delta$ of **23b** on the electrophilic site C$_\gamma$ of [A] leads to the intermediate [B]. The formation of **22b** would then result from an allylic hydrogen transfer.

These four new bimetallic compounds can be oxidized and reduced by contrast with the previous bimetallic complexes **11** and **14** (Table 4.3). They show a reversible one-electron oxidation wave followed by an almost reversible or irreversible second oxidation wave. To a first approximation, these two oxidation steps could be viewed as essentially involving the two Ru^{II}/Ru^{III} couples. The large separation of the redox processes ($\Delta E° = 650$ mV, $K_c = \exp(\Delta E° F/RT) = 1.50 \times 10^{11}$ for **22b**) establishes that all the monooxidized species are stable in solution with respect to disproportionation (mixed valence species) [5a,65] and supports the idea that the mixed valence (MV) forms of all of these complexes enjoy considerable stabilization due to delocalization. They also undergo a well-defined one-electron reduction process attributable to the reduction of the unsaturated carbon chain. This attribution is supported by ESR spectroscopy and by the fact that the reduction potential is highly influenced by the introduction of the phenyl group. All complexes are harder

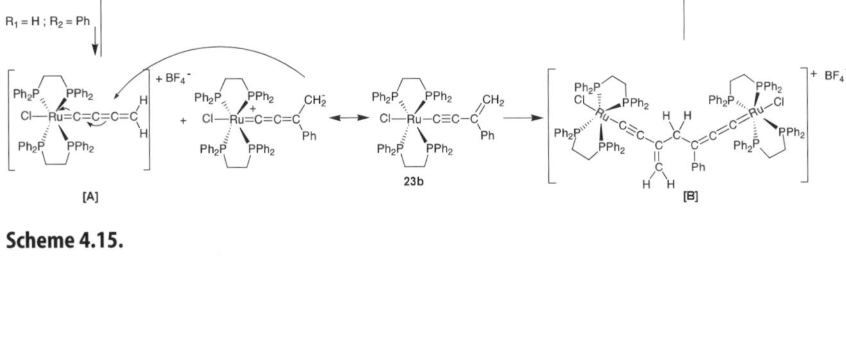

Scheme 4.15.

Table 4.3. Cyclic voltammetry and UV-vis data for complexes **20** and **22a–c**

	Electrochemistry[a]			UV-Vis
	E°_{red} (V)[b]	E°_{ox1} (V)[b]	E°_{ox2} (V)	λ_{max} [nm] (ε[mol^{-1} L cm^{-1}][f])
20	−1.48	0.42	0.91[c]	633 (141 000)
22a	−1.38	0.31	0.99[c]	710 (127 600)
22b	−1.24	0.32	0.97[d]	746 (98 000)
22c	−1.25	0.23	1.06[c]	764 (109 000)
24	−1.170	0.140	–	764 (63 000)
25	−1.240	0.120	–	782 (84 000)

[a] Sample, 1 mM; Bu$_4$NPF$_6$ (0.1 M) in CH$_2$Cl$_2$; v = 100 mV s^{-1}; potentials are reported in V vs. ferrocene as an internal standard. [b]Reversible redox processes $\Delta E_p \approx 60$ mV, $I_{pc}/I_{pa} \approx 1$. [c]Peak potential of an irreversible process. [d]Partially reversible peak $\Delta E_p \approx$ 100 mV, $I_{pc}/I_{pa} < 1$. [e]$\Delta E_p \approx 80$ mV. [f]In CH$_2$Cl$_2$

to oxidize than a neutral acetylide complex such as **3a** (E° = 0.130 V vs. ferrocene), easier to oxidize than a cationic allenylidene complex such as trans-[ClRu(dppe)$_2$=C=C=CPh$_2$]PF$_6$ (**1a**), and also harder to reduce than the latter. These observations are all consistent with a highly delocalized structure giving a formal half-positive charge on each ruthenium. Each complex displays a strong absorption band with λ_{max} values highly influenced by the nature of the bridge. For example, the band presents a batochromic shift from **20** to **22a** that displays a more effective conjugated path. The nature of the transition might be attributed to admixing of some charge transfer character of the acetylide and of the allenylidene moieties. In all cases the absorption bands are very broad, certainly including several transitions.

Finally, **22a–c** can be compared to the more localized bimetallic **24** and trimetallic **25** mixed acetylide-allenylidene complexes [23]. These complexes can be considered as bis-ruthenium species with a delocalized bridge allowing nine carbon atoms to connect a metal acetylide and one or two metal allenylidenes. They both show a one-electron ligand-centered reduction wave, and **24** displays a one-electron oxidation wave while **25** is oxidized twice at the same potential (two-electron wave) showing the lack of interaction between the two metal acetylides in an oxidation process. The reduction and oxidation potentials are (1) more favorable than those of **20** and **22** and (2) closer to those of the monometallic species **1a** and **3a**. Hence, as observed for the bis-allenylidenes, the aromatic cycle in the delocalized pathway and the slightly longer bridge lead to a decrease in the electronic coupling between the redox centers. The UV-vis spectra also present very high λ_{max} values with a large ε attributed to admixing of some charge transfer character of the acetylide to the allenylidene moieties.

All these studies led us to obtain new families of carbon-rich bimetallic complexes with novel alternative topologies of great interest for the building of nanoscaled molecular wires. The bis-allenylidene species along with unique annelated and W-shaped C_7 bridge arrangements are obtained under very mild conditions and display remarkable stabilities. The general trends observed with polymetallic bis-acetylides prevail with those species: (1) the existence of electronic communication between the metallic centers through the linker in a reduction or in an oxidation process if they are conjugated and (2) the influence of the structural nature of the bridge and of the efficiency of the delocalized path on the electronic properties of the complexes. These observations support the necessity of tuning a general system to find the best physical properties.

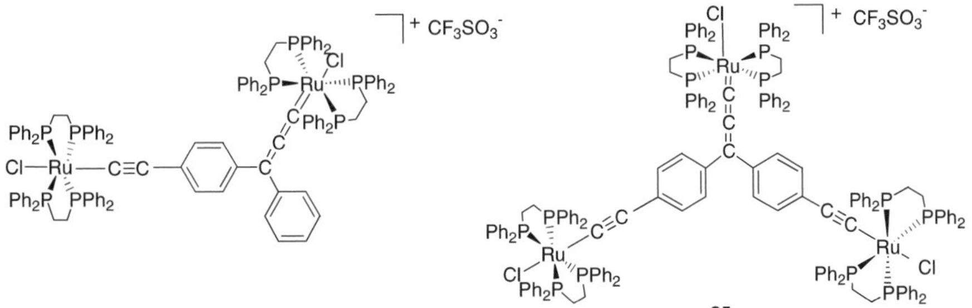

Scheme 4.16.

4.4
Connection of Two Carbon-Rich Chains with the Ruthenium System

Several studies indicate that mixed valence forms of various long bimetallic complexes are not chemically stable, owing to a higher reactivity of these forms in intra/interchain reactions or reactions with the solvent [5a,11a]. Therefore, the extension of a naked chain is not suitable to obtain long conductive molecular wires. An interesting alternative consists in the building of oligomeric systems in which metal centers connect carbon-rich chains as depicted in Scheme 4.17. The chain length can be accurately chosen in order to retain the metal protection, and connection of the systems to surfaces such as electrodes can also be further performed.

Much attention has been directed to the preparation of rod-shaped organometallic compounds containing two unsaturated chains [26–31]. However, many metals such as Pt, Pd, Hg, or Cu have been found to be mostly insulators as they disrupt the π-conjugation between two acetylide chains [31, 66]. By contrast, a *trans*-bis(alkynyl) ruthenium system increases communication between two ferrocene units in an oxidation process [27c,d,28a]. Indeed, for **26**, in which two ferrocene units are connected via a ruthenium atom, the second

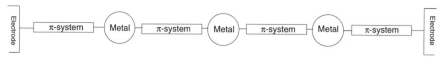

Scheme 4.17.

ferrocene oxidation occurs 200 mV after the first one. The difference is only 100 mV in the molecule without the metal **27** (Scheme 4.18) [27c].

To confirm the efficiency of the ruthenium system to act as an electron transmitter, our aim was to demonstrate that, in a reduction process, an electron can be delocalized on two carbon-rich chains in a *trans* arrangement. As the reduction of allenylidene complexes occurs on the carbon chain, an innovative solution to prove this communication was to study the reduction of a bis-allenylidene metal moiety $R_2C=C=C=[Ru]^{2+}=C=C=CR_2$.

An initial attempt consisted in the activation of a bis-diyne **28** complex with a leaving group at C_5 to provide the violet symmetric bis-alkenylallenylidene ruthenium complex *trans*-[(dppm)$_2$Ru(=C=C=C(OMe)(CH=CPh$_2$))$_2$](BF$_4$)$_2$ (**29**) via methanol addition to the metallacumulene intermediate (Scheme 4.19). However, spectroscopic studies and the crystal structure show that **29** has a strong bis-alkynyl character.

The successful preparation of the first real bis-allenylidene metal complex *trans*-[Ph$_2$C=C=C=Ru=C=C=CPh$_2$(dppe)$_2$]$^{2+}$ (**31**$^{2+}$) was achieved with an alternative method based on the chemical oxidation of [(dppe)$_2$Ru-C≡C-CHPh$_2$]PF$_6$ (**8a**) (vide infra). The allenylidene-acetylide *trans*-Ph$_2$C=C=C= (dppe)$_2$Ru-C≡C-CHPh$_2$]PF$_6$ (**30**) ($E° = 0.82$ vs. ferrocene) was oxidized with

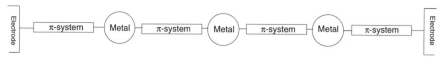

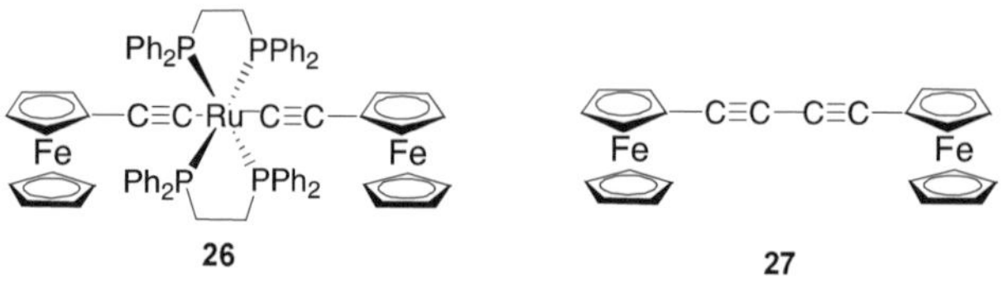

Scheme 4.18.

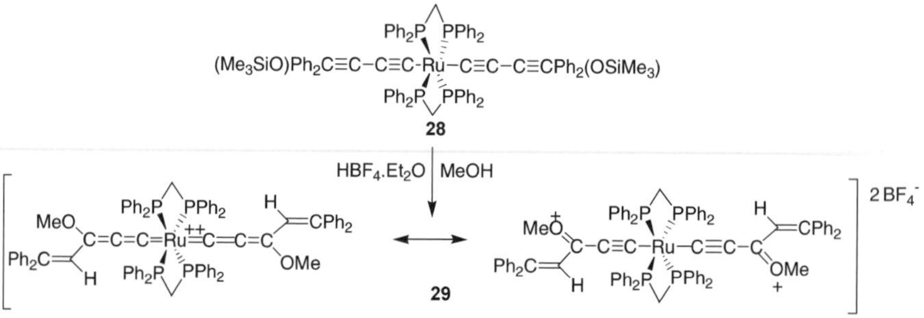

Scheme 4.19.

Ce^{IV} ammonium nitrate to induce the elimination of the γ proton that provided the bis-allenylidene complex 31^{2+} (Scheme 4.20) [28]. The spectroscopic features of this complex are characteristic of allenylidene ligands and two hydrid transfers on the C_γ atoms to form 32 confirmed the structure. Furthermore, full QM(DFT)/MM geometry optimizations led to a similar structure with consistent allenylidene bond lengths.

Two reversible one-electron reduction waves were observed at $E_1^\circ = -0.30$ V and $E_2^\circ = -0.93$ V vs. ferrocene. The singly reduced species was generated in situ with the addition of decamethylferrocene ($E^\circ = -0.59$ vs. ferrocene). Quenching of this radical 31^+ with Ph_3SnH at the C_γ carbon atom supports the partial localization of the single electron on the trisubstituted carbon atoms. The ESR quintet observed for 31^{2+} with $g = 1.9972$, and a coupling constant of 13.5 G due to the coupling of the single electron with the four equivalent phosphorus atoms, suggest its localization closer to the metal than in the reduced ruthenium monoallenylidene. In addition, IR studies of 31^+ show the vanishing of the allenylidene band and the formation of a new broad absorption at 1751 cm^{-1}, which is not consistent with the reduction of only one chain. Furthermore, QM(DFT)/MM calculations show that in the reduced state, 31^+ is symmetric with identical spin densities on the two C_α and the two C_γ carbon atoms (Fig. 4.3). Thus, all of these results are consistent with a novel radical with one *unpaired electron equally delocalized over both trans carbon-rich chains* linked by the ruthenium atom. This last point demonstrates the potential of ruthenium systems to mediate electron conduction, which is a key requirement for molecular wires.

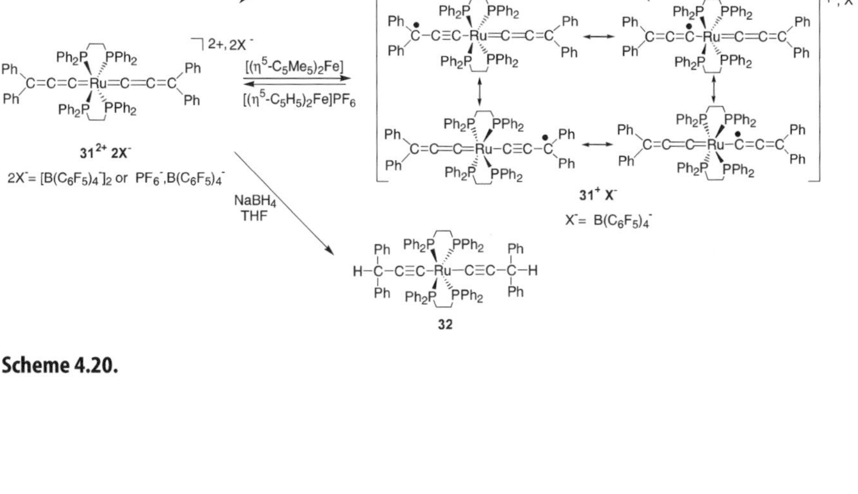

Scheme 4.20.

Fig. 4.3. Geometry of reduced bis-allenylidene 31^+ (QM(DFT)/MM), bond lengths (Å) (*left chain*) and spin density (*right chain*)

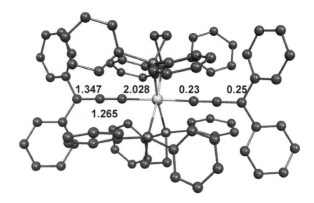

4.5
Trimetallic and Oligomeric Metal Complexes with Carbon-Rich Bridges

Conjugated organic polymers are currently used as organic light-emitting diodes (OLED) as they offer useful mechanical and processability properties. They have been produced by combination of dibromoaryl and terminal diyne derivatives using palladium-catalyzed Sonogashira reactions, and high-throughput screenings have revealed the most active combinations leading to efficient light-emitting polymers [67]. Similar approaches could be used for the production of carbon-rich metal oligomers in the search for new redox systems as illustrated in Scheme 4.21.

The rigid unit 1,4-diethylnylbenzene has been used as a bridge for building homometallic polymers (Pd, Pt or Ni) [68] and heterobimetallic Ru/Pd and Fe/Ni oligomers [69]. This rigid diynyl unit was first used to bridge two identical ruthenium moieties in **33** and ferrocenyl stable units in **34** as models [66] (Scheme 4.22).

Heterotrimetallic complexes Ru-Pd-Ru **35** and Fe-Pd-Fe **36** containing the same carbon-rich bridge were built using copper(I) catalyst for the activation of the terminal alkyne bond of organometallic alkynes **37** and **38** in the reaction with $PdCl_2(PBu_3)_2$ as shown in Scheme 4.23. The cyclic voltammetry of these carbon-rich complexes **33–38** gave useful information on the role of the bridge between metals (Table 4.4) [66].

The complex **33** showed two reversible waves corresponding to the systems Ru(III)/Ru(II) and Ru(III)/Ru(III), and its first oxidation is much easier than the mononuclear related complex **37** ($E° = 0.49$ V). The $E°$ value difference clearly shows the bridge communication ability in **33**. By contrast the

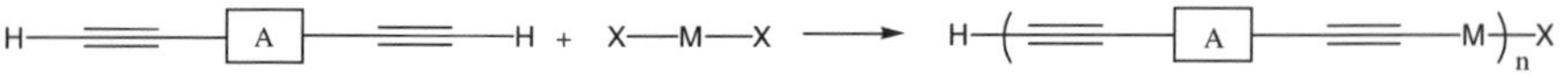

H——≡≡——[A]——≡≡——H + X—M—X $\longrightarrow$ H—(≡≡—[A]——≡≡—M)$_n$—X

Scheme 4.21.

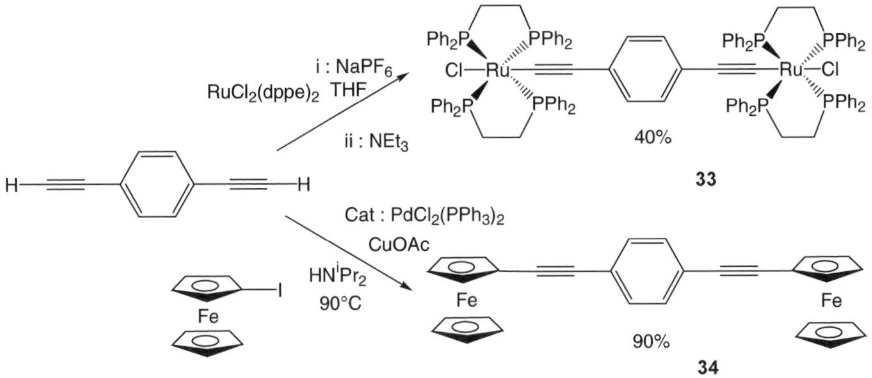

Scheme 4.22.

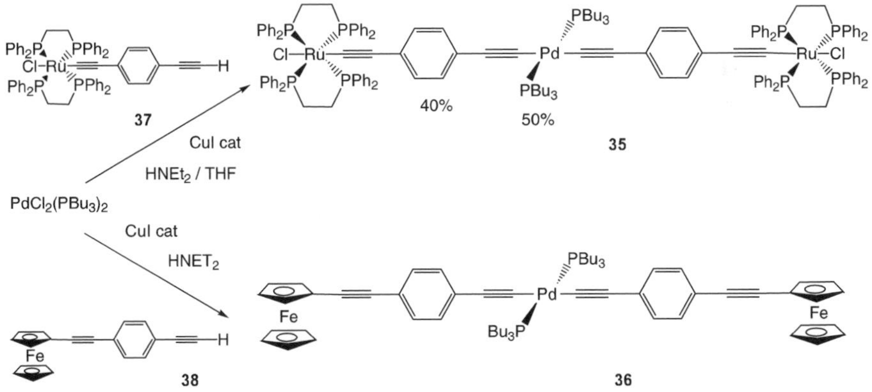

Scheme 4.23.

bis-ferrocenyl derivative **34** shows only one oxidation wave not significantly different from that of the monoferrocenyl derivative **38**. The Fe-Pd-Fe complex **36** behaves similarly to **34**. Thus, the diynyl bridge needs to be directly connected to the metal to show communication ability.

In the Ru-Pd-Ru complex **35** only one reversible wave ($E° = 0.33$ V$_{SCE}$) was observed for the oxidation of the two Ru sites (Ru(III)/Ru(II)), along with two irreversible waves for the Ru(IV)/Ru(III) and Pd(III)/Pd(II) oxidations. By comparison with **33**, the redox properties of **35** show that, surprisingly, the Pd unit inhibits the communication between the two ruthenium ends.

Based on these trimetallic models **35** and **36**, mixed metal oligomers could be produced from organometallic diynes on catalytic conditions. Thus the ruthenium diyne **39** ($E° = +0.45$ V vs. SCE) and PdCl$_2$(PBu$_3^n$)$_2$ were reacted with Cu(I) catalyst in the presence of amine, and the rigid rod oligomer **40** ($M_W = 15\,100$, $n = 10$) was obtained (Scheme 4.24) [69]. The cyclic voltammetry of **40** showed only one reversible Ru(III)/Ru(II) wave ($E° = +0.36$

vs. SCE) indicating that the palladium unit, as in model **35**, inhibits communication between the carbon-rich chains. Similarly, the ferrocene containing diyne **41** reacted with $PdCl_2(PBu_3^n)_2$ and Cu(I) catalyst to give the dark orange oligomer **42** (M_W = 26 000; n = 28) (Scheme 4.25) [69]. The diyne **41** and $Ni(C\equiv CH)_2(PBu_3^n)_2$ afforded the dark red mixed Fe/Ni oligomer

Table 4.4. Cyclic voltammetry data for complexes **33–38**

Complex[a,b]	E_{ox}° (V)	ΔE° (mV)	$E'_{ox p,a}$ (V)
[Ru]-C≡CC$_6$H$_4$C≡C-[Ru] (33)	0.15	80	1.6[c]
	0.51	80	
[Ru]-C≡CC$_6$H$_4$C≡C-H (37)	0.49	90	1.38[c]
[Pd](-C≡CC$_6$H$_4$C≡C-[Ru]$_2$) (35)	0.33	110	1.14[e]
			1.38[e]
[Fc]-C≡CC$_6$H$_4$C≡C-[Fc] (34)	0.56	160	
[Fc]-C≡CC$_6$H$_4$C≡CH (38)	0.58	120	
[Pd](-C≡CC$_6$H$_4$C≡C-[Fc])$_2$ (36)	0.55	110	1.42[d]

[a]Sample, 1 mM; Bu_4NPF_6 (0.1 M) in CH_2Cl_2; v = 100 mV s^{-1}; potentials are reported in V vs. SCE with ferrocene as an internal standard. [b]Abbreviations: [Ru] = ClRu(dppe)$_2$; [Pd] = Pd(P^{n}Bu$_3$)$_2$. [c]Peak potential of an irreversible process (Ru^{IV}/Ru^{III}). [d]Peak potential of an irreversible process (Pd^{III}/Pd^{II}). [e]Irreversible oxidations (Ru^{IV}/Ru^{III} and Pd^{III}/Pd^{II})

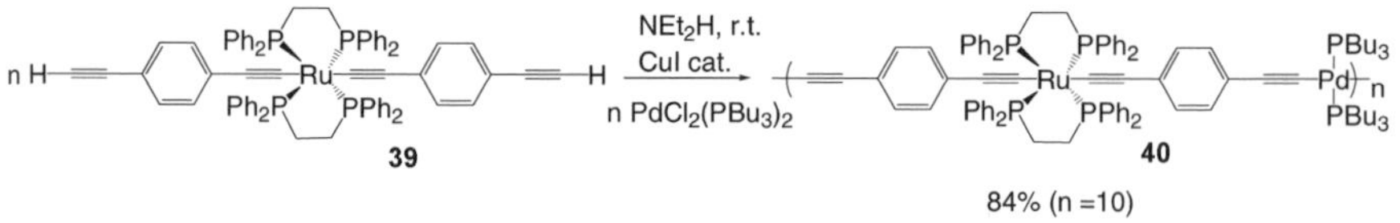

Scheme 4.24.

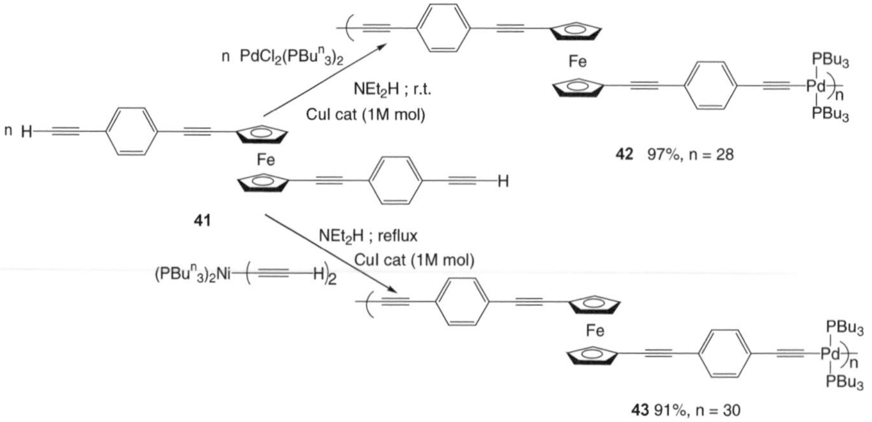

Scheme 4.25.

43 (M_W = 27 000, n = 30). Oligomers **42** and **43** with higher molecular weight than **40** are more flexible due to the rotating ferrocene moieties leading to zigzag-shape oligomers.

This study suggests that the building of communicating metal-containing oligomers should be better achieved by insertion of only Ru(dppe)$_2$ units in the carbon-rich systems.

4.6
Star Organometallic-Containing Multiple Identical Metal Sites

Star molecules constitute simple models for the production of multisite catalysts or stable redox systems. The latter have potential for applications as electron storage systems [70] or as modified electrodes [71]. The former allows recovery of catalysts and access to both polyfunctional molecules and star oligomers.

On the basis of known activation processes of terminal alkynes with 16-electron metal intermediates [38, 72], triynes containing a rigid core were first used for the access to star organometallics containing three identical stable or mixed redox systems [32a]. Thus the triyne 1,3,5-(HC≡C-C$_6$H$_4$-C≡C-)$_3$C$_6$H$_3$ **44** reacts with RuCl$_2$(dppe)$_2$ to give a trisvinylidene ruthenium complex **45**. Complex **45** is easily deprotonated into a trisalkynyl ruthenium compound **46** or used for the activation of ferrocenylacetylene to give complex **47** containing three identical Ru sites and three identical Fe sites (Scheme 4.26).

A cyclic voltammetry study of star organometallics **45–47** reveals that in each molecule the three ruthenium moieties are oxidized at the same potential for (V vs. ferrocene): +0.72 V for **45**, +0.02 V for **46**, and −0.28 V (Fe) and

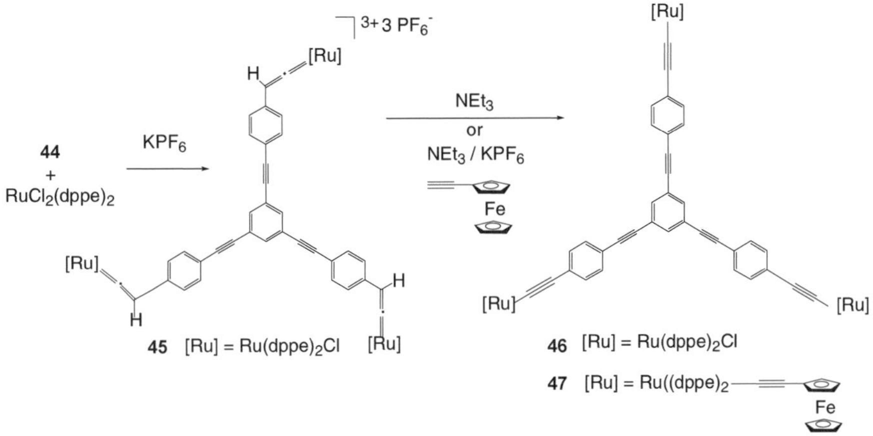

Scheme 4.26.

+0.30 V (Ru) for **47**. Thus the carbon-rich C_{3V} bridge does not allow significant communication between the three ruthenium centers and the three ferrocenyl groups. The vinylidene ligand is an electron withdrawing group in **45** with respect to tris-alkynyl bridge in **46**. The comparison of **46** and **47** with model compounds *trans*-Ru(CC-R)$_2$(dppe)$_2$ (R = Ph, Fc) indicates that in **47** the first oxidation is due to that of the three ferrocenyl groups that are thus electron enriched by the carbon-rich tris-ruthenium tripodal bridge [32a]. The electron donation of the ruthenium moieties is reflected in their higher oxidation potential in **47** as compared to **46**. Thus the organometallic star **47** can behave as an electron reservoir offering six electrons three by three.

The activation of tritopic terminal polyynes was explored in order to generate multiple identical catalyst sites. Indeed vinylidene- and allenylidene-ruthenium complexes have been used as catalyst precursors [34]. Triynes with nonrigid cores were activated by suitable ruthenium complexes. Thus, the triyne **48** was transformed into tris-vinylidene-ruthenium complexes **49** and **50** (Scheme 4.27) [32b]. By activation of the trispropargyl alcohol molecule **51** the new tris-allenylidene-ruthenium complexes **52** and **53** were selectively produced (Scheme 4.28) [32b].

These cationic tris-ruthenium complexes **49–50** and **52–53**, although they do not possess stable redox systems, contain identical ruthenium sites presenting potential as multiple catalysts with identical catalytic sites to promote alkene metathesis or ROMP polymerisation, thus offering route to star functional molecules and star polymers.

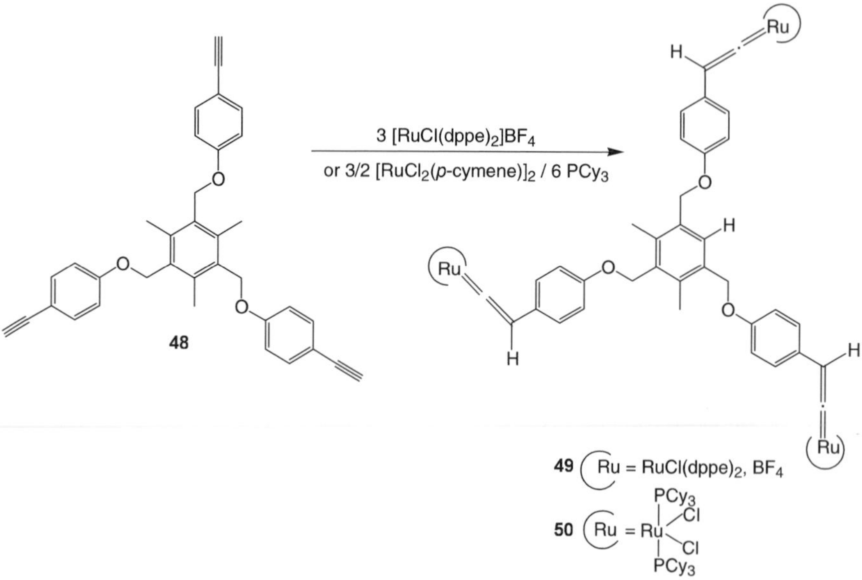

Scheme 4.27.

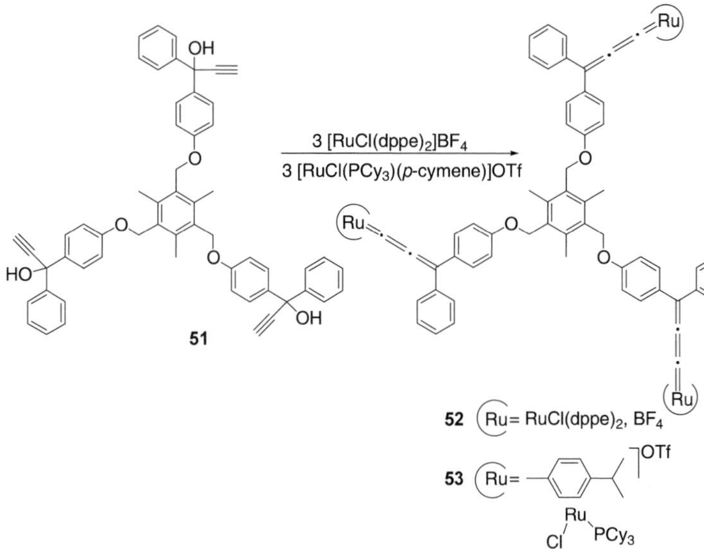

Scheme 4.28.

4.7
Conclusion

The above results show that controlled activation of carbon-rich molecules by suitable metal complexes led to formation of a wide range of ruthenium systems with potential for material science. Indeed, these ruthenium compounds are attractive precursors for the synthesis of bimetallic complexes as models for redox active molecular wires, i.e., bis-acetylides, bis-allenylidenes, or mixed acetylide-allenylidene ligands, with properties highly dependent on the nature of the chain. Furthermore, the preparation of a monometallic bis-allenylidene and its reduction showed the possibility of electronic communication between two carbon-rich chains through a ruthenium atom, which is a crucial issue for the development of extended molecular wires. Owing to the potential ability of these systems for chloride substitution with an acetylide or a cumulenic chain, they appear to be interesting building blocks for the construction of long one-dimensional nanoscaled organometallic networks to mediate electron transfer processes, i.e., metal-containing unsaturated oligomers and polymers, and for the building of star carbon-rich compounds with potential for enabling access to multiple identical redox systems or catalyst precursors.

4.8
References

1. (a) Aviram A, Ratner M, Chem Phys Lett (1974) 29:277; (b) Lehn JM (1995) Supramolecular Chemistry: Concepts and Perspectives, VCH, Weinheim Ch 8; (c) Long NJ (1999) In:

Roundhill DM, Fackler JP (eds) Optoelectronic Properties of Inorganic Compounds, Plenum, New York; (d) Yam VWW, Acc Chem Res (2002) 35:555; (e) Tour JM, Acc Chem Res (2000) 33:791; (f) Robertson N, McGowan CA, Chem Soc Rev (2003) 32:96; (g) Joachim C, Gimzewski JK, Aviram A, Nature (2000) 408:541; (h) Lloyd Carroll R, Gorman CB, Angew Chem Int Ed (2002) 41:4378; (i) McClecverty JA, Ward MD, Acc Chem Res (1998) 31:842

 2. Stang PJ, Chem Eur J (1998) 4:19
 3. (a) Baranof E, Collin J-P, Flamini L, Sauvage J-P, Coord Chem Rev (2004) 33:147; (b) Welter B, Brunner K, Hofstraat JW, De Cola L, Nature (2003) 21:54–57; (c) Ziessel R, Hissler M, El-ghaoury A, Harriman A, Chem Soc Rev (1998) 178–180:1251; (d) Hofmeier H, Shubert US, Chem Soc Rev (2004) 33:373; (e) Boyde S, Strouse GF, Jones WE, Meyer TJ, J Am Chem Soc (1990) 112:7395; Balzani V, Juris A, Venturi M, Campagna S, Serroni S, Chem Rev (1996) 96:759
 4. Nguyen P, Gómez-Elipe P, Manners I, Chem Rev (1999) 99:1515
 5. (a) Paul F, Lapinte C, Coord Chem Rev (1998) 178-180:431; (b) Szafert S, Gladysz JA, Chem Rev (2003) 103:4175; (c) Schwab RFH, Levin MD, Michl J Chem Rev (1999) 99:186; (d) Bruce MI, Low PJ, Adv Organomet Chem (2004) 50:179; (e) Paul F, Lapinte C (2002) In: Gielen M, Willem R, Wrackmeyer B (eds) Unusual Structures and Physical Properties in Organometallic Chemistry, Wiley, New York, p 220
 6. (a) Cotton FA, Liu CY, Murillo CA, Villagrán D, Wang X, J Am Chem Soc (2004) 126:14822; (b) Cotton FA, Donahue JP, Murillo CA, J Am Chem Soc (2003) 125:5436; (c) Diketo R, Hoshino Y, Higushi S, Fielder J, Su C-Y, Knölder A, Schwederski B, Sarkar B, Hartmann H, Kaim W, Angew Chem Int Ed (2003) 43:674; (d) Meacham AP, Druce KL, Bell ZR, Ward MD, Keister JB, Lever ABP, Inorg Chem (2003) 42:7887
 7. (a) Brunschwig BS, Creutz C, Sutin N, Chem Soc Rev (2002) 31:168; (b) Nelsen SF (2001) In: Balzani V (ed) Electron Transfer in Chemistry, Wiley-VCH, Weinheim, Germany, Vol 1, Chap 10; (c) Launay J-P, Coudret C (2001) In: Balzani V (ed) Electron Transfer in Chemistry, Wiley-VCH, Weinheim, Germany, Vol 5, Chap 1; (d) Launay J-P, Chem Soc Rev (2001) 30:386; (e) Demadis KD, Hartshorn CM, Meyer TJ, Chem Rev (2001) 101:2655
 8. (a) Antonova AB, Bruce MI, Ellis BG, Gaudio M, Humphrey PA, Jevric M, Melino G, Nicholson BK, Perkins GJ, Skelton BW, Stapleton B, White AH, Zaitseva NN, Chem Commun (2004) 960; (b) Bruce MI, Ellis BG, Low PJ, Skelton BW, White AH, Organometallics (2003) 22:3184; (c) Bruce MI, Low PJ, Costuas K, Halet J-F, Best SP, Heath GA, J Am Chem Soc (2000) 122:1949
 9. (a) Le Narvor N, Toupet L, Lapinte C, J Am Chem Soc (1995) 117:7129; (b) Roué S, Lapinte C, Organometallics (2004) 23:2558–2567; (c) Le Stang S, Paul F, Lapinte C, Organometallics (2000) 19:1035; (d) Paul F, Meyer WE, Toupet L, Jiao H, Gladysz JA, Lapinte C, J Am Chem Soc (2000) 122:9405
10. Sakurai A, Akita M, Moro-oka Y, Organometallics (1999) 18:3241
11. (a) Dembinski R, Bartik T, Bartik B, Jaeger M, Gladysz JA, J Am Chem Soc (2000) 122:810; (b) Brady M, Weng W, Zhou Y, Seyler JW, Amoroso AJ, Arif AM, Böhme M, Frenking G, Gladysz JA, J Am Chem Soc (1997) 119:775; (c) Mohr W, Stahl J, Hampel F, Gladysz JA, Chem Eur J (2003) 9:3324; (d) Stahl J, Bohling JC, Bauer EB, Peters TB, Mohr W, Martin-Alvarez JM, Hampel F, Gladysz JA, Angew Chem Int Ed (2002) 41:1872; (e) Owen GR, Stahl J, Hampel F, Gladysz JA, Organometallics (2004) 23:5889; (f) Owen GR, Hampel F, Gladysz JA, Organometallics (2004) 23:5893; (g) Owen GR, Stahl J, Hampel F, Gladysz JA, Organometallics (2004) 23:5889; (h) Zhuravlev F, Gladysz J, Chem Eur J (2004) 10:6510
12. (a) Fernandez F, Blacque O, Alfonso M, Berke H, Chem Commun (2000) 1266; (b) Kheradmandan S, Heinze K, Schmalle HW, Berke H, Angew Chem Int Ed (1999) 35:2270; (c)

Kheradmandan S, Venkatesan K, Blacque O, Schmalle HW, Berke H, Chem Eur J (2004) 10:4872; (d) Fernandez FJ, Kheradmandan S, Venkatesan K, Blacque O, Alfonso M, Schmalle HW, Berke H, Chem Eur J (2003) 9:6192

13. Xu G-L, Zou G, Ni Y-H, De Rosa MC, Crutchley RJ, Ren T, J Am Chem Soc (2003) 125:10057
14. Rigaut S, Touchard D, Dixneuf PH, Coord Chem Rev (2004) 248:1586
15. (a) Rigaut S, Le Pichon L, Daran J-C, Touchard D, Dixneuf PH, Chem Commun (2001) 1206; (b) Rigaut S, Massue J, Touchard D, Fillaut J-L, Golhen S, Dixneuf PH, Angew Chem Int Ed (2002) 41:4513
16. Rigaut S, Perruchon J, Le Pichon L, Touchard D, Dixneuf PH, J Organomet Chem (2003) 670:37
17. Rigaut S, Perruchon J, Guesmi S, Fave C, Touchard D, Dixneuf PH, Eur J Inorg Chem (2005) 447
18. Gil-Rubio J, Werberndörfer B, Werner H, Angew Chem Int Ed (2000) 39:786
19. Mantovani N, Brugnati M, Gonsalvi L, Grigiotti E, Laschi F, Marvelli L, Peruzzini M, Reginato G, Rossi R, Zanello P, Organometallics (2005) 24:405
20. Hartbaum C, Mauz E, Roth G, Weissenbach K, Fischer H, Organometallics (1999) 18:2619
21. (a) Etzenhouser BA, DiNiase Cavanaugh M, Spurgeon HN, Sponsler MB, J Am Chem Soc (1994) 116:2221; (b) Etzenhouser BA, Chen Q, Sponsler MB, Organometallics (1994) 13:4176; (c) Chung M-C, Gu X, Etzenhouser BA, Spuches AM, Rye PT, Seetharaman SK, Rose DJ, Zubieta J, Sponsler MB, Organometallics (2003) 22:3485; (d) Liu SH, Xia H, Wen TB, Zhou Z, Jia G, Organometallics (2003) 22:737; Liu SH, Hu QY, Xue P, Wen TB, Williams ID, Jia G, Organometallics (2005) 24:769
22. (a) Wong KMC, Lam SCF, Ko CC, Zhu N, Yam VWW, Roué S, Lapinte C, Fathallah S, Costuas K, Kahlal S, Halet J-F, Inorg Chem (2003) 42:7086; (b) Yam VWW, Wong KMC, Zhu N, Angew Chem Int Ed (2003) 43:1400; (c) Lu W, Xiang H-F, Zhu N, Che C-M, Organometallics (2002) 21:2343; (d) Fillaut JL, Price M, Johnson AL, Perruchon J, Chem Commun (2001) 739; (e) Yam VWW, Tang RP-L, Wong KM-C, Cheung K-K, Organometallics (2001) 20:4476
23. Ung A, Cargill Thompson AMW, Bardwell DA, Gatteschi D, Jeffery JC, McCleverty JA, Tolti F, Ward MD, Inorg Chem (1997) 36:3447
24. (a) Powell CE, Humphrey MG, Coord Chem Rev (2004) 248:725; (b) Weyland T, Ledoux I, Brasselet S, Zyss J, Lapinte C, Organometallics (2000) 19:5235; (c) Senechal K, Maury O, Le Bozec H, Ledoux I, Zyss J, J Am Chem Soc (2002) 124:4560; (d) Long NJ, Angew Chem, Int Ed Engl (1995) 34:21; (e) Fillaut J-L, Perruchon J, Blanchard P, Roncali J, Golhen S, Allain M, Migalsaka-Zalas A, Kityk IV, Sahraoui B, Organometallics (2005) 24:687
25. Astruc D (1995) Electron Transfer and Radical Processes in Transition Metal Chemistry VCH, New York
26. (a) Wong K-T, Lehn J-M, Peng S-M, Lee G-H, Chem Commun (2000) 2259; (b) Xu G, Ren T, Organometallics (2001) 20:2400; (c) Gil-Rubio J, Weberndörfer B, Werner H, Angew Chem Int Ed (2000) 39:786; (d) Qi H, Sharma S, Li Z, Snider S, Orlov AO, Lent CS, Fehlner TP, J Am Chem Soc (2003) 125:15251
27. (a) Touchard D, Haquette P, Daridor A, Romero A, Dixneuf PH, Organometallics (1998) 17:3844; (b) Touchard D, Haquette S, Guesmi P, Le Pichon L, Daridor A, Toupet L, Dixneuf PH, Organometallics (1997) 16:3640; (c) Lebreton C, Touchard D, Le Pichon L, Daridor A, Toupet L, Dixneuf PH, Inorg Chem Acta (1998) 272:188; (d) Zhu Y, Clot O, Wolf MO, Yap GPA, J Am Chem Soc (1998) 120:1812; (b) Lavastre O, Plass J, Bachmann P, Guesmi S, Moinet C, Dixneuf PH, Organometallics (1997) 16:184
28. Rigaut S, Costuas K, Touchard D, Saillard J-Y, Golhen S, Dixneuf PH, J Am Chem Soc (2004) 126:4072

29. (a) Xu G-L, De Rosa MC, Crutchley RJ, Ren T, J Am Chem Soc (2004) 126:3728; (b) Jones SC, Coropceanu V, Barlow S, Kinnibrugh T, Timofeeva T, Marder SR J Am Chem Soc (2004) 126:11782; (c) Schull TL, Kushmerick JG, Patterson CH, George C, Moore MH, Pollack SK, Shashidhar R, J Am Chem Soc (2003) 125:3202; (e) Stroh C, Mayor M, Von Hänisch C, Turek P, Chem Commun (2004) 2050; (f) Mayor M, von Hänisch C, Weber HB, Reichert J, Beckmann D, Angew Chem Int Ed (2002) 41:1183

30. (a) Yip JH, Wu J, Wong K-Y, Ho KP, Pun CS-N, Vittal JJ, Organometallics (2002) 21:5292; (b) Sheng T, Varenkamp H, Eur J Inorg Chem (2004) 41:1198; (c) Berry JF, Cotton FA, Murillo CA, Organometallics (2004) 23:2503; (d) Dewhurst RD, Hill AF, Willis AC, Organometallics (2004) 23:1646; (e) Zuo JL, Herdtweck E, de Biane FF, Santos AM, Kühn FE, New J Chem (2002) 26:883; (f) Weng W, Bartik T, Brady M, Bartik B, Ramsden JA, Arif AM, Gladysz JA, J Am Chem Soc (1995) 117:11922; (g) Zheng Q, Hampel F, Gladysz JA, Organometallics (2004) 23:5896

31. (a) Mayor M, Von Hänish C, Webr HB, Reichert J, Beckmann D, Angew Chem Int Ed (2002) 41:1183; (b) Bruce MI, Halet J-F, Le Guennic B, Skelton BW, Smith ME, White AH, J Inorg Chim Acta (2003) 350:175; (c) Yip HK, Wu J, Wong K-Y, Yeunh K-W, Vitall JJ, Organometallics (2002) 21:1612

32. (a) Uno M, Dixneuf PH, Angew Chem Int Ed (1998) 37:1714; (b) Weiss D, Dixneuf PH, Organometallics (2003) 22:2209

33. (a) Tamm M, Jentsche T, Werncke W, Organometallics (1997) 16:1418; (b) Roth G, Fischer H, Meyer-Friedrichsen T, Heck J, Houbrechts S, Persoons A, Organometallics (1998) 17:1511

34. Castarlenas R, Fichmeister C, Bruneau C, Dixneuf PH, J Mol Catal Chem (2004) 213:31

35. Castarlenas R, Dixneuf PH, Angew Chem Int Ed (2003) 114:5476

36. Fürstner A, Liebl M, Lehmann CW, Picquet M, Kunz R, Bruneau C, Touchard D, Dixneuf PH, Chem Eur J (2000) 6:1847

37. (a) Nishibayashi Y, Wakiji I, Hidai M, J Am Chem Soc (2000) 122:11019; (b) Nishibayashi Y, Yoshikawa M, Inada Y, Hidai M, Uemura S, J Am Chem Soc (2002) 124:11846

38. Touchard D, Dixneuf PH, Coord Chem Rev (1998) 178–180:409

39. (a) Bruce MI, Chem Rev (1998) 98:2797; (b) Bruce MI, Chem Rev (1991) 91:197; (c) Cardieno V, Gamasa MP, Gimeno J, Eur J Inorg Chem (2001) 571; (d) Werner H, Chem Commun (1997) 903; (e) Cardierno V, Diez J, Gamasa M-P, Gimeno J, Coord Chem Rev (1999) 193–195:147

40. Selegue JP, Organometallics (1982) 1:217

41. Pilette D, Ouzzine K, Le Bozec H, Dixneuf PH, Rickard CEF, Roper WR, Organometallics (1992) 11:809

42. Cadierno V, Gamasa MP, Gimeno J, González-Cueva M, Lastra E, Borge J, García-Granda S, Pérez-Carreño E, Organometallics (1996) 15:2137

43. Rigaut S, Maury O, Touchard D, Dixneuf PH, Chem Commun (2001) 373

44. (a) Gallop MA, Roper WR, Adv Organomet Chem (1986) 25:121; (b) Kim HP, Angelici RJ, Adv Organomet Chem (1987) 27:51; (c) Buhro WE, Chisholm MH, Adv Organomet Chem (1987) 27:311; (d) Fischer H, Hofmann P, Kreissl FR, Schrock RR, Schubert U, Weiss K (1988) Carbyne Complexes, VCH, Weinheim, Germany; (e) Mayr A, Hoffmeister H, Adv Organomet Chem (1991) 32:227; (f) Engel PF, Pfeffer M, Chem Rev (1995) 95:2281; (g) Hill AF (1995) In: Abel EW, Stone FGA, Wilkinson G (eds): Comprehensive Organometallic Chemistry II, Pergamon, Oxford, Vol 7; (h) Herdon JW, Coord Chem Rev (2002) 227:1

45. (a) Luecke HF, Bergman RG, J Am Chem Soc (1998) 120:11008; (b) Tang Y, Sun J, Chen J, Organometallics (1999) 18:2459; (c) Zhang L, Gamasa MP, Gimeno J, Carbajo RJ,

López-Ortiz F, Guedes da Silva MFC, Pombeiro AJL, Eur J Inorg Chem (2000) 341; (d) Bannwart E, Jacobsen H, Furno F, Berke H, Organometallics (2000) 19:3605; (e) Enriquez A, Templeton J Organometallics (2002) 21:852

46. (a) Baker LJ, Clark GR, Rickard CEF, Roper WR, Woodgate SD, Wright LJ, J Organomet Chem (1998) 551:247; (b) Coalter III JN, Bollinger JC, Eisentsein O, Caulton KG, New J Chem (2000) 24:925

47. (a) Gonzalez-Herrero P, Weberndörfer B, Ilg K, Wolf J, Werner H, Angew Chem Int Ed (2000) 39:3266; (b) Gonzalez-Herrero P, Weberndörfer B, Ilg K, Wolf J, Werner H, Organometallics (2001) 20:3672; (c) Stüer W, Wolf J, Werner H, Schwab P, Shultz M, Angew Chem Int Ed (1998) 37:3421

48. (a) Jung S, Brandt CD, Werner H, New J Chem (2001) 25:1101; (b) Jung S, Ilg K, Brandt CD, Wolf J, Werner H, J Chem Soc Dalton Trans (2002) 318; (c) Bustelo E, Jimenez-Tenorio M, Mereiter K, Puerta MC, Valerga P, Organometallics (2002) 21:1903

49. (a) Castarlenas R, Esteruelas MA, Onate E, Organometallics (2001) 20:3283; (b) Wen TB, Cheung YK, Yao J, Wong W-T, Zhou ZY, Jia G, Organometallics (2000) 19:3803; (c) Werner H, Jung S, Weberndorfer B, Wolf J, Eur J Inorg Chem (1999) 951; (d) Lapointe AM, Schrock RR, Davis WM, J Am Chem Soc (1995) 117:4802; (e) Clarck GR, Marsden K, Roper WR, Wright LJJ, Am Chem Soc (1980) 102:6570; (f) Espuelas J, Esteruelas MA, Lahoz FJ, Oro LA, Ruiz NJ, Am Chem Soc (1993) 115:4683; (g) Caulton KG, J Organomet Chem (2001) 617–618:56

50. Rigaut S, Touchard D, Dixneuf PH, Organometallics (2003) 22:3980

51. Koentjoro OF, Rousseau R, Low PJ, Organometallics (2001) 20:4502

52. (a) Winter RF, Hornung FM, Organometallics (1999) 18:4005; (b) Winter RF, Klinkhammer KW, Zalis S, Organometallics (2001) 20:1317; (c) Hartmann S, Winter RF, Sarkar B, Lissner F, J Chem Soc Dalton Trans (2004) 3273; (d) McGrady JE, Lovell T, Stranger R, Humphrey MG, Organometallics (1997) 16:4004

53. (a) Esteruelas MA, Gomez AV, Lopez A, Modrego J, Oñate E, Organometallics (1997) 16:5826; (b) Cadierno V, Gamasa MP, Gimeno J, González-Cueva M, Lastra E, Borge J, García-Granda S, Pérez-Carreño E, Organometallics (1996) 15:2137

54. Rigaut S, Monnier F, Mousset F, Touchard D, Dixneuf PH, Organometallics (2002) 21:2657

55. Amir-Ebrahimi V, Hamilton JG, Nelson J, Rooney JJ, Thompson JM, Beaumont AJ, Rooney AD, Harding CJ, Chem Commun (1999) 1621

56. (a) Winter RF, Eur J Inorg Chem (1999) 2121; (b) Winter RF, Hornung FM, Organometallics (1999) 18:4005; (c) Winter RF, Klinkhammer K-W, Zalis S, Organometallics (2001) 20:1317; (d) Winter RF, Hartmann S, Zalis S, Klinkhammer K-W, J Chem Soc Dalton Trans (2003) 2342

57. Auger N, Touchard D, Rigaut S, Halet J-F, Saillard J-Y, Organometallics (2003) 22:1638

58. Peters TB, Bohling JC, Arif AM, Gladysz JA, Organometallics (1999) 18:3261; (b) Mohr W, Stahl J, Hampel F, Gladysz JA, Inorg Chem (2001) 40:3263

59. Guesmi S, Touchard D, Dixneuf PH, Chem Commun (1996) 2773

60. Callejas-Gaspar B, Laubender M, Werner H, J Organomet Chem (2003) 684:144; Gil-Rubio J, Weberndörfer B, Werner H, Angew Chem Int Ed (2000) 39:786; Laubender M, Werner H, Chem Eur J (1999) 5:2937

61. (a) Bartik T, Johnson MT, Arif AM, Gladysz JA, Organometallics (1995) 14:889; (b) Fischer H, Leroux F, Stumpf R, Roth G, Chem Ber (1996) 129:1475; (c) Leroux F, Stumph R, Fischer H, Eur J Inorg Chem (1998) 1225; (d) Kolokova NY, Skripkin VV, Alexandrov GG, Struchkov YT, J Organomet Chem (1979) 169:293; (e) Bullock RM, J Am Chem Soc (1987) 109:8087

62. (a) Bartik T, Weng W, Ramsden JA, Szafert S, Falloon SB, Arif AM, Gladysz JA, J Am Chem Soc (1998) 120:11071; (b) Dembinski R, Szafert S, Haquette P, Lis T, Gladysz JA, Organometallics (1999) 18:5438; (c) Fuss B, Dede M, Weibert B, Fischer H, Organometallics (2002) 21:4425; (d) Hartbaum C, Fischer H, J Organomet Chem (1999) 578:186

63. (a) Jia G, Xia HP, Wu WF, Ng WS, Organometallics (1997) 16:2940; (b) Xia HP, Ng WS, Ye JS, Li XY, Wong WT, Lin Z, Yang C, Jia G, Organometallics (1999) 18:4552

64. Bruce MI, Coord Chem Rev (2004) 248:1603

65. Barrière F, Camire N, Geiger WE, Mueller-Westerhoff UT, Sanders R, J Am Chem Soc (2002) 124:7262

66. Lavastre O, Plass J, Bachmann P, Guesmi S, Moinet C, Dixneuf PH, Organometallics (1997) 16:184

67. Lavastre O, Illitchev I, Jegou G, Dixneuf PH, J Am Chem Soc (2002) 124:5278

68. (a) Sonogashira K, Olga K, Takahashi S, Hagihara N, J Organomet Chem (1980) 188:237; (b) Takahashi S, Sonogashira K, Morimoto H, Murata E, Kataoka S, Hagihara N, J Polym Sci Polym Chem Ed (1982) 20:565

69. Lavastre O, Even M, Dixneuf PH, Pacreau A, Vairon JP, Organometallics (1996) 15:1530

70. (a) Jutzi P, Batz C, Neumann B, Stammler H-G, Angew Chem Int Ed Engl (1996) 35:2118; (b) Bard AJ, Nature (1995) 374:13; (c) Astruc D, New J Chem (1992) 16:305

71. Alonso B, Morain M, Casado CM, Lobete F, Losada J, Cuadrado I, Chem Mater (1995) 7:1440

72. Touchard D, Haquette P, Pirio N, Toupet L, Dixneuf PH, Organometallics (1993) 12:3132

Chapter 5

Molecular Metal Wires
Built from a Linear Metal Atom Chain Supported by Oligopyridylamido Ligands

CHEN-YU YEH[1] · CHIH-CHIEH WANG[2] · CHUN-HSIEN CHEN[3] ·
SHIE-MING PENG[4]

[1] Department of Chemistry, National Chung Hsing University, Taichung, Taiwan
  e-mail: cyyeh@dragon.nchu.edu.tw
[2] Department of Chemistry, Soochow University, Taipei, Taiwan
[3] Department of Chemistry, National Tsing Hua University, Hsinchu, Taiwan
[4] Department of Chemistry, National Taiwan University, Taipei, Taiwan
  e-mail: smpeng@ntu.edu.tw

Summary Molecular wires incorporating four oligopyridylamine ligands and a metal atom chain have been synthesized. The structures show that the metal atom chain is in a linear arrangement and is helically wrapped by four ligands. The metal atom chain may adopt a symmetrical or unsymmetrical arrangement depending on the crystallization environment, the nature of the metal ions, the oxidation states of the complexes, and the nature of the axial ligands. The strength of metal–metal interactions in the neutral form of linear multinuclear metal compounds is in the order Cr > Co > Ni. Upon oxidation of the complexes, the chromium and nickel complexes exhibit significant changes in the metal–metal interactions while the cobalt analogs show slight changes. Electrochemical studies show that these complexes have rich redox properties. Some of the complexes have more than two stable oxidation states. The oxidation of nickel and chromium complexes becomes easier as the number of metal atoms in the metal chains increases. The first oxidation potential increases with an increasing atomic number for the first-row transition metals. STM studies of multinuclear metal string complexes show significant differences in the conductivity of the oxidized and neutral forms. This scenario provides a paradigm for the use of molecular electronics such as molecular switches, molecular wires, and storage devices.

Abbreviations

bpyan	The dianion of 2,7-bis(2-pyridylamino)naphthyridine
depa	The anion of di(2-(4-ethylpyridyl))amine
dpa	The anion of di(2-pyridyl)amine
Fc	Ferrocenyl
Hdpa	Dipyridylamine
H_2tpda	Tripyridyldiamine
$H_3teptra$	Tetrapyridyatriamine
$H_4peptea$	Pentapyridyltetraamine
$H_5heppea$	Hexapyridylpentaamine
$H_6hpphea$	Heptapyridylhexaamine
$H_7ocphpa$	Octapyridylheptaamine
$H_8nopoca$	Nonapyridyloctaamine

H_9depnoa	Decapyridylnonaamine
H_{10}unpdea	Undecapyridyldecaamine
H_{11}dopuna	Dodecapyridylundecaamine
H_{12}trpdoa	Tridecapyridyldodecaamine
N_3Br	Bromodipyridylamine
N_4	Dipyridyldiamine
N_4Br	Bromodipyridyldiamine
N_5	Dipyridyltriamine
N_7Br	Bromotetrapyridyltriamine
N_8Br	Bromotetrapyridyltetraamine
N_9Br	Bromopentapyridyltetraamine
peptea	Tetraanion of pentapyridyltetraamine
pyan	Anion of 2-(2-pyridylamino)naphthyridine
STM	Scanning tunneling microscopy
teptra	Trianion of tetrapyridyltriamine
tpda	Dianion of tripyridyldiamine

5.1
Introduction

The studies on organic molecules used for molecular wires have attracted much attention. Several new classes of organic molecular wires have been constructed [1–10]. In contrast, the development of inorganic molecular wires has lagged [11–14]. Our approach to inorganic molecular wires is the use of oligopyridylamines for the construction of linear multinuclear complexes. The complexes consist of a linear metal atom chain supported by four oligopyridylamido ligands. It has been shown that di-(2-pyridyl)amine can stabilize linear trinuclear coordination complexes of various first-row and second-row transition metals. The first examples of the trinuclear copper and nickel complexes bridged by di-α-pyridylamido ligand were reported in 1990–1991 [15–17]. In 1994 we reported the first unsymmetrical structure of a tricobalt complex [Co_3(dpa)$_4$Cl$_2$] (dpa = the anion of di(2-pyridyl)amine), in which the central cobalt ion forms a metal–metal bond with one of the two terminal cobalt ions and leaves the other terminal cobalt ion isolated [18]. Three years later, the symmetrical structure of this complex was discovered by Cotton et al. [19]. Further work by the same group confirmed that both symmetrical and unsymmetrical structures could exist [20–26]. Similar to the tricobalt system, the trichromium complexes also display both symmetrical and unsymmetrical structures [27–30]. To see how metal–metal interactions behave in a linear chain of metal atoms, we set out to design and synthesize a series of oligopyridylamine ligands and their corresponding complexes. In the past decade, we have successfully synthesized numerous trimetal complexes [M_3(dpa)$_4$X$_2$] (M = Cr, Co, Ni, Cu, Rh, and Ru, X = axial ligands) [18, 31], pentametal complexes $[M_5(tpda)_4X_2]$ (M = Cr, Co, Ni, and Ru, tpda = dianion of tripyridyldiamine) [32–36], heptametal complexes $[M_7(teptra)_4X_2]$ (M = Cr, Co, and Ni, teptra = trianion of tetrapyridyltriamine) [37, 38], and nonametal complex $[M_9(peptea)_4X_2]$ (M = Cr and Ni, peptea = tetraanion of pentapyridyltetraamine) [39, 40], along with various multinuclear systems supported by modified ligands.

In this chapter, we will discuss the synthesis of oligopyridylamine and naphthyridine-containing ligands, along with the preparation, structures, and physical properties of their complexes. The strong metal–metal interactions observed in some of these linear multinuclear complexes indicate that the linear metal chain may provide a conductive pathway for electrons, i.e., they may find use in the application as molecular wires. The complexes bearing thiocyanate axial ligands have been self-assembled on metal substrate for the STM studies of their conductive behavior. These results will also be described.

5.2
Synthesis of Oligopyridylamine Ligands

The ligands used in our laboratory for the construction of linear multinuclear complexes are of precise length and constitution. The structures of these oligopyridylamine ligands synthesized are outlined in Scheme 5.1. The synthesis of the shorter ligands such as **Hdpa** and **H₂tpda** is straightforward. For example, **Hdpa** was synthesized by the amination reaction of commercially available 2-bromopyridine with 2-aminopyridine in the presence of strong base *tert*-BuOK. By employing a procedure similar to that used with **Hdpa**, **H₂tpda** was synthesized by reacting 2,6-diaminopyridine with 2-chloropyridine. For ligands having seven or more than seven nitrogen atoms, the amination was unsuccessful under the conditions similar to those for **Hdpa** or **H₂tpda**. Recently,

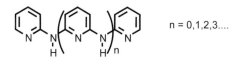

n = 0,1,2,3....

n = 0   Hdpa = dipyridylamine
n = 1   H₂tpda = tripyridyldiamine
n = 2   H₃teptra = tetrapyridyatriamine
n = 3   H₄peptea = pentapyridyltetraamine
n = 4   H₅heppea = hexapyridylpentaamine
n = 5   H₆hpphea = heptapyridylhexaamine
n = 6   H₇ocphpa = octapyridylheptaamine
n = 7   H₈nopoca = nonapyridyloctaamine
n = 8   H₉depnoa = decapyridylnonaamine
n = 9   H₁₀unpdea = undecapyridyldecaamine
n = 10   H₁₁dopuna = dodecapyridylundecaamine
n = 11   H₁₂trpdoa = tridecapyridyldodecaamine

Scheme 5.1. Structures of oligopyridylamine ligands

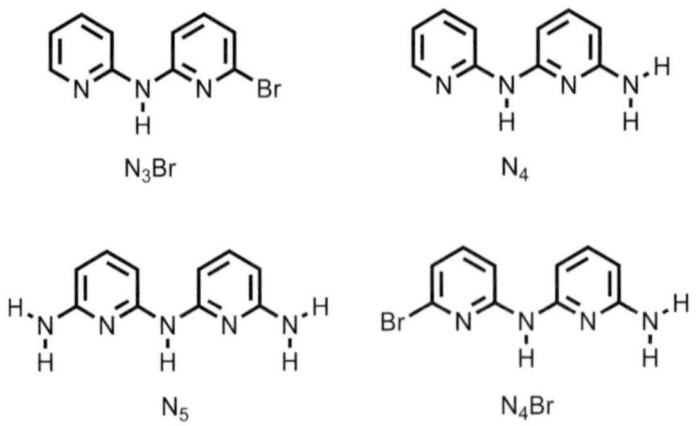

Scheme 5.2. Building blocks for synthesis of oligopyridylamine ligands

Buchwald et al. have developed a facial means for cross-coupling amination by the reaction of the corresponding aryl halide and aryl amine in the presence of palladium catalyst and the appropriate phosphine ligand [41–43]. Using Buchwald's conditions, a series of oligo-α-aminopyridines were successfully synthesized.

The key building blocks used for the synthesis of these oligo-α-aminopyridines are shown in Scheme 5.2. N_3Br was synthesized by treating 2-aminopyridine with 2,6-dibromopyridine [44]. The reaction between 2,6-diaminopyridine with 2-chloropyridine gave the building block N_4 along with tripyridyldiamine as the byproduct. However, the yield of the byproduct can be suppressed when excess 2,6-diaminopyridine is used. Building block N_5 was prepared from 2,6-diaminopyridine catalyzed by hydrochloride. The bifunctional N_4Br building block was prepared by amination of 2,6-diaminopyridine and 2,6-dibromopyridine using $NaNH_2$ as the base. Using the appropriate building blocks, the preparation of short ligands can be achieved by a one-step reaction. For example, the Pd-catalyzed coupling reaction between N_3Br and N_4 gave tetrapyridyltriamine ($H_3teptra$), and coupling of one equivalent of 2,6-diaminopyridine to two equivalents of N_3Br afforded pentapyridyltetraamine ($H_4peptea$) in the presence of palladium catalyst. In contrast, the synthesis of long ligands is time consuming and laborious. As an example, Scheme 5.3 depicts the synthetic routes for three long-chained ligands nonapyridyloctaamine ($H_8nopoca$), decapyridylnonaamine ($H_9depnoa$), and undecapyridyldecaamine ($H_{10}unpdea$) [45]. The starting material N_4Br was coupled to N_3Br using typical Buckwald's conditions to give intermediate N_7Br which was then dicoupled to 2,6-diaminopyridine under similar conditions to afford $H_8nopoca$. The ligand $H_9depnoa$ was synthesized by self-coupling of N_4Br followed by reaction with 2-bromopyridine to form N_9Br. The reaction of N_9Br with one equivalent of 2,6-diaminopyridine afforded N_{12}, which was then coupled to N_7Br and N_9Br to give $H_9depnoa$ and $H_{10}unpdea$, respectively. The cross-coupling amination step leading to $H_9depnoa$ or $H_{10}unpdea$ has low yield when commonly used solvents are employed as the media. However,

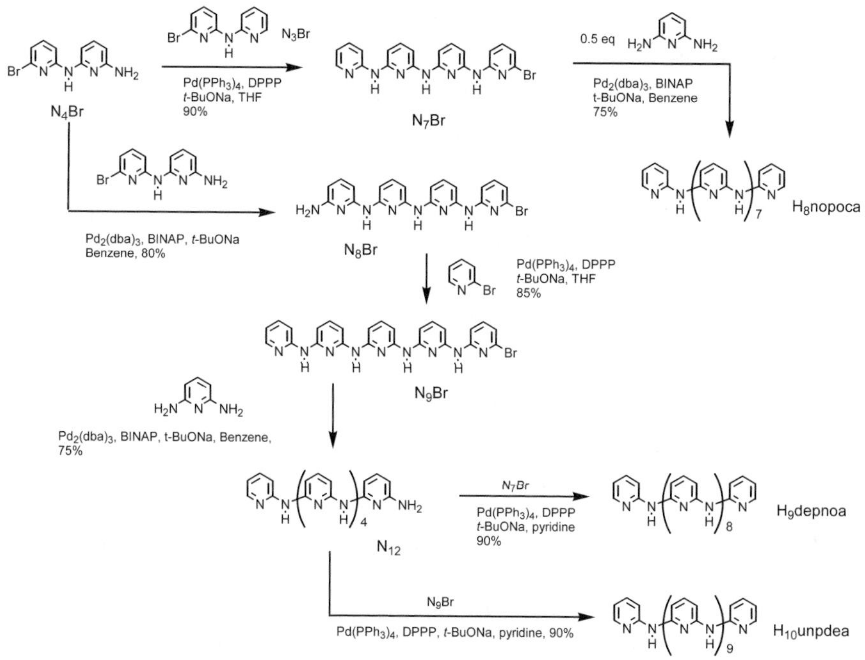

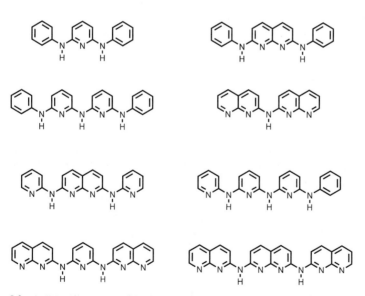

Scheme 5.3. Synthetic routes for long ligands **H₈nopoca**, **H₉depnoa**, and **H₁₀unpdea**

Scheme 5.4. Structures of naphthyridine-containing ligands

the reaction proceeds with good yield when pyridine is used as the solvent due to its high solvating ability. Using an approach similar to that shown in Scheme 5.3, we have successfully synthesized the long ligand that contains up to 25 nitrogen atoms, although the solubility of long ligands is moderately low in most organic solvents and their purification is tedious.

To fine-tune the properties of the multinuclear complexes, some new ligands with naphthyridyl and/or phenyl groups have been synthesized in our laboratory. Their structures are shown in Scheme 5.4. Likewise, these new ligands can be prepared under Buchwald's amination conditions.

5.3
Dimerization by Self-Complementary Hydrogen Bonding

It should be noted that this new class of ligands contains complementary donor and acceptor units that can undergo self-dimerization by hydrogen bonding interactions [46].

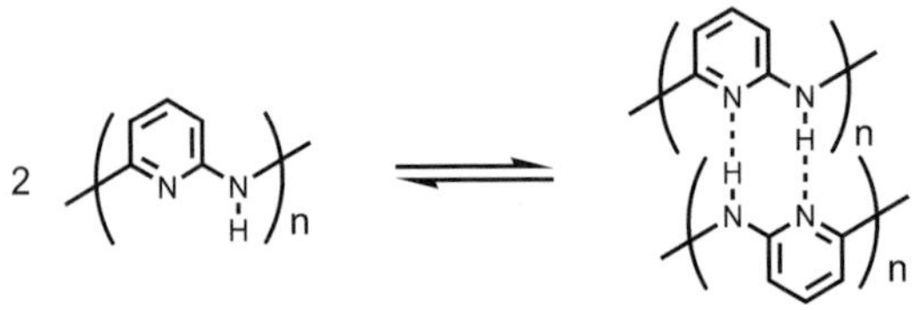

Figure 5.1 shows the crystal structure of an oligopyridylamine in the dimeric form. The structure adopts a spiral conformation with six pairs of N–H·····N hydrogen bonds. The N–H·····N distances vary in a range of 3.02 to 3.12 Å, suggesting strong intermolecular hydrogen-bond interactions. The ligands dimerize not only in the solid state but also in solution. The formation of

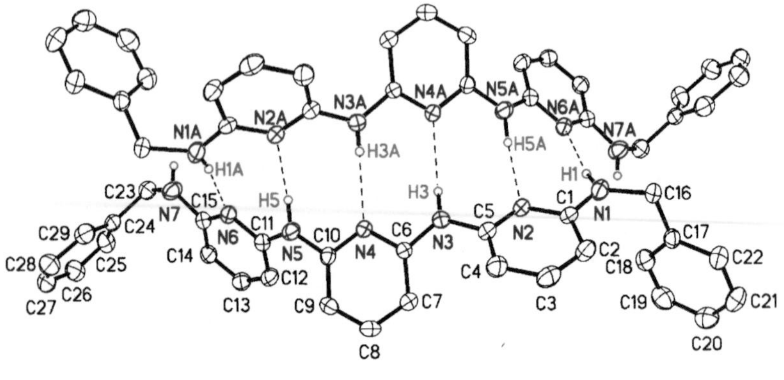

Fig. 5.1. Crystal structure of dimeric form of oligopyridylamine

the hydrogen-bonded dimers is confirmed by ^{1}H NMR and UV-vis studies. The dimerization constant increases as the number of hydrogen-bonded pairs increases. For example, the ligands with two and ten pairs of N–H ... N hydrogen bonds have dimerization constants in the orders of 10^{-2} and 10^4 M^{-1}, respectively. Thermodynamic studies suggest that about 54% of the enthalpy gain from dimerization interactions would be sacrificed to compensate for the entropic free-energy loss.

5.4
Complexation of Oligopyridylamine Ligands

Oligopyridylamines have two types of nitrogen atoms, i.e., the pyridyl nitrogens and the amido nitrogens, which provide binding capability with metal ions. The possible complexation of these oligopyridylamines can be catagorized into three types, *syn–syn*, *syn–anti*, and *anti–anti* [47–50].

syn-syn   anti-anti   syn-anti

5.5
Mono- and Dinuclear Complexes

Numerous all-syn and all-anti forms of complexes have been synthesized in our laboratory. Figure 5.2 shows the structures of [Fe(H$_2$tpda)$_2$]$^{2+}$ and [Cu(H$_3$teptra)]$^{2+}$ with the ligands in an all-anti conformation [51]. The mononuclear complex [Fe(H$_2$tpda)$_2$]Cl$_2$ was synthesized by treating iron(II) chloride with two equivalents of **H$_2$tpda** in methanol. In the structure of [Fe(H$_2$tpda)]$^{2+}$, the three pyridyl groups of each **H$_2$tpda** ligand are essentially planar and the pyridyl rings are twisted with a dihedral angle of 38° between the neighboring pyridyls. The Fe–N distances range from 1.981 to 2.005 Å. Under similar conditions, [Cu(H$_3$teptra)](ClO$_4$)$_2$ was synthesized by reacting Cu(ClO$_4$)$_2$ and one equivalent of **H$_3$teptra** ligand in acetonitrile. The four pyridyl groups of the **H$_3$teptra** ligand are planar, and the dihedral angles between the neighboring pyridyl rings are 25.3°, 15.0°, and 20.4°, respectively. When long ligands having more than eight pyridyl groups are used, it is possible to construct their dinuclear complexes. As an example, Fig. 5.3 shows the structures of the dication of two dinuclear complexes [Ni$_2$(H$_{10}$unpdea)](ClO$_4$)$_4$ and [Cu$_2$(H$_{10}$unpdea)](ClO$_4$)$_4$, which were prepared by reacting **H$_{10}$unpdea** with

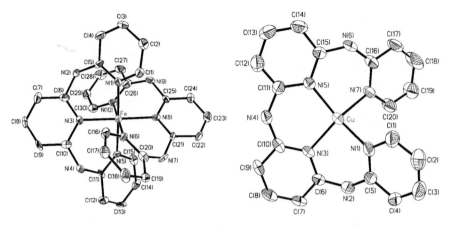

Fig. 5.2. Crystal structures of $[Fe(H_2tpda)]^{2+}$ (*left*) and $[Cu(H_3teptra)]^{2+}$ (*right*) with ligands in all-anti conformation

2.1 equivalents of $Ni(ClO_4)_2$ and $Cu(ClO_4)_2$, respectively [45]. In the structure of $[Ni_2(H_{10}unpdea)]^{4+}$, the ligand helically wraps the two nickel ions. The six-coordinated nickel ion adopts an octahedral arrangement, whereas the five-coordinated nickel ion has a distorted square pyramid geometry. The Ni ... Ni distance is 5.900 Å. The structure of $[Cu_2(H_{10}unpdea)]^{4+}$ also exhibits a helical arrangement with two four-coordinated copper ions. Both copper ions have distorted square planar geometry, and the distance between these two copper ions is 7.420 Å. As some dinickel and dicopper complexes promote urease and oxygenase reactions, respectively [52, 53], our dinuclear complexes may find use in the catalytic reactions. Considering the helical arrangement of the complexes, they may also be used in molecular recognition.

5.6
Structures of Linear Multinuclear Nickel Complexes

In addition to the *all-anti* complexes, *all-syn* form complexes of these oligopyridylamines have been synthesized. In general conditions, the ligand is reacted with the stoichiometric amount of the metal salt in the presence of *tert*-BuOK using refluxing naphthalene as the solvent. In the past decade, numerous linear multinuclear metal complexes have been prepared. The longest multinuclear metal complex of oligopyridylamines with a crystal structure is the nonanickel complex $[Ni_9(peptea)_4Cl_2]$ [40]. Figure 5.4 shows the side view and the view along the Ni_9 axis of its crystal structure. The structure of this complex consists of a linear chain of nine nickel atoms that are supported by four peptea^{4-} ligands, with two chloride ions in the axial positions. As can be seen from the end view, the peptea^{4-} ligands helically wrap the linear metal chain with an overall twist angle of about 180° from one pyridyl end to the other pyridyl end

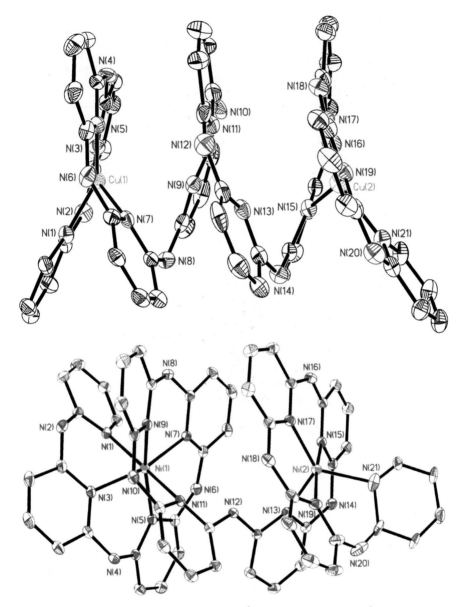

Fig. 5.3. Crystal structures of $[Ni_2(H_{10}unpdea)]^{4+}$ and $[Cu_2(H_{10}unpdea)]^{4+}$

due to the steric crowding in the pyridyl hydrogens. Similar to the nonanickel system, all the neutral molecules of the tri-, penta-, and heptanickel complexes adopt a symmetrical form with a linear metal atom chain.

Figure 5.5 shows the comparisons of Ni–Ni and Ni–N distances for these multinickel complexes. Though the Ni–Ni distances are short, there is no bond

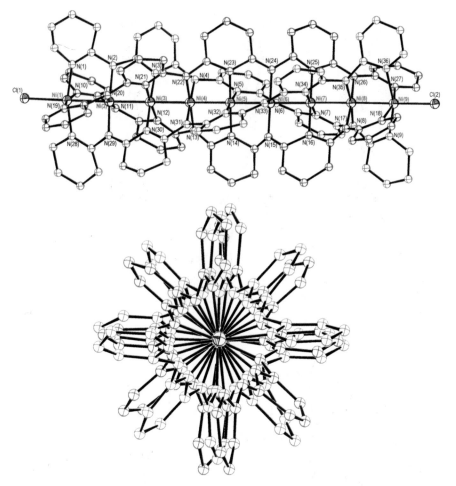

Fig. 5.4. Side view (*top*) and end view (*bottom*) of crystal structure of [Ni$_9$(peptea)$_4$Cl$_2$]

between the nickel ions. Generally, the terminal Ni–Ni distances are longer than the inner ones in these complexes. The outermost nickel ions have Ni–N lengths of about 2.1 Å, which are longer by about 0.2 Å than the internal ones. These results indicate that the terminal nickel ions are in a high-spin state and the internal ones in a low-spin state, which are in accordance with the magnetic measurements and qualitative M.O. calculation. The antiferro-magnetic coupling constants of the magnetic moments on the terminal nickel ions are −99, −8.3, −3.8, and −1.7 cm^{-1} for [Ni$_3$(dpa)$_4$Cl$_2$], [Ni$_5$(tpda)$_4$Cl$_2$], [Ni$_7$(teptra)$_4$Cl$_2$], and [Ni$_9$(peptea)$_4$Cl$_2$], respectively [40].

As mentioned, there is no metal–metal bonding between the neighboring nickel ions in the neutral forms of these multinickel complexes. In contrast, the molecules show a partial Ni–Ni bond in their oxidized forms. In 2002, Cot-

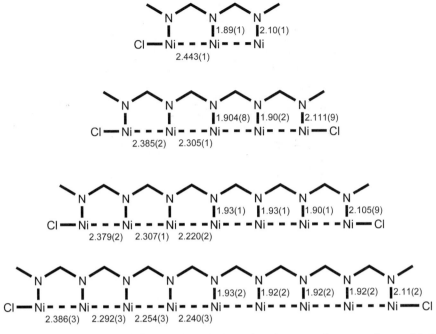

Fig. 5.5. Comparisons of Ni–Ni and Ni–N distances for tri-, penta-, hepta-, and nonanickel complexes

ton et al. reported the oxidation of [Ni$_3$(dpa)$_4$Cl$_2$] to [Ni$_3$(dpa)$_4$(PF$_6$)$_2$](PF$_6$) by the use of AgPF$_6$ and the crystal structure of [Ni$_3$(dpa)$_4$(PF$_6$)$_2$](PF$_6$) [54]. The trinickel core is symmetrical with two PF$_6^-$ anions as the axial ligand and the positive charge is compensated by another PF$_6^-$ anion. The most striking feature of the structure is the short Ni–Ni distance of 2.283 Å, which is shorter by about 0.15 Å than that in the neutral molecule [Ni$_3$(dpa)$_4$Cl$_2$]. The electronic structure of the one-electron oxidation complex has been significantly changed. Instead of two terminal nickel ions being in a high-spin state in the neutral molecule, there is only one unpaired electron delocalized over the trinickel unit in the oxidized form.

At the time Cotton et al.'s work on the oxidized trinickel complex was reported, we had synthesized the one-electron oxidation counterparts of the pentanickel complexes [35]. The reaction of [Ni$_5$(tpda)$_4$Cl$_2$] with excess AgBF$_4$ and AgCF$_3$SO$_3$ afforded [Ni$_5$(tpda)$_4$(H$_2$O)(BF$_4$)](BF$_4$)$_2$ and [Ni$_5$(tpda)$_4$(CF$_3$SO$_3$)$_2$] (CF$_3$SO$_3$), respectively. Figure 5.6 shows the crystal structure of [Ni$_5$(tpda)$_4$ (CF$_3$SO$_3$)$_2$] (CF$_3$SO$_3$). The most striking feature of their structures is the unsymmetrical Ni$_5$ chain. The unsymmetrical Ni–Ni bond distances in [Ni$_5$(tpda)$_4$(CF$_3$SO$_3$)$_2$](CF$_3$SO$_3$)   are   2.358(2), 2.276(2), 2.245(2), and 2.304(2) Å. Compared to the Ni–Ni bond lengths in [Ni$_5$(tpda)$_4$Cl$_2$], both the internal and external Ni–Ni bond distances in the oxidized compounds are

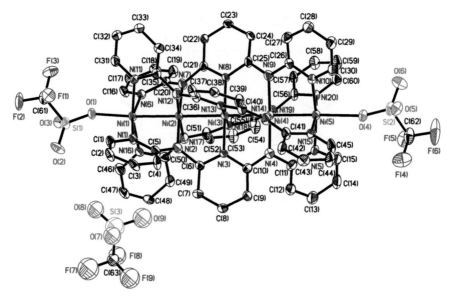

Fig. 5.6. Crystal structure of $[Ni_5(tpda)_4(CF_3SO_3)_2](CF_3SO_3)$

shorter by 0.04–0.08 Å. The significant differences between the two terminal nickel ions can also be viewed from the different metal–ligand bond distances of 2.059(7) and 2.338(8) Å for Ni(1)–O(1) and Ni(5)–O(4), respectively. Unlike the case of compound $[Ni_5(tpda)_4Cl_2]$ where the Ni–N distances for the terminal Ni ions are considerably longer than those for the internal Ni ions, the average Ni–N distances in $[Ni_5(tpda)_4(CF_3SO_3)_2](CF_3SO_3)$ are 2.081(10), 1.874(10), 1.904(9), 1.885(11), and 1.908(9) Å for Ni(1), Ni(2), Ni(3), Ni(4), and Ni(5), respectively. These results indicate that Ni(1) is in a high-spin state while the other four Ni ions are in a low-spin state, and the one-electron oxidation occurs at the terminal Ni(5) ion.

Theoretical calculations show that there are no metal–metal bonds in the neutral molecules of the pentanickel complexes and the bond order for Co–Co in neutral pentacobalt complexes is estimated to be 0.5 [36]. Comparing $[Ni_5(tpda)_4(CF_3SO_3)_2](CF_3SO_3)$ to $[Ni_5(tpda)_4Cl_2]$ and $[Co_5(tpda)_4Cl_2]$, the Ni–Ni bond distances are in between the corresponding metal–metal distances in $[Ni_5(tpda)_4Cl_2]$ and $[Co_5(tpda)_4Cl_2]$ (Fig. 5.7). We believe that considerable unsymmetrical Ni–Ni bonding interactions, especially on the Ni(5) site, exist in the oxidized pentanickel complexes. The magnetic data suggest that the terminal Ni(1) ion with $S = 1$ is antiferromagnetically coupled with the terminal Ni(5) ion with $S = 1/2$ or the remaining Ni$_4$ unit with a delocalized unpaired electron. The coupling constants of the antiferromagnetic interactions in the pentanickel core are estimated to be −555 and −318 cm^{-1} for $[Ni_5(tpda)_4(H_2O)(BF_4)](BF_4)_2$ and $[Ni_5(tpda)_4(CF_3SO_3)_2](CF_3SO_3)$, respectively, which are much higher than those for $[Ni_5(tpda)_4Cl_2]$ (−8.3 cm^{-1}) and

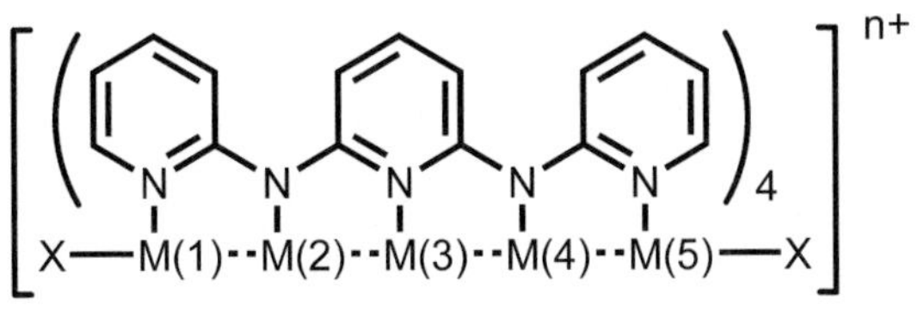

Co—Co		2.281	2.236	2.233	2.277	n = 0	X = NCS⁻
Ni—Ni		2.385	2.306	2.306	2.385	n = 0	X = Cl⁻
Ni—Ni		2.358	2.276	2.245	2.304	n = 1	X = CF₃SO₃⁻

Fig. 5.7. Comparisons of metal–metal bond distances in $[Ni_5(tpda)_4Cl_2]$, $[Co_5(tpda)_4Cl_2]$, and $[Ni_5(tpda)_4(CF_3SO_3)_2](CF_3SO_3)$

$[Ni_3(dpa)_4Cl_2]$ (-99 cm^{-1}), indicating strong metal–metal interactions in these oxidized complexes. The magnetic results are in agreement with the X-ray crystal structural analyses.

Using a procedure similar to that for the electron oxidation pentanickel complexes, the one-electron oxidation heptanickel complexes, $[Ni_7(teptra)_4(OH)(NO_3)](NO_3)$ and $[Ni_7(teptra)_4(OH)(H_2O)](CF_3SO_3)_2$, were prepared by reacting $[Ni_7(teptra)_4Cl_2]$ with excess AgNO$_3$ and AgCF$_3$SO$_3$, respectively. The structure of $[Ni_7(teptra)_4(OH)(NO_3)](NO_3)$ shows that the Ni–Ni distances are 2.242(2), 2.278(2), and 2.360(3) Å for the innermost, intermediate, and terminal Ni–Ni bonds, respectively. The average Ni$_{terminal}$–N bond distance of 2.098(12) Å is about 0.20 Å longer than the average Ni$_{internal}$–N one. One might expect a decrease in the Ni–Ni bond distances upon one-electron oxidation. However, the innermost average Ni–Ni bond distances of 2.242(2) Å for $[Ni_7(teptra)_4(OH)(NO_3)](NO_3)$ and 2.248(2) Å for $[Ni_7(teptra)_4(OH)(H_2O)]$ $(CF_3SO_3)_2$ are slightly longer by 0.02 to 0.03 Å when compared to those in $[Ni_7(teptra)_4Cl_2]$. The average Ni–Ni and Ni–N bond distances in $[Ni_7(teptra)_4$ $(OH)(NO_3)](NO_3)$ and $[Ni_7(teptra)_4(OH)(H_2O)](CF_3SO_3)_2$ are not significantly different from those in $[Ni_7(teptra)_4Cl_2]$. It is difficult to describe the electronic configurations for the nickel atoms in $[Ni_7(teptra)_4(OH)(NO_3)]$ (NO_3) and $[Ni_7(teptra)_4(OH)(H_2O)](CF_3SO_3)_2$ on the basis of crystal data alone. However, one-electron oxidation on one of the terminal nickel atoms, which will result in a spin-state change from high spin to low spin, can be ruled out since significant changes on the Ni–Ni and Ni–N bond lengths for the terminal nickel atoms are not found. Based on the measured μ_{eff} value of 4.69 B.M. for $[Ni_7(teptra)_4(OH)(NO_3)](NO_3)$ at 300 K, there should be five unpaired electrons, two on each terminal Ni(II) ion and one on the internal pentanickel unit. Thus the electronic configuration of $[Ni_7(teptra)_4(OH)(NO_3)](NO_3)$ is best described as the terminal nickel ions in a high-spin state and the internal pentanickel unit, which carries an unpaired electron, in a low-spin state.

5.7
Structures of Linear Multinuclear Cobalt Complexes

It has been found that trinuclear cobalt complexes exhibit both symmetrical and unsymmetrical structures depending on the nature of the axial ligands, the crystal environment, and the oxidation states [18–26]. We reported the first unsymmetrical structure of a tricobalt complex $[Co_3(dpa)_4Cl_2]$, in which the cobalt ions group up to form a Co_2 unit and an isolated Co(II) ion [18]. As shown in Fig. 5.8, the short and long Co–Co distances are 2.290(3) and 2.472(3) Å, respectively. The average Co–N distances in the Co_2 unit are 1.96 Å for the pyridyl nitrogen and 1.90 Å for the amido nitrogen. The average Co–N distance of 2.121 Å for the isolated cobalt ion is considerably longer than those in the Co_2 unit. The long Co–N distance for the isolated cobalt ion is partly compensated by the stronger Co–Cl interaction. Shortly after our report, the symmetrical structure of this complex was discovered by Cotton et al. [10]. Further work by the same group confirmed that both symmetrical and unsymmetrical structures can exist. In the symmetrical structure, the Co–Co bond distances range from 2.25 to 2.32 Å from one compound to another depending on the crystallization conditions and the nature of axial ligands. The metal–metal distances are shorter in the symmetrical tricobalt system than in the trinickel analogs, indicating that a partial metal–metal bond exists in the Co_3 unit. The Co–N distances for the pyridyl nitrogens are in the range 1.95–2.03 Å, and those for the amido nitrogen are in the range 1.89–1.91 Å. The magnetic measurements show that both symmetrical and unsymmetrical tricobalt complexes exhibit spin crossover behavior.

Unlike the case of $[Co_3(dpa)_4Cl_2]$, wherein both unsymmetrical and symmetrical structures can exist, the five cobalt atoms in pentacobalt complexes can only exist in a symmetrical arrangement [36]. The crystal structures of the neutral pentacobalt complexes with only the axial ligands being different have a great deal in common. In the complexes, two types of Co–Co bond distances are observed. The average internal Co–Co lengths are in the range 2.22–2.24 Å, whereas the external bond distances are in the range 2.27–2.29 Å. All the Co–N lengths vary in the range 1.90–1.99 Å. The shorter metal–metal bond distances as compared to those in the pentanickel complexes suggest Co–Co interactions in the Co_5 core. Theoretical calculations show that the bond order delocalized over the Co_5 unit is two and there is an unpaired electron in the HOMO. Magnetic studies indicate that the molecules are paramagnetic and there is an unpaired electron in the Co_5 core. As the tricobalt complexes exhibit spin crossover, the behavior of spin crossover is not observed in the pentacobalt system.

The structure of the oxidized tricobalt complex $[Co_3(dpa)_4Cl_2](BF_4)$ is essentially symmetrical with a linear Co_3 chain [23]. The Co–Co distances are 2.3168(8) and 2.3289(8) Å, which are slightly longer than in the neutral molecules. This can be ascribed to the larger positive-charge repulsion between

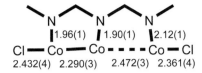

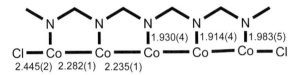

Fig. 5.8. Comparisons of Co–Co and Co–N bond distances in unsymmetrical and symmetrical forms of [Co$_3$(dpa)$_4$Cl$_2$], and in [Co$_5$(tpda)$_4$Cl$_2$]

the cobalt ions upon oxidation. As expected for a metal cation with more positive charge, the average terminal Co–N distance of 1.977 Å and the average central Co–N distance of 1.869 Å are shorter by 0.011 and 0.028 Å, respectively, relative to that of the neutral counterpart. The most significant change in bond distances of the oxidized molecule is found in the Co–Cl distance, which is about 0.15 Å shorter than that in [Co$_3$(dpa)$_4$Cl$_2$]. The significant decrease in the Co–Cl distance is consistent with the theoretical calculation, which suggests that the singly occupied HOMO of [Co$_3$(dpa)$_4$Cl$_2$] is of Co–Co nonbonding and Co–Cl antibonding character. Similar to the neutral molecule, the oxidized form also exhibits spin crossover behavior.

One might expect that some of the interatomic bond distances would be shorter as the neutral pentacobalt compounds are oxidized. However, the Co–N distances for all cobalt atoms remain essentially unchanged upon one-electron oxidation [36]. Moreover, all of the Co–Co bond lengths are slightly longer (by about 0.01–0.02 Å) than those in the neutral analogs. As discussed in the case of the tricobalt system, the average Co–Cl (X = axial ligand) bond distance in the one-electron oxidation species is 0.15 Å shorter than that of the neutral one. Similar to the case of the tricobalt compounds, theoretical calculations for the pentacobalt complexes reveal that the singly occupied HOMO also characterizes metal–metal nonbonding and metal–ligand antibonding. A large decrease in Co–Cl bond distances is expected upon removal of an electron from the neutral molecule. However, the average Co–Cl bond distance of 2.412(3) Å is only slightly shorter (by 0.033 Å) than that in the neutral compound. Most likely, the electron is removed not from the singly occupied HOMO (σ nonbonding) but from the HOMO-1 (δ-nonbonding) to form a paramagnetic complex with $S = 1$, which is in accordance with the measured μ_{eff} value of 3.18 B.M. for $[Co_5(tpda)_4Cl_2](ClO_4)$ at 300 K.

5.8
Structures of Linear Multinuclear Chromium Complexes

In 1997, Cotton et al. reported the structure of the first trichromium complex $[Cr_3(dpa)_4Cl_2]$ with a linear Cr_3 chain in a symmetrical arrangement [27]. The Cr–Cr distance is 2.365(1) Å. In subsequent work on the trichromium complexes, they found that both symmetrical and unsymmetrical Cr_3 chains exist for complexes with identical axial ligands while only unsymmetrical arrangement is observed for complexes with different ligands in the axial positions. For the unsymmetrical structure, the Cr_3 chain consists of a diamagnetic quadruply bonded Cr_2 unit and an isolated Cr(II) ion. In 2004, the same group reinvestigated their previously reported structures of $[Cr_3(dpa)_4Cl_2]$ that crystallize in various solvents [30]. They found that the Cr_3 unit has one short Cr–Cr distance ranging from 2.227 to 2.254 Å and a longer one in the range 2.469–2.483 Å. It should be mentioned that the structure of the ethyl-substituted $[Cr_3(depa)_4Cl_2]$ (depa = anion of di(4-ethylpyridyl)amine) has a symmetrical Cr_3 core with a Cr–Cr distance of 2.3780(5) Å. Our work on the trichromium complex $[Cr_3(dpa)_4(NCS)_2]$ shows that the Cr_3 unit is symmetrical, while Cotton's group described the structure for the same complex as unsymmetrical. Though conclusive evidence for the exact structure has not been obtained, the symmetrical structure of $[Cr_3(dpa)_4(NCS)_2]$ is consistent with our STM studies, as will be discussed later on [55].

As compared to trinickel and tricobalt complexes, the unsymmetrical structure of trichromium complexes has a shortest M–M distance ascribed to the quadruple bond of the Cr_2 unit and the symmetrical form shows an intermediate M–M distance. Similar to the case of trichromium complexes, we found that the neutral pentachromium complexes $[Cr_5(tpda)_4X_2]$ (X = Cl or NCS) have a symmetrical Cr_5 core while others described the structure as unsymmetrical with alternating short and long Cr–Cr distances [34]. In the symmetrical structure of $[Cr_5(tpda)_4Cl_2]$, two types of Cr–Cr distance are observed, with the outer one being 2.284(1) Å and the inner one 2.2405(8) Å. The Cr–N distances are 2.119(3), 2.021, and 2.050 Å for the terminal Cr–N_{py}, terminal Cr–N_{amido}, and central Cr–N_{py}, respectively. The Cr–Cr and Cr–N distances do not significantly change when the axial ligands are replaced by thiocyanate ions. The M.O. calculation shows that pentachromium complexes have an electronic configuration of $\sigma_1^2\sigma_2^2\pi_1^4\pi_2^4\delta_{n1}^2\delta_{n2}^2\delta_{n3}^2\pi_{n(dxz)}^1\pi_{n(dyz)}^1$, suggesting that there are two unpaired electrons delocalized over the Cr_5 unit, and a total bond order of six is delocalized over the five chromium ions. The measured μ_{eff} value of 4.0 B.M. is larger than the expected spin-only value (2.82 B.M.), and this can be ascribed to the orbital contribution of π-nonbonding orbitals. In the unsymmetrical structure of $[Cr_5(tpda)_4Cl_2]$ reported by Cotton et al., the Cr_5 unit has alternating Cr–Cr distances of 2.587(7), 1.901(6), 2.587(6), 2.031(6), and 2.604(5) Å. The short Cr–Cr distances correspond to the quadruply bonded Cr–Cr units and there is no metal–metal bonding between the chromium ions with the long metal·····metal interatomic distances.

Simply based on the results of trichromium and pentachromium complexes without any structural evidence for the heptachromium complexes, Cotton et al. predict that the neutral heptachromium analogs should be unsymmetrical with alternating long and short Cr–Cr bond distances [56]. However, our work on the heptachromium complexes shows that the neutral molecule $[Cr_7(teptra)_4Cl_2]$ adopts a symmetrical structure with three types of Cr–Cr bond lengths [37]. The average Cr–Cr distances are 2.286(2), 2.243(2), and 2.213(2) Å corresponding to the terminal, intermediate, and outermost Cr–Cr bonds, respectively. The comparisons of Cr–Cr and Cr–N distances for tri-, penta-, and heptachromium complexes are given in Fig. 5.9.

To date, all the oxidized forms of tri-, penta-, and heptachromium complexes have been found to be unsymmetrical whether the axial ligands are identical or different. Numerous one-electron oxidation trichromium complexes of the type $[Cr_3(dpa)_4Cl_2](X)$ (X = counter anion) have been structurally characterized [30]. In the structures, two of three chromium ions group up to form a quadruple Cr–Cr bond with bond distances varying in the range 2.01–2.12 Å.

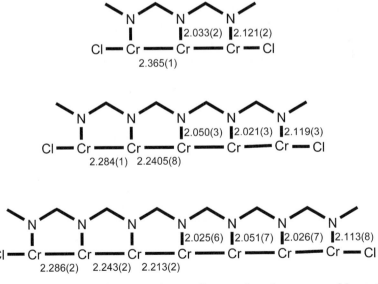

Fig. 5.9. Comparison of Cr–Cr and Cr–N distances for tri-, penta-, and heptachromium complexes

The long Cr·····Cr distances range from 2.47 to 2.56 Å. For one-electron oxidation complexes with different ligands occupying the axial positions, the
difference between the long and short Cr–Cr distances is quite large. The short
Cr–Cr length lies within the range 1.90–2.07 Å and the long ones range from
2.48 to 2.58 Å. The electronic configuration of these oxidized complexes is described as the quadruply bonded Cr_2 unit being diamagnetic and the isolated
$Cr^{(III)}$ being paramagnetic ($S = 3/2$).

In our laboratory, the one- and two-electron oxidation pentachromium complexes have been synthesized by reacting the neutral complex $[Cr_5(dpa)_4Cl_2]$
with two and ten equivalents of silver salts, respectively. Under similar conditions, the one-and two-electron oxidation heptachromium complexes have
also been obtained. Some selected bond distances for the oxidized complexes
are given in Table 5.1. In the structures of these oxidized complexes, one of
the terminal chromium ions is isolated and the other chromium ions pair
up to form Cr_2 units. Again, the alternating long and short Cr–Cr distances
are observed. The long Cr·····Cr distances vary in the range 2.377–2.596 Å,
and the short Cr–Cr distances lie within the range of a quadruple Cr–Cr
bond. It should be mentioned that the Cr(2)–Cr(3) distance of 1.8377(14) Å
in $[Cr_5(tpda)_4(BF_4)(F)](BF_4)$ is comparable to the supershort Cr–Cr distances
(<1.90 Å). The first electron removed from the neutral molecules definitely
occurs at the isolated chromium ion to form a high-spin $Cr^{(III)}$ ion ($S = 3/2$),
while the exact oxidation center is unclear as the second electron is removed.

Table 5.1. Selected bond distances for oxidized penta- and heptachromium complexes

	1	2	3	4
Cr(1)–X	1.840(4)	1.809(4)	1.887(6)	1.871(5)
Cr(5 or 7)–Y	1.931(4)	2.184(4)	1.923(6)	1.937(5)
Cr(1)–Cr(2)	2.487(2)	2.5963(14)	2.377(5)	2.455(2)
Cr(2)–Cr(3)	1.969(2)	1.8377(14)	1.948(10)	1.996(2)
Cr(3)–Cr(4)	2.419(2)	2.5719(14)	2.462(10)	2.402(2)
Cr(4)–Cr(5)	2.138(2)	1.9614(14)	1.963(10)	2.007(2)
Cr(5)–Cr(6)			2.454(10)	2.395(2)
Cr(6)–Cr(7)			2.220(6)	2.143(2)
Cr(1)–N	2.070(7)	2.059(6)	2.068(8)	2.053(7)
Cr(2)–N	2.004(6)	2.007(6)	2.004(9)	2.007(7)
Cr(3)–N	2.034(6)	2.037(5)	2.056(9)	2.042(7)
Cr(4)–N	1.992(6)	1.993(6)	2.019(9)	2.011(7)
Cr(5)–N	2.079(6)	2.080(6)	2.055(9)	2.037(7)
Cr(6)–N			2.010(10)	1.999(7)
Cr(7)–N			2.069(9)	2.073(8)

1 = $[Cr_5(tpda)_4F_2](BF_4)$, 2 = $[Cr_5(tpda)_4(BF_4)(F)](BF_4)$, 3 = $[Cr_7(teptra)_4F_2](BF_4)$,
4 = $[Cr_7(teptra)_4F_2](BF_4)_2$; X and Y are the axial ligands

5.9
Structures of Triruthenium and Trirhodium Complexes

The structures of $[Ru_3(dpa)_4Cl_2]$ and $[Rh_3(dpa)_4Cl_2]$ are symmetrical with delocalized metal–metal bond distances of 2.2537(5) Å for the Ru–Ru bond and 2.3920(5) Å for the Rh–Rh bond, respectively [31]. Table 5.2 lists the structural comparison of symmetrical trinuclear metal complexes. According to M.O. analysis, there is an overall bond order of three delocalized over the Ru_3 unit and the overall bond order is 1.5 for the Rh_3 unit. The Ru–Ru bond distance of 2.2537(5) Å is slightly shorter than that of $[Ru_2(OAc)_4Cl] \cdot 2H_2O$ [57], 2.267(1) Å, which has a bond order of 2.5 and that of $[Ru_2(OAc)_4(thf)_2]$ [58] is 2.261(3) Å, which has a bond order of 2.0. The Rh–Rh bond distance of 2.3920(5) is comparable to that of the dinuclear complex $[Rh_2(OAc)_4(H_2O)_2]$ [59], 2.3855(5) Å, which has a bond order of 1.0. It should be mentioned that the Ru_3 unit deviates slightly from linearity with $\angle Ru(1)-Ru(2)-Ru(3)$ of 171.17(4)° and $\angle Ru(2)-Ru(3)-Cl$ of 174.10(8)°.

Recently, we successfully synthesized the one- and two-electron oxidation triruthenium complexes $[Ru_3(dpa)_4Cl_2](BF_4)$ and $[Ru_3(dpa)_4F_2](BF_4)_2$ by reacting $[Ru_3(dpa)_4Cl_2]$ with ferrocenium tetraflouroborate and $AgBF_4$, respectively. In the structure of $[Ru_3(dpa)_4Cl_2](BF_4)$, the Ru_3 unit is essentially symmetrical. The Ru–N distances lie in the range 2.04–2.12 Å, which are comparable with those of $[Ru_3(dpa)_4Cl_2]$. The average Ru–Ru bond length is 2.291 Å, which is slightly longer (by 0.035 Å) than that of the neutral analog $[Ru_3(dpa)_4Cl_2]$. The lengthening of the Ru–Ru bond may be ascribed to the stronger repulsion between the ruthenium ions in the Ru_3^{7+} unit. The most striking feature is that the average Ru–Cl distance exhibits a significant change from 2.574 to 2.466 Å as an electron is removed from the neutral compound $[Ru_3(dpa)_4Cl_2]$. The significant change in the Ru–Cl distances is consistent with the metal–metal nonbonding and metal–axial ligand antibonding character of the HOMO. It is noteworthy that the Ru_3 unit is not in a linear arrangement ($\angle Ru(1)-Ru(2)-Ru(3) = 166.66(3)°$). The reason for the nonlinearity of the Ru_3 unit remains unclear. Table 5.3 gives a comparison of bond distances and angles of these neutral and oxidized triruthenium complexes.

Table 5.2. Structural comparison of symmetrical trinuclear metal complexes $[M_3(dpa)_4Cl_2]$ (M = Cr(II), Co(II), Ni(II), Cu(II), Ru(II), and Rh(II))

	Cr	Co	Ni	Cu	Ru	Rh
M–Cl	2.561(4)	2.520(2)	2.325(3)	2.465(1)	2.596(2)	2.586(1)
M–M	2.365(1)	2.3178(9)	2.443(1)	2.471(1)	2.2537(5)	2.3920(5)
M–N_{py}	2.121(2)	1.997(3)	2.10(1)	2.05(2)	2.11(4)	2.08(1)
M–N_{amido}	2.033(2)	1.907(3)	1.89(1)	1.96(2)	2.07(3)	2.01(1)

Table 5.3. Selected bond distances (Å) and angles (°) for $[Ru_3(dpa)_4Cl_2](BF_4)$ (5), $[Ru_3(dpa)_4F_2](BF_4)_2$ (6), and $[Ru_3(dpa)_4Cl_2]$ (7)

	5	6	7
Ru(1)–X[a]	2.4656(13)	1.884(4)	2.596(2)
Ru(3)–X	2.4656(13)	1.918(3)	2.596(2)
Ru(1)–Ru(2)	2.2939(6)	2.2727(6)	2.2537(5)
Ru(2)–Ru(3)	2.2882(6)	2.2694(6)	2.2537(5)
Ru(1)–N	2.108(4)	2.086(5)	2.108(6)
Ru(2)–N	2.041(5)	2.033(5)	2.066(6)
Ru(3)–N	2.118(5)	2.093(5)	2.108(6)
∠Ru(1)–Ru(2)–Ru(3)	166.66(3)	179.35(3)	171.15(4)

[a]X = axial ligands

5.10
Complexes of Modified Ligands

Some multinuclear metal complexes of naphthyridine-modified ligands have been synthesized and structurally characterized in our laboratory. Figure 5.10 shows the structures of the dications of the hexanickel and hexacobalt complexes. All the metal ions in these complexes have an oxidation state of two. In the structure of $[Ni_6(bpyan)_4(NCS)_2]^{2+}$ [60], the Ni_6 core adopts a symmetrical arrangement with distances of 2.403(1), 2.314(1), and 2.295(1) Å for the outermost, intermediate, and inner Ni–Ni separations, respectively. The Ni–Ni distances are comparable to those observed in tri-, penta-, hepta-, and nonanickel complexes. The Ni–N_{py} distance is 2.11(2), which is longer by 0.20 Å than Ni–N_{amido} and Ni–N_{naph} distances, suggesting that the terminal Ni(II) ions are in a high-spin state and the internal ones in a low-spin state. The structure of $[Co_6(bpyan)_4(NCS)_2]^{2+}$ is essentially symmetrical [61]. There are two sets of Co–Co distances: the average terminal distance being 2.294(2) Å and the internal ones being 2.251 Å. These Co–Co distances are slightly shorter (by 0.01–0.02 Å) than the corresponding Co–Co distances in the pentacobalt complex $[Co_5(tpda)_4(NCS)_2]$. The average Co–N distances are 1.984(10), 1.911(10), and 1.927(10) Å for Co–N_{py}, Co–N_{amido}, and Co–N_{naph}, respectively, which are shorter by 0.01–0.02 Å than those in the pentacobalt analog.

We have also synthesized the tetranickel complex of the pyridylaminonaphthyridine ligand. The crystal consists of two types of molecules: one with the four ligands orienting in the same direction and the other one with three of the four ligands orienting in one direction and the other in the opposite direction. Figure 5.11 shows the structure of $[Ni_4(pyan)_4(NCS)_2]^{2+}$ with the four ligands positioned in the same direction. The separation between metal ions are 2.4177(16), 2.3335(16), and 2.4044(17) Å for Ni(1)–Ni(2), Ni(2)–Ni(3), and Ni(3)–Ni(4), respectively. The Ni–N distances of 1.911(7) and 1.885(7) Å for

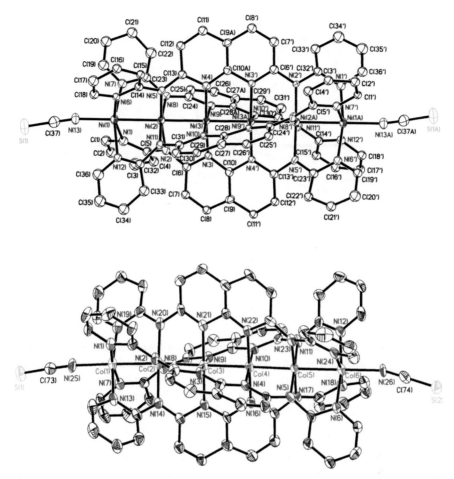

Fig. 5.10. Crystal structures of $[Ni_6(bpyan)_4(NCS)_2]^{2+}$ (*top*) and $[Co_6(bpyan)_4(NCS)_2]^{2+}$ (*bottom*)

the internal nickel ions are shorter by about 0.2 Å than those for the terminal nickel ions, indicative of the terminal Ni(II) being in a high-spin state and the internal Ni(II) in a low-spin state.

5.11
Electrochemical Properties of the Complexes

It has long been known that dinuclear metal complexes exhibit rich redox chemistry and some of them have two stable oxidation states [62,63]. Remarkable richness in the electrochemical reactions of these multinuclear metal complexes can be expected. Intuitively, their electrochemical properties will strongly depend on the type and the number of the metal ions. The elec-

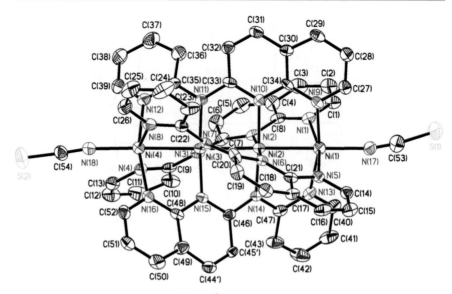

Fig. 5.11. Structure of $[Ni_4(pyan)_4(NCS)_2]^{2+}$ with three of the four ligands orienting in the same direction and the other in the opposite direction

trochemistry of these complexes was investigated by cyclic voltammetry. The electrochemical data are listed in Table 5.4. All the ligands show irreversible electrochemical oxidation reactions, and the first oxidative peak potential $E_p(ox1)$ shifts negatively as the ligand becomes longer. The electrochemical data reported in Table 5.4 for the metal complexes illustrate several general trends. First, the first oxidation potential increases with an increase in the atomic number for the first-row transition metals. For example, in the series of trinuclear complexes the first oxidations occur at +0.18, +0.43, and +1.05 V for $[Cr_3(dpa)_4Cl_2]$, $[Co_3(dpa)_4Cl_2]$, and $[Ni_3(dpa)_4Cl_2]$, respectively. This trend is also observed in the penta- and heptanuclear systems. Comparison of the data for the multinickel complexes in Table 5.4 illustrates a second trend: the longer the string complex, the lower the oxidation potential. From the viewpoint of M.O., this is because the HOMO energy increases as the number of metal ions in the linear chain increases. In the multinickel series, the first oxidation occurs at +1.05, +0.65, +0.18, and −0.12 V for the $[Ni_3(dpa)_4Cl_2]$, $[Ni_5(tpda)_4Cl_2]$, $[Ni_7(teptra)_4Cl_2]$, and $[Ni_9(peptea)_4Cl_2]$, respectively. The effect of the axial ligands reflects on the redox potentials. As shown in Table 5.4, the nature of the ligands shows a slight influence on their electrochemical behavior and the stability of the oxidized species. As an example, Fig. 5.12 shows the cyclic voltammograms of pentacobalt complexes with a variety of axial ligands. All these complexes exhibit two reversible redox couples at about $E_{1/2} = +0.35$ and +0.85 V, respectively [36]. Both electrochemical reactions involve one electron transfer as ascertained by spectroelectrochemistry. For $[Co_5(tpda)_4Cl_2]$, another two oxidative waves at potentials of about +1.25 and +1.40 V, which are

Table 5.4. Selected redox potentials for multinuclear metal complexes[a]

Compounds	Oxidation				Reduction	
	4th	3rd	2nd	1st	1st	2nd
Hdpa				+1.40*		
H$_2$tpda				+1.03*		
H$_3$teptra				+0.95*		
[Cr$_3$(dpa)$_4$Cl$_2$]		+1.75*	+0.95	+0.18		
[Cr$_5$(tpda)$_4$Cl$_2$]		+1.25*	+0.63	+0.29	−1.30	
[Cr$_7$(teptra)$_4$Cl$_2$]	+1.30*	+0.94	+0.48	+0.27		
[Co$_3$(dpa)$_4$Cl$_2$]		+1.60*	+1.34	+0.43	−0.85*	
[Co$_3$(dpa)$_4$(NCS)$_2$]				+0.52		
[Co$_5$(tpda)$_4$Cl$_2$]	+1.34	+1.21	+0.86	+0.34	−0.55*	
[Co$_5$(tpda)$_4$(NCS)$_2$]			+0.88	+0.38	−0.53	
[Co$_5$(tpda)$_4$(N$_3$)$_2$]		+1.16	+0.82	+0.33	−0.54	
[Co$_5$(tpda)$_4$(CN)$_2$]		+0.19	+0.73	+0.27	−0.66	
[Ni$_3$(dpa)$_4$Cl$_2$]			+1.63*	+1.05	−1.15*	−1.30*
[Ni$_5$(tpda)$_4$Cl$_2$]		+1.37	+1.25	+0.65		
[Ni$_5$(tpda)$_4$(NCS)$_2$]			+1.20*	+0.68		
[Ni$_7$(teptra)$_4$Cl$_2$]			+0.73*	+0.18		
[Ni$_7$(teptra)$_4$(NCS)$_2$]		+1.02*	+0.79*	+0.17	−0.95	
[Ni$_9$(peptea)$_4$Cl$_2$]			+0.66	−0.12		
[Ni$_9$(peptea)$_4$(NCS)$_2$]			+0.77	−0.06		
[Ru$_3$(dpa)$_4$Cl$_2$]		+1.41	+0.80	+0.01		

[a] Cyclic voltammetry was performed in CH$_2$Cl$_2$ with 0.1 M n-Bu$_4$NClO$_4$ using Pt as the working electrode. Potentials are reported vs. Ag/AgCl (sat'd) and referenced to the ferrocene/ferrocenium (Fc/Fc$^+$) couple that occurs at $E_{1/2}$ = +0.54 V vs. Ag/AgCl (sat'd)
* irreversible reaction

well resolved using differential pulse techniques, are observed and each step involves one-electron abstraction as judged by the peak current maximum. The first reduction near −0.50 V is reversible for [Co$_5$(tpda)$_4$(NCS)$_2$] and irreversible for [Co$_5$(tpda)$_4$Cl$_2$], [Co$_5$(tpda)$_4$(N$_3$)$_2$], and [Co$_5$(tpda)$_4$(CN)$_2$]. It should be noted that the cyclic voltammogram of [Co$_5$(tpda)$_4$(CN)$_2$] shows a small reduction peak near −0.35 V upon the first reduction.

As shown in Fig. 5.13, this peak progressively increases as the potential is scanned in cycles between 0.00 and −0.92 V, indicating that an absorbed species forms when [Co$_5$(tpda)$_4$(CN)$_2$] is electrochemically reduced. The insoluble film exhibits four newly developed redox couples at −0.36, +0.47, +0.84, and +1.18 V. Most likely, the electrochemical reduction reaction leading to [Co$_5$(tpda)$_4$(CN)$_2$]$^-$ is followed by the dissociation of one of the cyanide ligands to form an intermediate [Co$_5$(tpda)$_4$CN], and subsequent reactions with [Co$_5$(tpda)$_4$(CN)$_2$] afford the insoluble polymer {−(Co$_5$(tpda)$_4$CN)$_n$−} or dimer {NCCo$_5$(tpda)$_4$CNCo$_5$(tpda)$_4$CN} film. To see the stability of the film, the applied potentials consecutively scan 20 cycles between +1.33 and

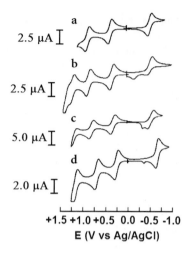

Fig. 5.12. Cyclic voltammograms of linear pentanuclear string complexes in CH$_2$Cl$_2$ containing 0.1 M TBAP. Compounds [Co$_5$(tpda)$_4$(NCS)$_2$] (**a**); [Co$_5$(tpda)$_4$Cl$_2$] (**b**); [Co$_5$(tpda)$_4$(N$_3$)$_2$] (**c**); [Co$_5$(tpda)$_4$(CN)$_2$] (**d**)

−0.92 V for the film adsorbed on the electrode surface. The peak currents slightly decrease, suggesting the decomposition of the film. However, the peak currents remain unchanged when the potentials consecutively scan between +1.05 and −0.75 V (Fig. 5.13c). The insoluble film is electrochemically robust and exhibits three well-resolved, reversible redox couples. These electrochem-

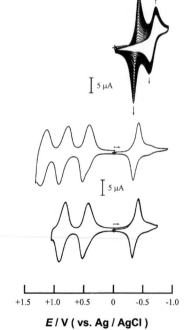

Fig. 5.13. Cyclic voltammograms in CH$_2$Cl$_2$ containing 0.1 M TBAP at 0.10 V s^{-1} using a glassy carbon electrode. **a** 20 consecutive cycles between 0 V and −0.92 V in 1.0 mM [Co$_5$(tpda)$_4$(CN)$_2$] solution. **b** First cycle between +1.33 V and −0.92 V for insoluble film adsorbed on electrode surface. **c** 10 consecutive cycles between +1.05 V and −0.75 V for insoluble film adsorbed on electrode surface

ical characteristics suggest that this insoluble material is well suited for the use of multibit storage devices.

The electrochemical reactions of these complexes have also been investigated by coupled thin-layer spectroelectrochemical technique, which is especially useful for evaluating the electrochemical reaction center and the number of electron transfers and provides definitive information about the electrochemical reactions [64]. As an example, Fig. 5.14 shows a spectral changes of $[Co_5(tpda)_4Cl_2]$ at applied potentials from +0.18 to +0.44 V in CH_2Cl_2 containing 0.1 M TBAP. There are clear isosbestic points during the electrolysis process, indicating that no intermediates are produced during the oxidation reaction. The resulting spectrum is similar to that of the one-electron oxidation products of $[Co_5(tpda)_4Cl_2]$, obtained by chemical method, suggesting that the oxidation is a metal-centered reaction. Based on the spectral changes of $[Co_5(tpda)_4Cl_2]$ at various applied potentials, the number of electrons transferred is calculated to be one.

An electrochemical process involving the oxidation of the neutral multinuclear metal complex is often accompanied by the appearance of a new broad band in the near-IR region, which may be ascribed to the intervalence charge transfer. Figure 5.15 shows the spectral changes for the electrochemical oxidation reactions of $[Ni_3(dpa)_4Cl_2]$ and $[Ni_5(tpda)_4Cl_2]$, and Fig. 5.16 shows the changes for $[Ni_7(teptra)_4Cl_2]$ and $[Ni_9(peptea)_4Cl_2]$ at various applied potentials. As can be seen, a new broad band in the near-IR region corresponding to the intervalence charge transfer evolves during the electrochemical oxidation process. In general, the broad band in the near-IR region undergoes bathochromic shift as the number of metal ions in the metal chain increases.

As was mentioned, some of the multinuclear metal complexes may find application in molecular wires. To be considered as molecular wires, the

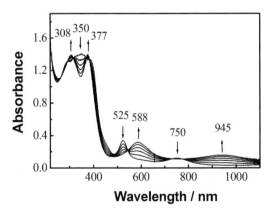

Fig. 5.14. Spectral changes for first oxidation of $[Co_5(tpda)_4Cl_2]$ in CH_2Cl_2 with 0.1 M TBAP at various applied potentials from +0.18 to +0.44 V

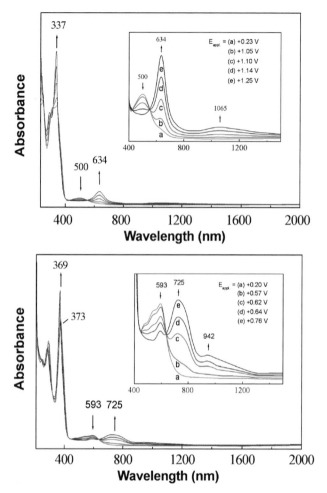

Fig. 5.15. Spectral changes for first oxidation of [Ni$_3$(dpa)$_4$Cl$_2$] (*top*) and [Ni$_5$(tpda)$_4$Cl$_2$] (*bottom*) in CH$_2$Cl$_2$ with 0.1 M TBAP at various applied potentials from +0.23 to +1.05 V for [Ni$_3$(dpa)$_4$Cl$_2$] and from +0.20 to +0.76 V for [Ni$_5$(tpda)$_4$Cl$_2$]

molecules must be able to provide a pathway for electrons. A convenient way to evaluate the electron conduction capability is to look at the electronic communication between the redox-active termini bonded to the ends of a molecular wire. Our studies on the triruthenium complexes show that there is an overall bond order of three over the Ru$_3$ unit. Thus, the triruthenium complex is a suitable candidate used for a molecular wire. We have synthesized the triruthenium complex [Ru$_3$(dpa)$_4$(CCFc)$_2$] with two ferrocenylacetylide axial ligands [65].

The electronic communication between the two ferrocenyl units can be studied using electrochemical techniques. As shown in Fig. 5.17, the cyclic

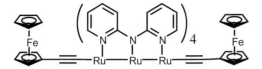

voltammagram of compound [Ru$_3$(dpa)$_4$(CCFc)$_2$] displays three reversible redox couples. The unresolved waves at potentials between +0.30 and +0.60 V
for the two consecutive redox processes correspond to the ferrocenyl units.
The Ru$_3$-centered ([Ru$_3$]$^{7+}$/[Ru$_3$]$^{6+}$) redox reaction occurs at $E_{1/2}$ = +0.76 V,
which is comparable to those of analogs [Ru$_3$(dpa)$_4$(CCAr)$_2$] ($E_{1/2}$ = +0.63–
+0.70 V). The stepwise one-electron oxidations instead of a two-electron oxidation are indicative of electronic coupling between the two iron centers,
suggesting that the bridge may provide a passive path for electrons. These
reversible waves for the ferrocenyl units are resolved using a differential-

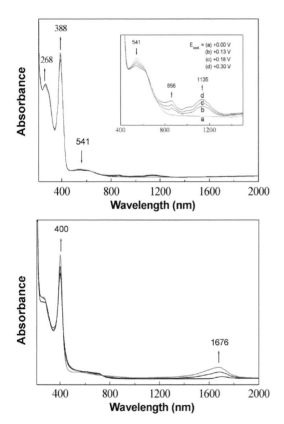

Fig. 5.16. Spectral changes for first oxidation of [Ni$_7$(teptra)$_4$Cl$_2$] (*top*) and
[Ni$_9$(peptea)$_4$Cl$_2$] (*bottom*) in CH$_2$Cl$_2$ with 0.1 M TBAP at various applied potentials from
+0.00 to +0.30 V for [Ni$_7$(teptra)$_4$Cl$_2$] and from −0.35 to +0.00 V for [Ni$_9$(peptea)$_4$Cl$_2$]

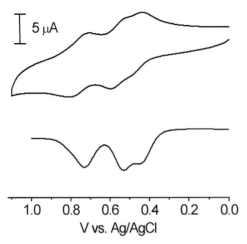

Fig. 5.17. Cyclic voltammogram (*top*) and differential pulse voltammogram (*bottom*) of [Ru$_3$(dpa)$_4$(CCFc)$_2$] in CH$_2$Cl$_2$ with 0.1 M TBAP

pulse technique. The comproportionation constant K_c is calculated to be about 33 [66]. This is larger than the corresponding tricobalt complex ($K_c = 16$) in which the Co$_3$ unit is unsymmetrical, and an overall bond order of 1.5 over the Co$_3$ unit is estimated [67]. As compared to the diruthenium complex with ferrocenylacetylide axial ligands [68], the electronic coupling of our system is weaker due to the longer Fe–Fe distance (ca. 16.3 Å). This complex can be classified as a weakly coupled (class II) mixed-valence system and is comparable to the α, ω-dipyridylpolyene molybdenum complex [Mo(4,4'-NC$_5$H$_4$(CH=CH)$_4$H$_4$C$_5$N)Mo] and the polyyne rhenium complex [ReC$_{16}$Re] [69,70].

5.12
Scanning Tunneling Microscopy Studies

Structural analysis and magnetic measurements show that the strength of metal–metal interactions of these multinuclear metal string complexes is in the order Cr > Co > Ni. The complexes with strong metal–metal interactions may show good electrical conductivity. In some of the multinuclear metal complexes, the structures exhibit significant changes upon oxidation. For example, the neutral complex of trichromium complex is symmetrical with a bond order of 1.5 between adjacent chromium ions, whereas upon one-electron oxidation the Cr$_3$ unit is comprised of a Cr–Cr quadruple bond and an isolated chromium ion at the end. The penta- and heptachromium complexes also exhibit a behavior similar to that observed in the trichromium system. Therefore, an on/off-switch-like behavior associated with the neutral oxidized states of the complexes can be expected. To test the possible use of these complexes in molecular electronics such as molecular conducting wires and molecular switches, scanning tunneling microscopy (STM) studies of these molecules have been carried out [55]. Using STM techniques, the molecules can be individually imaged and addressed as shown in Fig. 5.18.

Figure 5.19 shows typical STM images of $[Ni_3(dpa)_4(NCS)_2]$, $[Co_3(dpa)_4(NCS)_2]$, and $[Cr_3(dpa)_4(NCS)_2]$ molecules inserted into the self-assembled monolayer (SAM) of dodecanethiolate on gold [55]. The STM measurements showing the heights of the protrusions for $[Ni_3(dpa)_4(NCS)_2]$, $[Co_3(dpa)_4(NCS)_2]$, and $[Cr_3(dpa)_4(NCS)_2]$ embedded within dodecanethiolate SAM are 7.8 ± 1.0, 9.8 ± 0.6, and 12.8 ± 0.7 Å, respectively, suggesting that the tunneling current increases with increasing metal–metal interactions in these multinuclear complexes. The heights for the pentacobalt and the pentachromium complexes are 2.9 and 5.5 Å higher than that of the pentanickel complex (11.3 ± 0.3 Å) when the molecules are embedded within the dodecanethiolate SAM. The results are consistent with the trend observed in that of the trimetal system.

To further probe the effect of metal–metal interactions on the conducting properties, the one-electron oxidation complexes are also inserted into the dodecanthiolate SAM for the STM studies. As expected, the oxidized pentanickel complex exhibits better conductivity than the neutral analog. The protrusions are, remarkably, raised 2.6 Å, from 11.3 ± 0.3 to 13.9 ± 0.4 Å. In the case of the pentacobalt complex, the height of the protrusions is 13.2 ± 0.4 Å, which drops slightly as compared to the neutral one. The height for the pentachromium complex exhibits a considerable decrease from 16.8 ± 1.2 to 13.6 ± 0.8 Å upon oxidation.

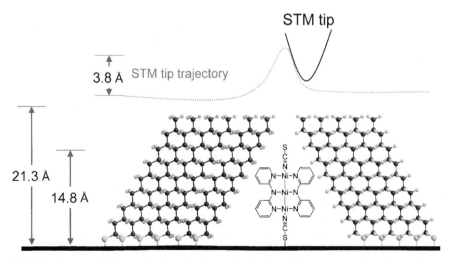

Fig. 5.18. Protocol for inserting $[Ni_3(dpa)_4(NCS)_2]$ molecular wires into a hexadecanethiolate self-assemble monolayer. Relative conductance was measured by a STM tip

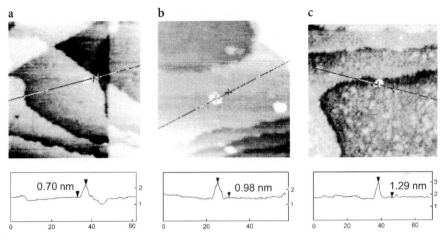

Fig. 5.19. STM images of [Ni₃(dpa)₄(NCS)₂] (**a**), [Co₃(dpa)₄(NCS)₂] (**b**), and [Cr₃(dpa)₄(NCS)₂] (**c**) embedded within dodecanethiolate SAMs

5.13
Summary

Enormous progress has been made in synthesizing oligopyridylamine and modified ligands and in constructing molecular wires with a linear metal atom chain supported by ligands. These complexes exhibit interesting physical properties and provide us with the chance to better understand metal–metal interactions in the linear metal atom chain. We have shown that the neutral and oxidized forms of the multichromium complexes exhibit significant differences in their structures, the strength of metal–metal interactions, and the electronic configuration. The oxidized multinickel complexes have a partial metal–metal bond in the metal atom chain, whereas the neutral molecules do not show a metal–metal bond between the neighboring nickel ions. In the multicobalt system, slight changes in the metal–metal distances are observed.

Since current downsizing of silicon-based chips will reach its physical limits, an alternative approach needs to be found. Our multinuclear metal string complexes with a linear metal chain in a defined length have been shown to be electron conducting. This type of molecules may be used as molecular wires. Based on structural analysis and STM studies, it is possible to make molecular switches using these molecules. Figure 5.20 shows the design of molecular switches in which [Cr₅(tpda)₄(NCS)₂] or [Ni₅(tpda)₄(NCS)₂] molecules are positioned between two gold electrodes. The Cr₅ unit of the neutral molecule may provide a conductive path for electrons, and the conductivity of the Cr₅ unit is expected to decrease upon oxidation due to weak or no metal–metal interactions between the chromium ions with long distances. The molecular switches made by these multinuclear complexes will function in an electrochemically

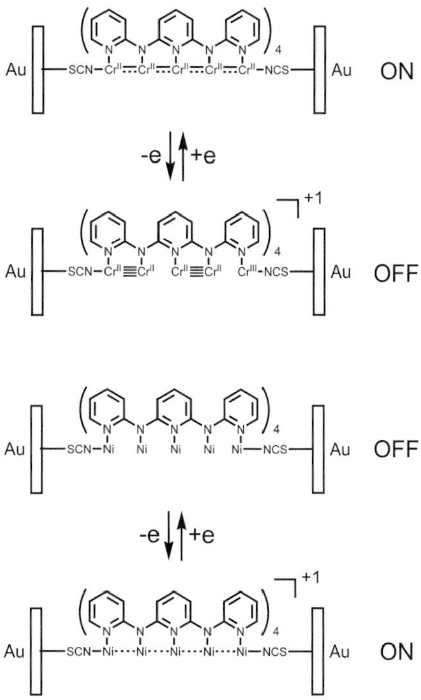

Fig. 5.20. [Cr$_5$(tpda)$_4$(NCS)$_2$]-based (*top*) and [Ni$_5$(tpda)$_4$(NCS)$_2$]-based (*bottom*) molecular switches

controllable fashion. The multinickel complexes are expected to work in a similar way. In contrast to the chromium system, the oxidized form of multinickel complexes will have better conductivity since it shows stronger Ni–Ni interactions as compared to the neutral form. In the case of the multicobalt complexes, neither neutral nor oxidized complexes show significant differences in the metal–metal interactions, and they exhibit comparable conductivity. Thus, the multicobalt complexes may be used as conducting wires, but they are not suited for use as molecular switches. Significant progress in nanotechnology has allowed us to address a single molecule, but there is still a long way to go before molecule-based electronics becomes practical. The next key step is to find methods of integrating these molecules into electronic devices.

Acknowledgement This work has been financially supported by the National Science Council and Ministry of Education of Taiwan. The authors would like to express special appreciation to the graduate students, postdoctoral fellows, and collaborators.

5.14
References

1. Davies WB, Svec WA, Ratner MA, Wasielewski MR (1998) Nature 396:60
2. Tour M (2000) Acc Chem Res 33:791

3. Boldi AM, Anthony J, Gramlich V, Boudon C, Gisselbrecht J, Gross M, Diederich F (1995) Helv Chim Acta 78:779
4. Ziener U, Godt A (1997) J Org Chem 62:6137
5. Tsuda A, Osuka A (2001) Science 293:79
6. Crossley MJ, Burn PL (1991) J Chem Soc Chem Commun 1569
7. Sakai M, Lizuka M, Nakamura M, Kudo K (2005) J Appl Phys 97:053509
8. Bildstein B, Loza O, Chizhov Y (2004) Organometallics 23:1825
9. Gonzalez C, Morales RGE (1999) Chem Phys 250:279
10. Cygan MT, Dunbar TD, Arnold JJ, Bumm LA, Shedlock NF, Burgin TP, Jones II L, Allara DL, Tour JM, Weiss PS (1998) J Am Chem Soc 120:2721
11. Tejel C, Ciriano MA, Villarroya BE, López JA, Lahoz FJ, Oro LA (2003) Angew Chem Int Ed 42:509
12. Murahashi T, Mochizuki E, Kai Y, Kurosawa H (1999) J Am Chem Soc 121:10660
13. Matsumoto K, Sakai K, Nishio K, Tokisue Y, Ito R, Nishide T, Shichi Y (1992) J Am Chem Soc 114:8110
14. Chern S-S, Lee G-H, Peng S-M (1994) J Chem Soc Chem Commun 1645
15. Saduldecha S, Hathaway B (1991) J Chem Soc Dalton Trans 993
16. Pyrka GJ, El-Mekki M, Pinkerton AA (1991) J Chem Soc Chem Commun 84
17. Wu L-P, Field P, Morrissey T, Murphy C, Nagle P, Hathaway B, Simmons C, Thornton P (1990) J Chem Soc Dalton Trans 3835
18. Yang E-C, Cheng M-C, Tsai M-S, Peng S-M (1994) J Chem Soc Chem Commun 2377
19. Cotton FA, Daniels LM, Jordan IV GT (1997) Chem Commun 421
20. Cotton FA, Daniels LM, Jordan IV GT, Murillo CA (1997) J Am Chem Soc 119:10377
21. Cotton FA, Murillo CA, Wang X (1999) J Chem Soc Dalton Trans 3327
22. Cotton FA, Murillo CA, Wang X (1999) Inorg Chem 38:6294
23. Clérac R, Cotton FA, Daniels LM, Dunbar KR, Lu T, Murillo CA, Wang X (2000) J Am Chem Soc 122:2272
24. Clérac R, Cotton FA, Dunbar KR, Lu T, Murillo CA, Wang X (2000) Inorg Chem 39:3065
25. Clérac R, Cotton FA, Daniels LM, Dunbar KR, Kirschbaum K, Murillo CA, Pinkerton AA, Schultz AJ, Wang X (2000) J Am Chem Soc 122:6226
26. Clérac R, Cotton FA, Daniels LM, Dunbar KR, Murillo CA, Wang X (2001) J Am Chem Soc 123:1256
27. Cotton FA, Daniels LM, Murillo CA, Pascual I (1997) J Am Chem Soc 119:10223
28. Clérac R, Cotton FA, Danniels LM, Dunbar KR, Murillo CA, Pascual I (2000) Inorg Chem 39:748
29. Clérac R, Cotton FA, Daniels LM, Dunbar KR, Murillo CA, Pascual I (2000) Inorg Chem 39:752
30. Berry JF, Cotton FA, Lu T, Murillo CA, Roberts BK, Wang X (2004) J Am Chem Soc 126:7082
31. Sheu J-T, Lin C-C, Chao I, Wang C-C, Peng S-M (1996) Chem Commun 315
32. Wang C-C, Lo W-C, Chou C-C, Lee G-H, Chen J-M, Peng S-M (1998) Inorg Chem 37:4059
33. Shieh S-J, Chou C-C, Lee G-H, Wang C-C, Peng S-M (1997) Angew Chem Int Ed Engl 36:56
34. Chang H-C, Li J-T, Wang C-C, Lin T-W, Lee H-C, Lee G-H, Peng S-M (1999) Eur J Inorg Chem 1243
35. Yeh C-Y, Chiang Y-L, Lee G-H, Peng S-M (2002) Inorg Chem 41:4096
36. Yeh C-Y, Chou C-H, Pan K-C, Wang C-C, Lee G-H, Su YO, Peng S-M (2002) J Chem Soc Dalton Trans 2670
37. Chen Y-H, Lee C-C, Wang C-C, Lee G-H, Lai S-Y, Li F-Y, Mou C-Y, Peng S-M (1999) Chem Commun 1667

38. Lai S-Y, Lin T-W, Chen Y-H, Wang C-C, Lee G-H, Yang M-H, Leung M-K, Peng S-M (1999) J Am Chem Soc 121:250
39. Lai S-Y, Wang C-C, Chen Y-H, Lee C-C, Liu Y-H, Peng S-M (1999) J Chin Chem Soc 46:477
40. Peng S-M, Wang C-C, Jang Y-L, Chen Y-H, Li F-Y, Mou C-Y, Leung M-K (2000) J Magn Magn Mater 209:80
41. Wagaw S, Buchwald SL (1996) J Org Chem 61:7240
42. Wolfe JP, Buchwald SL (2000) J Org Chem 65:1144
43. Singh UK, Strieter ER, Blackmond DG, Buchwald SL (2002) J Chem Am Soc 124:14104
44. Hasan H, Tan U-K, Lin Y-S, Lee C-C, Lee G-H, Lin T-W, Peng S-M (2003) Inorg Chim Acta 351:369
45. Hasan H, Tan U-K, Wang R-R, Lee G-H, Peng S-M (2004) Tetrahedron Lett 45:7765
46. Leung M-K, Mandal AB, Wang C-C, Lee G-H, Peng S-M, Cheng H-L, Her G-R, Chao I, Lu H-F, Sun Y-C, Shiao M-Y, Chou P-T (2002) J Am Chem Soc 124:4287
47. Kar S, Chanda N, Mobin SM, Urbanos FA, Niemeyer M, Puranik VG, Jimenez-Aparicio R, Lahiri GK (2005) Inorg Chem 44:1571
48. Cabeza JA, Río ID, García-Granda S, Riera V, Suárez M (2002) Organometallics 21:2540
49. Haukka M, Costa PD, Luukkanen S (2003) Organometallics 22:5137
50. Blakley RL, DeArmond MK (1987) J Am Chem Soc 109:4895
51. Yang M-H, Lin T-W, Chou C-C, Lee H-C, Chang H-C, Lee G-H, Leung M-K, Peng S-M (1997) Chem Commun 2279
52. Day EP, Peterson J, Sendova MS, Todd MJ, Hausinger RP (1993) Inorg Chem 32:634
53. Karlin KD, Hayes JC, Gultneh Y, Cruse RW, McKown JW, Hutchinson JP, Zubieta J (1984) J Am Chem Soc 106:2121
54. Berry JF, Cotton FA, Daniels LM, Murillo CA (2002) J Am Chem Soc 124:3212
55. Lin S-Y, Chen I-W P, Chen C-H, Hsieh M-H, Yeh C-Y, Lin T-W, Chen Y-H, Peng S-M (2004) J Phys Chem B 108:959
56. Cotton FA, Daniels LM, Murillo CA, Wang X (1999) Chem Commun 2461
57. Bino A, Cotton FA, Felthouse TR (1979) Inorg Chem 18:2599
58. Lindsay AJ, Wilkinson G, Motevalli M, Hursthouse MB (1985) J Chem Soc Dalton Trans 2321
59. Cotton FA, DeBoer BG, LaPrade MD, Dipal JR, Ucko DA (1970) J Am Chem Soc 92:2926
60. Chieh C-H, Chang J-C, Yeh C-Y, Lee G-H, Fang J-M, Song Y, Peng S-M. Dalton Trans (submitted)
61. Chieh C-H, Chang J-C, Yeh C-Y, Lee G-H, Fang J-M, Peng S-M. Dalton Trans (submitted)
62. Cotton FA, Walton RA (1993) Multiple Bonds Between Metal Atoms, 2nd edn. Clarendon Press, Oxford
63. Cotton FA, Daniels LM, Hillard EA, Murrilo CA (2002) Inorg Chem 41:1639
64. Rohrbach DF, Deutsch E, Heineman WR, Pasternack RF (1977) Inorg Chem 16:2650
65. Kuo C-K, Chang J-C, Yeh C-Y, Lee G-H, Wang C-C, Peng S-M (2005) Dalton Trans 3696
66. Richardson DE, Taube H (1981) Inorg Chem 20:1278
67. Berry JF, Cotton FA, Murillo CA (2004) Organometallics 23:2503
68. Xu GL, DeRosa MC, Crutchley RJ, Ren T (2004) J Am Chem Soc 126:3728
69. McWhinnie SLW, Thomas JA, Hamor TA, Jones CJ, McCleverty JA, Collison D, Mabbs FE, Harding CJ, Yellowlees LJ, Hutchings MG (1996) Inorg Chem 35:760
70. Dembinski R, Bartik T, Bartik B, Jaeger M, Gladysz JA (2000) J Am Chem Soc 122:810

Chapter 6

Multielectron Redox Catalysts in Metal-Assembled Macromolecular Systems

Takane Imaoka · Kimihisa Yamamoto

Keio University, Faculty of Science and Technology, Department of Chemistry, 3-14-1 Hiyoshi, Kohoku-ku, Yokohama 223-8522, Japan

Summary Multielectron redox catalysis of small molecules has been researched as an ambitious objective, because it relates to chemical processes involved in the life reactions and is important for industrial applications. This review focuses on the multielectron transfer reaction facilitated by redox-assembled systems in macromolecular architectures. Recent advances in constructing a nano-redox system based on precise assembly of metal complexes in dendritic macromolecules are shown in particular.

Abbreviations

ET   Electron transfer
DPA   Dendritic phenylazomethine

6.1
Introduction

Reduction and oxidation (redox) reactions are some of the most important chemical processes in the biological molecular conversion system. In many cases, a redox reaction prefers to proceed as a multielectron transfer (multi-ET) rather than a single-electron transfer (Single-ET) due to the stability of the resulting products. Actually, the biological reductions of small molecules such as O_2, N_2, and CO_2 occur via the multielectron pathway without generating an unstable radical anion. However, in artificial systems, single-electron products are often observed or control the entire reaction process without appearing as a product. The difference is caused by the function of the biological enzymes and artificial catalysts.

In general, the main role of catalysts is to stabilize this intermediate species of chemical reactions. The reaction kinetics are accelerated by the stabilization effect because it results in a lower activation energy for the entire reaction process. It is enough to facilitate most chemical reactions with a practically available rate constant, but another role would be required when the direct electron transfer to/from the redox reagent or electrode surface is very slow. The second role of redox catalysts is mediation of the electron exchange between each substrate and reagent.

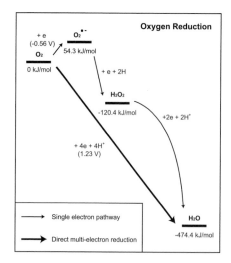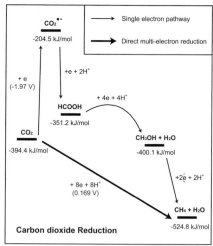

Scheme 6.1. Energy diagram of O_2 and CO_2 multielectron reduction, heat of formation (kJ/mol), and theoretical redox potential (V vs. NHE)

Conventional metal complexes having one redox center can mediate only a Single-ET process, while biological enzymes with multicored metalloproteins can promote a multi-ET process at the same time. The catalytic system with a mononuclear metal complex inevitably produces unstable radical species through the reaction. The chemical potentials of radical ions are very high for small molecules. For example, the heat of formation and theoretical redox potentials of CO_2 and O_2 are shown in Scheme 6.1. The differences between the required redox potential for a single or multielectron reduction are 2.14 V (CO_2) and 1.79 V (O_2), respectively. Therefore, an exceptionally strong coordination energy of radical anions to the catalysts are needed to lower the activation energy of a single-electron pathway to the level of a multielectron pathway. In short, multielectron redox catalysts are practically feasible for promoting such reductions by applying a relatively low overpotential.

This review focuses on the methodology to fabricate multielectron redox systems by assembling redox centers around a catalytic center using a macromolecular architecture.

6.2
Multielectron Redox Systems

In principle, the one-step multi-ET is favorable for many organic or inorganic molecules with covalent bonds. As mentioned above, small molecules are one of the most typical examples. The driving forces, which accelerate the multi-ET, is a combination of proton binding [1–3], formation of a new bond [4,5], or a structural transformation [6–12]. However, most multielectron reactions

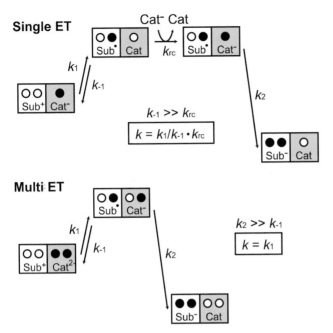

Scheme 6.2. Kinetic schemes of single-ET and multi-ET system for catalytic reduction of substrate

do not readily proceed except for a few examples. A key aspect of the multielectron reaction is the kinetic control. If the kinetics was the ideal rate, the reaction would proceed at the redox potential of the multielectron pathway. Traction, which pulls the first endothermic single-electron process, is the second exothermic process.

For example [13], the two-electron reduction of NAD^+ leads to NADH through $NAD^•$ connected to flavin, which is a two-electron mediator. The required overpotential to form the intermediate radical ($NAD^•$) is 0.7 V higher than that of the NADH formation, but, in practice, no overpotential is needed. In this system, the reaction should proceed as a multi-ET system. When the intrinsic electron exchange among the catalytic redox centers is very fast (close to the adiabatic limit), the entire electron transfer kinetics via the radical formation will be a practically available rate ($> 10^1$ s^{-1}) even when the production of the transition radical species is very hard. The close proximity of the redox centers (within 6 Angstrom) among flavin-NAD plays the very essential role of accelerating the entire reaction by the exothermic second ET. The fast kinetic of the second ET (k_2) results in a rate-limiting step by the first ET (k_1). Although the first ET is strongly endothermic, the rate constant is relatively fast when the electron mediator (Cat) is close to the substrate [13].

In contrast, if the electron mediator has only one electron, at least two mediators are required to start the multielectron reaction. Therefore, the oxidized catalyst (Cat) should be exchanged by another catalyst (Cat$^-$) before the second ET. The recomplexation (k_{rc}) should be the rate-limiting step of the second ET process because it is usually a second-order reaction. As a result, the entire reaction becomes a two-step process producing unstable radical species as the intermediate. The two-step reaction is very slow because the radical formation equilibrium constant (k_1/k_{-1}) under a low applied overpotential is small. This means that a radical formation pathway with a very fast electron exchange rate can be regarded as well as a multielectron pathway. In other words, a one-electron reduced/oxidized species is regarded not as an interme-diate state but as a transition state through the multielectron pathway. Based on this idea, the designed strategy for an efficient multielectron redox reaction is to construct the redox-assembled system in a very narrow space.

6.3
Multinuclear Complexes as Redox Catalysts

The redox potential of the mediating catalysts must be similar to that of the substrates. For reduction, if the potential of the catalysts was much more negative, the actually required negative potential would be higher than that required in the theoretical prediction. In contrast, if the potential was more

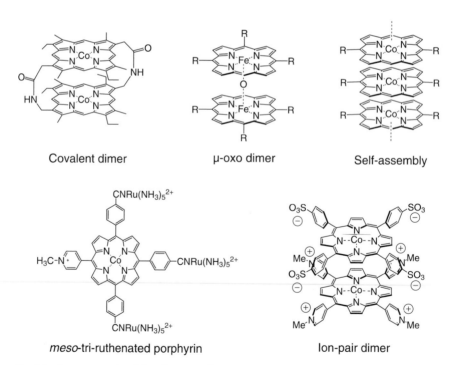

Covalent dimer   μ-oxo dimer   Self-assembly

meso-tri-ruthenated porphyrin   Ion-pair dimer

Fig. 6.1. Structures of multinuclear metalloporphyrins for O_2 four-electron reduction

positive, the electron relay will not work. For the catalytic reduction of O_2, a metalloporphyrin complex such as cobalt or iron is usually employed. Their redox potential, corresponding to the redox between the di- and trivalence states, is located in the region at 0.5 V vs. NHE. It is close to the theoretical redox potential of the two-electron O_2 reduction to hydrogen peroxide (H_2O_2), so that the main product of the catalytic reduction would be H_2O_2 using the mononuclear porphyrin complex. The role of the mononuclear catalyst is only to decrease the activation energy of the $O_2^{-\bullet}$ radical formation.

The first strategy, aimed at facile multielectron processes, was to construct a multinuclear cofacial porphyrin complex [14]. The multiplying process by covalent bonding [15, 16], μ-oxo bridge [17], ionic interaction [18], and self-assembly [19, 20], was successfully applied to the four-electron reduction of O_2 producing water (Fig. 6.1). The triruthenated cobalt porphyrin [21] also catalyzed the four-electron reduction. The complex has a four-metal center in which the redox potentials are close to each other.

6.4
Macromolecule-Metal Complexes

The multielectron process can be developed by concentrating the redox active center in the polymer matrix. We have reported the efficient production of H_2O from O_2 via a four-electron reduction process using redox active polyaniline derivatives [22, 23] (Fig. 6.2). During the reduction process, the key point is the matching of the redox potential for both the multielectron mediator (polyaniline) and catalytic center (cobalt porphyrin). The gap in the redox potentials between each redox center prevents a fast successive electron injection into the catalytic center. A similar multielectron reaction using the macromolecule-metal architecture was also successful for the O_2 reduction by concentrating the redox centers in a matrix such as Nafion [24, 25]. However, most of the macromolecule-metal complexes catalyze the O_2 reduction at a potential more negative than the theoretical potential of the H_2O_2 formation. This fact cannot completely rule out the quasi-

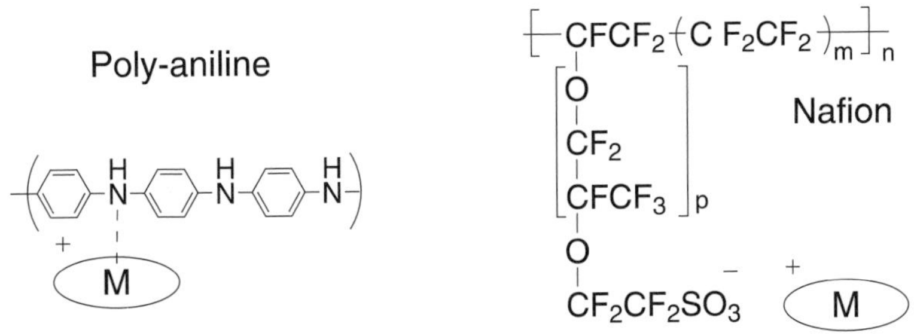

Fig. 6.2. Polymer matrices for multi-ET reduction of O_2

multi-ET mechanism involving two successive electron reductions via H_2O_2 formation.

6.5
Metal Ion Assembly on Dendritic Macromolecules

Metalloproteins [26] as biological catalysts have a well-defined structure of metal-organic (polypeptide) hybrids. A critical difference between the metalloproteins and conventional macromolecule-metal complexes [27] is the structural unity in the complexation. The structure of the macromolecular complexes by linear polymers cannot be predicted in advance. Previous research showed that a further precise hybridizing method is required in order to facilitate the fast electron exchange between the assembled metal complexes and a catalytic center.

The next step in the efficient multi-ET catalysis is to construct a finely defined metal-organic hybrid in the macromolecule with a single molecular weight and structure. For a scaffold of the metal-assembled nanocomposite, dendritic macromolecules (dendrimer) [28,29] bearing multicoordination sites on the branching units are used because of their structural unity. Several dendrimers, which can assemble metal ions, have been reported [30–34]. The phenylazomethine dendrimer (**DPA**) in particular showed a layer-by-layer stepwise complexation [35,36] with various metal ions from the core to the periphery (A → D in Fig. 6.3), whereas the other dendrimer complex could define the number and position of the metal ion. The stepwise complexation can be determined by the UV-vis titration method, transmission electron microscope (TEM), and electrochemical approach [37]. This unique property originates from repeating electronic donation of the phenylazomethine unit

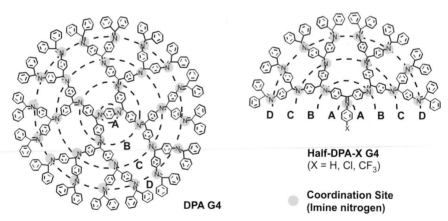

Fig. 6.3. Structures of phenylazomethine dendrimers with a *p*-phenylene (**DPA G4**) and *p*-substituted phenyl group (**Half-DPA-X G4**) as the core

to the focal point. The electronic negativity on inner units is amplified layer to layer because the inner coordination sites are affected much more by outer electron donating groups. Thus, the basicity of the inner imine nitrogen groups are enhanced.

The complexation behavior of the **half-DPA** clearly shows that the stepwise metal assembly depends on the electronic factor of the dendritic phenylazomethine (DPA) structure [38]. Similar spectral changes in the UV-vis absorption were also observed with the **half-DPA G4**. Not only the $SnCl_2$ complexation, but protonation with trifluoroacetic acid also occurred from the core-side imines. The coordination constants (K_N) of the protons to imines could be separately estimated for each layer of the dendrimer by a theoretical analysis of the isosbestic points of the UV-vis spectral change. The K_N values of each imine in the different layers (A~D) for various generation numbers of the dendrimer (Fig. 6.4) strongly indicate that the basicity of the imine units close to the core are enhanced. The coordination constants of $SnCl_2$ to the inner imines are two orders of magnitude higher than that to the neighboring outer imines. However, if the core is substituted by a strong electron withdrawing unit (CF_3), the basicity gradient is disordered. As a result, the complexation behavior is different from the nonsubstituted dendrimers.

The stepwise metal assembly can be applied to dendrimers having various functional core units [39–41] because it is based on the electronic property of

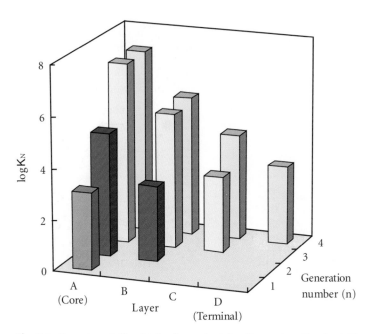

Fig. 6.4. Experimentally obtained constant for the protonation by trifluoroacetic acid to each imine site (A~D) of the **half-DPA** with various generation numbers [38]

the dendritic structure itself. Furthermore, a variety of metallic cations such as Fe^{3+}, Au^{3+}, and so on are also available for precise hybridization. The dendritic phenylazomethine-substituted porphyrins [39] have a structure similar to that of the heme-containing metalloproteins. For example, cytochrome c oxidase [42, 43], known to catalyze the four-electron O_2 reduction, has a Cu complex close to the heme center. The additional metal complex plays a synergetic role as an electron mediator and O_2 binding site [44, 45]. The porphyrin cored dendrimer also assembles a metal ion in a stepwise fashion [39]. The UV-vis spectrum changes during titration by $SnCl_2$ are shown in Fig. 6.5. Four isosbestic points were observed in turn by adding equimolar amounts of $SnCl_2$ to the dendrimer solution as well as to the previous phenylazomethine dendrimer [35, 36]. The number of coordination sites (imines) in the different layers agreed with the number of required equivalents for each shift in the isosbestic points. This means that a precise hybridization between metal ions and a macromolecular architecture is available around a functional porphyrin core. In this dendrimer, a metal ion is also assembled from the core-side imines, as shown in Scheme 6.3.

The observed shift in the isosbestic points toward the shorter wavelength region was stepwise. This fact represents a very large difference in the coordination constants between the neighboring imines in different layers. From a theoretical point of view, the required gradient of the coordination constants is two orders of magnitude per layer (Fig. 6.6a). If the gradient were smaller, the shift in the isosbestic point should be continuous (Fig. 6.6b).

The metal-assembling dendrimer with a porphyrin complex is expected to be used as a multielectron catalyst. In practice, the lanthanide metal ion

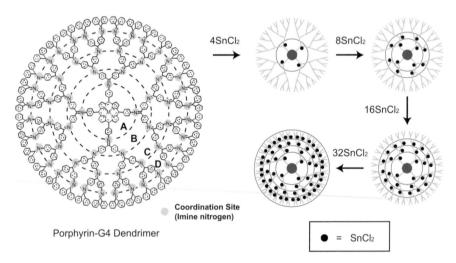

Scheme 6.3. Schematic representation of stepwise complexation to phenylazomethine dendrimer (generation 4) with metalloporphyrin core

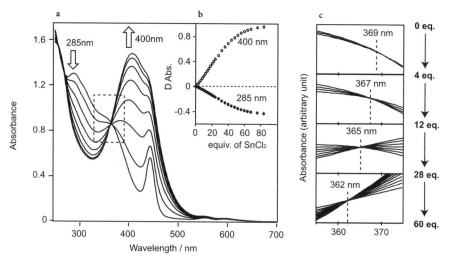

Fig. 6.5. **a** UV-vis spectra changes in the stepwise addition of $SnCl_2$ to the 2.5 µM phenyla-zomethine dendrimer (generation 4) with a cobalt porphyrin core. **b** Absorption changes vs. equivalents ratio of added $SnCl_2$. **c** Enlargement of isosbestic points

($La^{3+} = Y^{3+}$, Tb^{3+}, Er^{3+}, Nd^{3+}) assembling phenylazomethine dendrimer with a cobalt porphyrin core acts as an excellent catalyst for the CO_2 electrochemical reduction. Cobalt tetraphenylporphyrin (CoTPP) in the zero valent state is known as a CO_2 reduction catalyst [46]. In the cyclic voltammogram of CoTPP, a reductive current response by CO_2 was observed at the potential of the Co(0) formation (-1.8 V vs. NHE). It corresponds to a redox-mediated catalytic CO_2 reduction. The theoretical redox potential of the CO_2 multielectron reduction is

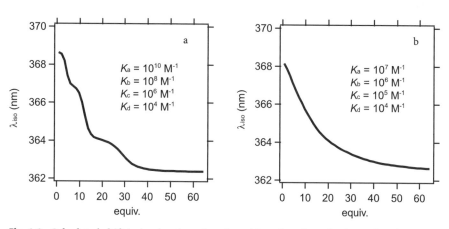

Fig. 6.6. Calculated shift in isosbestic points for arbitrarily selected values of each coordi-nation constant (K_a : K_N of intime $A \sim K_d$: K_N of intime D) based on equilibrium theory assuming the same experimental condition of Fig. 6.5

much higher than the observed reduction potential; thus the overpotential for the CO_2 reduction is very high. It is explained by a single-electron mechanism, which corresponds to the reduction via a single-ET process, once the CO_2 radical anion is generated as the intermediate. In contrast, as shown in Fig. 6.7, the metal (Tb^{3+})-assembling cobalt porphyrin dendrimer showed a catalytic response toward the CO_2 reduction at the potential of the Co(I) formation (−0.7 V vs. NHE), 1.1 V higher than that of Co(0). The mechanism of the reduction by the metal-assembling dendrimer cannot be explained by the single-electron process. The theoretical redox potential of the CO_2 one-electron reduction (−1.97 V vs. NHE) is 1.3 V lower than the observed redox potential. If the reduction proceeds via a one-electron process, the required stabilization energy of $CO_2^{-\bullet}$ should be 125 kJ/mol. This is unlikely for the CO_2-Co(I) binding energy [47, 48] because an exceptionally high coordination constant ($K = 10^{21}$ M^{-1}) is required.

The result is easier to understand as a multi-ET reaction. Phenylazomethine complexes with Tb^{3+} can act as an electron mediator to the cobalt porphyrin core. The electron-exchanging kinetics between the core and metal complexes should be very fast due to their close proximity. The dropping overpotential (1.3 V) can be distributed as a contribution to the stabilization energy and thermal activation energy. Moreover, the metal ion concentrated in the dendrimer may also coordinate to the oxygen terminal of CO_2, thus stabilizing the activation energy [49].

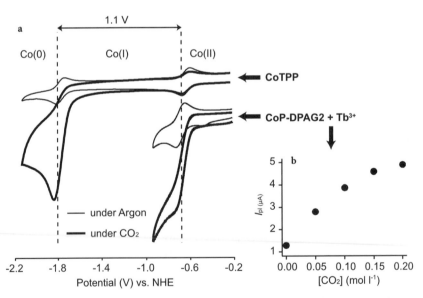

Fig. 6.7. a Cyclic voltammograms of 0.5 mM CoTPP and 0.5 mM CoP-DPA (G2) with (50 mM) $Tb(OTf)_3$ in argon or CO_2-saturated DMF solution. b CO_2 concentration dependence of plateau current for CoP-DPA-Tb^{3+}

6.6
Conclusion

A multi-ET catalyst, contributing the biological molecular conversions, has been a difficult target from a synthetic approach because it requires a very close proximity of the metal complexes. The most important requirement for the multi-ET catalysts is the fabrication of a nanosized redox-assembling structure with a designed flexibility. We have developed a macromolecule-metal hybridization technique. A novel method using the phenylazomethine dendrimer architecture as the framework of the metal ion assembly can provide us with a well-defined hybrid structure. The goal – ingenious hybrid structure in metalloproteins – is still not easy, but the method using the dendrimer should be an important tool for the precise hybridization of metals and organic macromolecules. Since the metal-assembling property does not depend on a specific dendrimer core or choice of the metallic elements, a variety of multi-ET system including catalysts and molecular electronic devices could be applied.

Acknowledgement   This work was partially supported by CREST from the Japan Science and Technology Agency, grants-in-aid for Scientific Research (No. 15350073), and the 21st COE Program (Keio-LCC) from the Ministry of Education, Science, Culture and Sports.

6.7
References

1. Pipes DW, Meyer TJ (1984) J Am Chem Soc 106:7653
2. Che CM, Wong KY, Anson FC (1987) J Electroanal Chem 226:211
3. Che CM, Wong KY, Poon CK, Anson FC (1985) Inorg Chem 24:1797
4. Hill MG, Rosenhein LD, Mann KR, Mu XH, Schults FA (1992) Inorg Chem 31:4108
5. Zhuang B, McDonald JW, Schults FA, Newton WE (1984) Organometallics 3:943
6. Miedaner A, Haltiwagner RC, DuBois DL (1991) Inorg Chem 30:417
7. Wang PW, Fox MA (1994) Inorg Chem 33:2938
8. Wang PW, Fox MA (1998) Chem Eur J 9:4
9. Pierce DT, Geiger WE (1992) J Am Chem Soc 114:6063
10. Bowyer WJ, Geiger WE (1988) J Electroanal Chem 239:253
11. Umakoshi K, Kinoshita I, Ichimura A, Ooi S (1987) Inorg Chem 26:3551
12. Nishiumi T, Chimoto Y, Hagiwara Y, Higuchi M, Yamamoto K (2004) Macromolecules 37:2661
13. Page CC, Moser CC, Chen X, Dutton PL (1999) Nature 402:47
14. Collman JP, Wagenknecht PS, Hutchison JE (1994) Angew Chem Int Ed Engl 33:1357
15. Collman JP, Marrocco M, Denisevich P, Koval C, Anson FC (1979) J Electroanal Chem Interfacial Electrochem 101:117
16. Collman JP, Denisevich P, Konai Y, Marrocco M, Koval C, Anson FC (1980) J Am Chem Soc 102:6027
17. Haryono A, Oyaizu K, Yamamoto K, Natori J, Tsuchida E (1998) Chem Lett 1998:233
18. D'Souza F, Hsieh Y-Y, Deviprasad GR (1998) Chem Commun 1998:1027

19. Shi C, Steiger B, Yuasa M, Anson FC (1997) Inorg Chem 36:4294
20. Yuasa M, Oyaizu K, Yamaguchi A, Kuwakado M (2004) J Am Chem Soc 126:11128
21. Steiger B, Anson FC (1994) Inorg Chem 33:5767
22. Yamamoto K, Taneichi D (2000) Chem Lett 2000:4
23. Matsufuji A, Nakazawa S, Yamamoto K (2001) J Inorg Organometal Polym 11:47
24. Imaoka T, Yamamoto K (2001) Phys Chem Chem Phys 3:4462
25. Abe T, Shoji K, Tajiri A (2004) J Mol Catal A 208:11
26. Messerschmidt A, Huber R, Poulos T, Wieghardt K, Bode W, Gygler M (2002) (eds) Handbook of Metalloproteins: A. Wiley, New York
27. Ciardelli F, Tsuchida E, Wohrle D (1996) (eds) Macromolecule-metal Complexes. Springer, Berlin Heidelberg New York
28. Fréchet JMJ, Tomalia DA (2002) Dendrimers and Other Dendritic Polymers. Wiley, New York
29. Tomalia DA, Naylor AM, Goddard III WA (1990) Angew Chem Int Ed Engl 29:138
30. Ottaviani MF, Bossmann S, Turro NJ, Tomalia DA (1994) J Am Chem Soc 116:661
31. Klein Gebbink RJM, Bosman AW, Feiters MC, Meijer EW, Nolte RJM (1999) Chem Eur J 5:65
32. Tominaga M, Hosogi J, Konishi K, Aida T (2000) Chem Commun 2000:719
33. Crooks RM, Zhao M, Sun L, Chechik V, Yeung LK (2001) Acc Chem Res 34:181
34. Zhao M, Sun L, Crooks RM (1998) J Am Chem Soc 120:4877
35. Yamamoto K, Higuchi M, Shiki S, Tsuruta M, Chiba H (2002) Nature 415:509
36. Higuchi M, Tsuruta M, Chiba H, Shiki S, Yamamoto K (2003) J Am Chem Soc 125:9988
37. Nakajima R, Tsuruta M, Higuchi M, Yamamoto K (2004) J Am Chem Soc 126:1630
38. Yamamoto K, Higuchi M, Kimoto A, Imaoka T, Masachika K (2005) Bull Chem Soc Jpn 78:349
39. Imaoka T, Horiguchi H, Yamamoto K (2003) J Am Chem Soc 125:340
40. Enoki O, Imaoka T, Yamamoto K (2003) Org Lett 5:2547
41. Satoh N, Cho J-S, Higuchi M, Yamamoto K (2003) J Am Chem Soc 125:8104
42. Babcock GT, Wikström M (1992) Nature 356:301
43. Malmström BG (1990) Chem Rev 90:1247
44. Collman JP, Fu L, Herrmann PC, Zhang X (1997) Science 275:949
45. Collman JP, Rapta M, Bröring M, Raptova L, Schwenninger R, Boitrel B, Fu L, L'Her M (1999) J Am Chem Soc 121:1387
46. Behar D, Dhanasekaran T, Neta P, Hosten CM, Ejeh D, Hambright P, Fujita E (1998) J Phys Chem A 102:2870
47. Fujita E, Creutz C, Sutin N, Szalda DJ (1991) J Am Chem Soc 113:343
48. Fujita E, Furenlid LR, Renner MW (1997) J Am Chem Soc 119:4549
49. Bhugun I, Lexa D, Savéant J-M (1996) J Phys Chem 100:19981

Part II

Redox Systems via Coordination Control

Chapter 7

Triruthenium Cluster Oligomers that Show Multistep/Multielectron Redox Behavior

Tomohiko Hamaguchi[1] · Tadashi Yamaguchi[2] · Tasuku Ito[3]

[1] Department of Chemistry, Faculty of Science, Fukuoka University, Nanakuma 8-19-1, Jonan-ku, Fukuoka 814-0180, Japan
[2] Department of Chemistry, Waseda University, Okubo, Shinjuku-ku, Tokyo 169-8555, Japan
[3] Department of Chemistry, Graduate School of Science, Tohoku University, Sendai 980-8578, Japan
e-mail: ito@agnus.chem.tohoku.ac.jp

7.1
Introduction

Nanosized molecules having plural redox active sites, and thus showing multiple redox processes, are interesting and attractive not only from the viewpoint of pure science but also for their use in electronic devices [1]. However, single molecules usually have a limited degree of multiplicity. For example, although fullerenes C_{60} and C_{70} exhibit six reversible single-electron reductions, these processes are spread over a large and extremely negative potential window (-0.98 V to -3.26 V vs. Fc/Fc$^+$) [2]. A few examples of multinuclear complexes having redox-active ligands that possess a large number of redox processes are known [3–5]. Balzani and coworkers reported 26 successive redox processes for a hexanuclear ruthenium complex with bridging polypyridyl ligands, but their redox processes are highly overlapped [3]. Recently, Lehn and coworkers reported a $[2 \times 2]$ grid-type Co_4^{II} complex that can be electrochemically reduced by 11 electrons in ten well-resolved reversible steps [4]. Metal-assembled complexes with redox-active metal centers and their oligomers should be useful for making multiredox systems.

In this chapter, we report the design, syntheses, and electrochemical properties of a pyrazine-bridged triruthenium cluster trimer and tetramer that show well-separated multistep one-electron redox waves [6]. The triruthenium clusters that we used are a class of oxocentered triruthenium clusters bridged by six acetate ligands with the general formulae $[Ru_3O(CH_3CO_2)_6(L)_3]^+$ and $[Ru_3O(CH_3CO_2)_6(CO)(L)_2]$. In general, the ancillary ligand L is a neutral monodentate ligand such as a pyridine derivative, triphenylphosphine, isocyanide, and/or a solvent molecule. In the isolated state, $[Ru_3O(CH_3CO_2)_6(L)_3]^+$ without a carbonyl or isocyanide ligand formally has three Ru(III) centers and an overall charge of +1 ($Ru_3^{III,III,III}$), whereas $[Ru_3O(CH_3CO_2)_6(CO)(L)_2]$ formally contains one Ru(II) and two Ru(III) centers and thus has an overall charge of 0 ($Ru_3^{III,III,II}$). The noncarbonyl cluster $[Ru_3O(CH_3CO_2)_6(L)_3]^+$ has, in addition to four reversible cluster-based single-electron redox waves from +3 ($Ru_3^{IV,IV,III}$) to −1 ($Ru_3^{III,II,II}$) state, a quasireversible ligand-based one-electron wave at the most negative potentials. On the other hand, the carbonyl cluster $[Ru_3O(CH_3CO_2)_6(CO)(L)_2]$ shows four reversible single-electron redox waves from a +2 ($Ru_3^{IV,III,III}$) to a −2 (possibly $Ru_3^{II,II,II}$) state in a normally accessible potential region [7, 8]. In addition, the redox potentials of the carbonyl Ru_3 complexes are 300 to 500 mV more positive than those of corresponding waves of the noncarbonyl Ru_3 complexes, and the redox potential of Ru_3 complexes strongly depends on the basicity, or pK_a, of the ancillary ligand L. For example, redox potentials shift toward negative potential by 200–500 mV upon changing L from 4-cyanopyridine (pK_a = 1.7) to 4-dimethylaminopyridine (9.7) [8e]. Moreover, oligomers of Ru_3 clusters are easily obtained when a linear didentate bridging ligand such as pyrazine (pz) [8c,9,10] or 4,4′-bipyridine [8c,10,11] is used and each Ru_3 unit in the oligomers maintains a redox-potential-controllable ligand site. Thus, we can control the redox potential of the Ru_3 cluster oligomers by choosing the appropriate ligands, L and/or CO, for each Ru_3 cluster unit. Using these unique properties, we designed the pyrazine-bridged Ru_3 trimer 1 and tetramer 2 that show multiple one-electron redox waves. The molecular structures of 1 and 2 are shown in Fig. 7.1.

To reduce the complexity of chemical formulae, we have abbreviated the pyrazine-bridged Ru_3 oligomers such that the ancillary ligand set, excluding the bridging pyrazine, for each Ru_3 unit is in parentheses and hyphens represent the bridging pyrazines, e.g., the trimer $[\{Ru_3O(CH_3CO_2)_6(L^1)(L^2)\}$-$(\mu$-pz$)$-$\{Ru_3O(CH_3CO_2)_6(L^3)\}$-$(\mu$-pz$)$-$\{Ru_3O(CH_3CO_2)_6(L^4)(L^5)\}]$ is abbreviated as $[(L^1,L^2)$-(L^3)-$(L^4,L^5)]$. When the oxidation state of Ru or the ligand set in an oligomer unit must be shown, abbreviations such as $Ru_3^{III,III,II}$ or $\{Ru_3(L^1,L^2)\}$, in which μ_3-oxo, six acetate ligands, and pyrazine ligand are omitted, are used. In this study, pyridine and 4-dimethyaminopyridine are used as ancillary ligands L and are abbreviated as py and dmap, respectively. Meyer and coworkers reported a similar pyrazine-bridged Ru_3 trimer $[(py,py)$-(CO)-$(py,py)]^{2+}$, in which the onset of bandlike electronic properties was observed [9b]. However, oligomers 1 and 2 are the first examples of this type that have well- separated multiple one-electron redox waves.

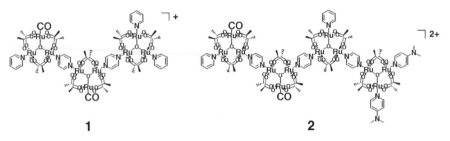

Fig. 7.1. Molecular structures of Ru_3 trimer 1 and Ru_3 tetramer 2

7.2
Syntheses of Oligomers 1 and 2

The objective of this study is to prepare Ru_3-based compounds that show many reversible well-separated one-electron redox waves. To accomplish this, the selection of the ancillary ligands and their location within the oligomer, i.e., arrangement of the Ru_3 cluster units with appropriate ancillary ligand sets in the oligomers, is extremely important. The syntheses of oligomers 1 and 2 were easily carried out using a "*complexes as metal and complexes as ligand*" strategy [12], as shown in Schemes 7.1 and 7.2.

The coordinated solvent molecules in the Ru_3 complexes ("*complex as metal*") were easily replaced by those containing pyrazine with a free donor nitrogen atom ("*complex as ligand*"). This strategy worked nicely for coupling Ru_3 unit(s) one at a time to the growing oligomers via a μ-pz ligand. Compounds 1 and 2 were obtained as PF_6^- salts in relatively good yields. De-

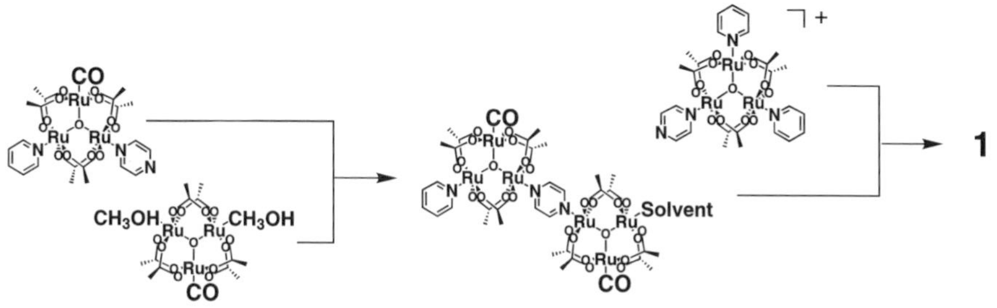

Scheme 7.1.

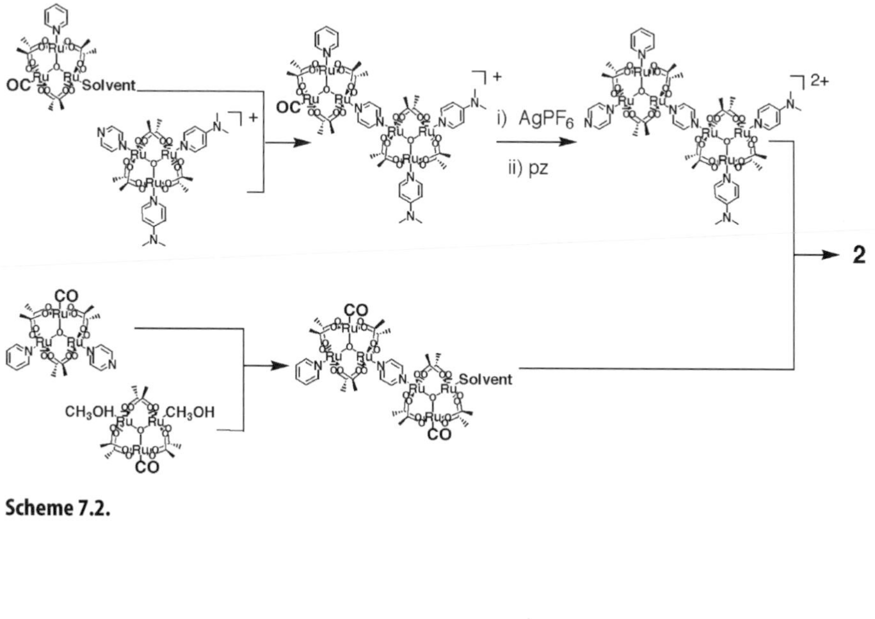

Scheme 7.2.

tailed synthetic methods and characterization of these compounds have been reported elsewhere [6].

7.3
Redox Behavior of 1 and 2

In **1** and **2**, all the Ru$_3$ units in the oligomers have different ancillary ligand sets and thus each unit should have different redox potentials. In actual fact, such redox behavior has been observed in the cyclic voltammograms of **1** and **2** in acetonitrile (Figs. 7.2 and 7.3).

The cyclic voltammogram of **1** in acetonitrile shows 11 reversible redox waves involving 12 electrons overall (Fig. 7.2, Table 7.1).

Each wave is a single-electron process and is well separated from the other waves except for two waves at $E_{1/2}$ = ca. 1.4 and 0.70 V, where two single-electron processes overlap. All of the redox waves can be easily assigned by comparing the Ru$_3$ monomers bearing the same ligand set ([Ru$_3^{III,III,II}$O(CH$_3$CO$_2$)$_6$(CO)(py)(pz)] (corresponding to unit A in Fig. 7.2): $E_{1/2}$ = 1.23, 0.58, −0.83, −1.64; [Ru$_3^{III,III,II}$O(CH$_3$CO$_2$)$_6$(CO)(pz)$_2$] (correspond-

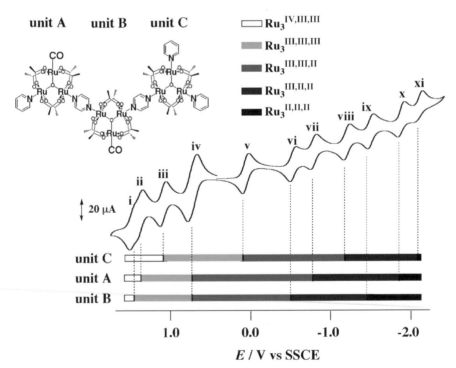

Fig. 7.2. Cyclic voltammogram and redox wave assignments of Ru$_3$ trimer **1**. The oxidation state of each unit is shown in the differently grayed regions

Table 7.1. Electrochemical data for Ru$_3$ trimer **1** in 0.1 M [(n-C$_4$H$_9$)$_4$N]PF$_6$-CH$_3$CN

Redox wave	Assignment	$E_{1/2}$ [a] V vs. SSCE (ne) [b]
i	(IV,III,III)/(III,III,III) in unit B	1.44 (1e)
ii	(IV,III,III)/(III,III,III) in unit A	1.36 (1e)
iii	(IV,III,III)/(III,III,III) in unit C	1.07 (1e)
iv	(III,III,III)/(III,III,II) in unit B and A	0.70 (2e)
v	(III,III,III)/(III,III,II) in unit C	0.09 (1e)
vi	(III,III,II)/(III,II,II) in unit B	−0.50 (1e)
vii	(III,III,II)/(III,II,II) in unit A	−0.76 (1e)
viii	(III,III,II)/(III,II,II) in unit C	−1.16 (1e)
ix	(III,II,II)/(II,II,II) in unit B [c]	−1.46 (1e)
x	(III,II,II)/(II,II,II) in unit A [c]	−1.84 (1e)
xi	(III,II,II)/(II,II,II) in unit C [c]	−2.07 (1e)

[a] $E_{1/2}$s are determined from the differential pulse voltammetry
[b] ne = number of electrons exchanged
[c] See text for the Ru$_3^{II,II,II}$ state

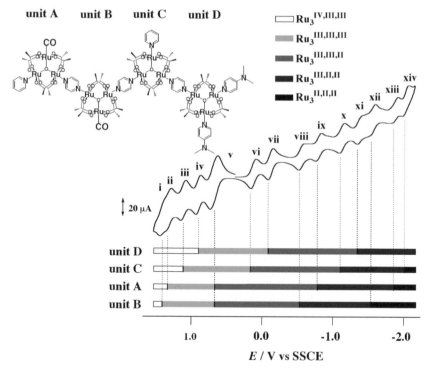

Fig. 7.3. Cyclic voltammogram and redox wave assignments of Ru$_3$ tetramer **2**. The oxidation state of each unit is shown in the differently grayed regions

ing to unit B): $E_{1/2} = 1.37, 0.70, -0.65, -1.30$; $[Ru_3^{II,III,III}O(CH_3CO_2)_6(py)_2(pz)]^+$ (corresponding to unit C): $E_{1/2} = 0.96, -0.03, -1.17, -1.74$ V). In assigning the waves based on the comparison with the component Ru_3 monomers, two important points were considered: (1) the electron-donating ability of the bridging pyrazine is much lower than that of the coordinated monodentate pyrazine, and (2) in the negative potential region, electronic interaction through the bridging pyrazine between adjacent Ru_3 units makes the difference in redox potentials of each unit larger [8c,9a,10]. The redox wave assignments are shown in Fig. 7.2. On going from the most positive potential to the most negative potential, the successive waves were assigned to $(Ru_3^{IV,III,III}/Ru_3^{III,III,III})$, $(Ru_3^{III,III,III}/Ru_3^{III,III,II})$, $(Ru_3^{III,III,II}/Ru_3^{III,II,II})$, and $(Ru_3^{III,II,II}/Ru_3^{II,II,II})$ processes in each cluster unit, where the reduction of the cluster units occurred in the order B to A to C for each redox couple (Fig. 7.2). Waves ix–xi were tentatively assigned to the $(Ru_3^{III,II,II}/Ru_3^{II,II,II})$ processes (Fig. 7.2 and Table 7.1). However, they might be ligand-based redox waves (see below). The cyclic voltammogram of **2** in acetonitrile consists of 14 reversible step waves involving 15 electrons in all (Fig. 7.3, Table 7.2).

Except for the wave at 0.70 V, which corresponds to two overlapping single-electron redox processes, all of the waves involve single-electron processes and are well separated. Tetramer **2** was designed in such a way that a $\{Ru_3^{III,III,III}(dmap,dmap)\}$ unit is linked to the right-hand side of trimer **1** to give $[Ru_3^{III,III,II}(CO,py)-Ru_3^{III,III,II}(CO)-Ru_3^{III,III,III}(py)-Ru_3^{III,III,III}$

Table 7.2. Electrochemical Data for Ru_3 tetramer 2 in 0.1 M $[(n\text{-}C_4H_9)_4N]PF_6\text{-}CH_3CN$

Redox wave	Assignment	$E_{1/2}{}^a$ V vs SSCE $(ne)^b$
i	(IV,III,III)/(III,III,III) in unit B	1.44 (1e)
ii	(IV,III,III)/(III,III,III) in unit A	1.36 (1e)
iii	(IV,III,III)/(III,III,III) in unit C	1.15 (1e)
iv	(IV,III,III)/(III,III,III) in unit D	0.93 (1e)
v	(III,III,III)/(III,III,II) in unit B and A	0.70 (2e)
vi	(III,III,III)/(III,III,II) in unit C	0.19 (1e)
vii	(III,III,III)/(III,III,II) in unit D	−0.06 (1e)
viii	(III,III,II)/(III,II,II) in unit B	−0.50 (1e)
ix	(III,III,II)/(III,II,II) in unit A	−0.75 (1e)
x	(III,III,II)/(III,II,II) in unit C	−1.08 (1e)
xi	(III,III,II)/(III,II,II) in unit D	−1.32 (1e)
xii	(III,II,II)/(II,II,II) in unit B ^{c}	−1.50 (1e)
xiii	(III,II,II)/(II,II,II) in unit A ^{c}	−1.83 (1e)
xiv	(III,II,II)/(II,II,II) in unit C ^{c}	−1.99 (1e)

$^a E_{1/2}$s are determined from the differential pulse voltammetry
$^b ne$ = number of electrons exchanged
^{c} See text for $Ru_3^{II,II,II}$ state

(dmap,dmap)]$^{2+}$. The strong electron-donating dmap ligands cause the redox waves from the {Ru$_3$(dmap,dmap)} unit to shift toward more negative potentials and thus these waves do not overlap with other waves (Fig. 7.3). Redox wave assignment of **2** was made in the following manner. Cyclic voltammogram pattern of **2** can be understood as a sum of CVs of trimer **1** and the Ru$_3$ monomer [Ru$_3$O(CH$_3$CO$_2$)$_6$(dmap)$_2$(pz)]$^+$ added to trimer **1** (unit D, Fig. 7.3). As is seen in Fig. 7.3, the redox potentials of the (Ru$_3^{IV,III,III}$/Ru$_3^{III,III,III}$/Ru$_3^{III,III,II}$/Ru$_3^{III,II,II}$/Ru$_3^{II,II,II}$) couples for unit D appear at more negative potentials than than those of units A, B, and C. Note that a wave corresponding to the Ru$_3^{III,II,II}$/Ru$_3^{II,II,II}$ process of unit D is at a potential more negative than the accessible potential window and thus does not appear in the CV. On going from the most positive potential to the most negative potential, the reduction of the cluster units occurs in the order B to A to C to D for each redox couple (Fig. 7.3). Waves xii–xiv are tentatively assigned to the (Ru$_3^{III,II,II}$/Ru$_3^{II,II,II}$) processes (Fig. 7.3 and Table 7.2). However, these might be ligand-based redox waves (see below). As mentioned above, the assignments to the (Ru$_3^{III,II,II}$/Ru$_3^{II,II,II}$) processes in **1** and **2** are tentative. A redox wave based likely on the bridging pyrazine is often observed in these potential regions; in actual fact, Meyer and coworkers assigned the wave that corresponds to our tentative assignment to the (Ru$_3^{III,II,II}$/Ru$_3^{II,II,II}$) process to a ligand-based process for similar Ru$_3$ complexes [8c]. There is a possibility that the waves observed at potentials more negative than −1.45 V are due to reduction to an electronic state created by mixing of the reduced states of pyrazine and Ru$_3$ cluster core. This study deals with compounds showing multistep one-electron redox waves and not the detailed assignments for the observed waves. Therefore, waves observed at extremely negative potentials are tentatively assigned to the (Ru$_3^{III,II,II}$/Ru$_3^{II,II,II}$) processes in this paper.

7.4
Conclusion

We synthesized triruthenium cluster oligomers which showed well-separated multistep one-electron redox waves. It has been accomplished by assembling cluster units with different redox potentials. To the best of our knowledge, the present compounds show the largest number of well-characterized and well-separated single-electron redox processes reported thus far. As an extension of this sudy, we are now making dendrimers comprised of the present Ru$_3$ clusters.

7.5
References

1. (a) Lehn JM (1995) Supramolecular Chemistry. VHC, Weinheim; (b) Boulas PL, Gòmez-Kaifer M, Echegoyen L (1998) Angew Chem Int Ed 37:216; (c) Balzani V, Scandola F

(1996) In: Atwood JL, Davies JED, MacNicol DD, Vögtle F, Lehn JM (eds) Comprehensive Supramolecular Chemistry. Elsevier, Oxford, vol 10, p 687; (d) Balzani V, Scandola F (1991) Supramolecular Photochemistry. Ellis Harwood, Chichester; (e) Balzani V, Credi A, Raymo FM, Stoddart JF (2000) Angew Chem Int Ed 39:3348

2. Xie Q, Pérez-Cordero E, Echegoyen L (1992) J Am Chem Soc 114:3978
3. Marcaccio M, Paolucci F, Paradisi C, Roffia S, Fontanesi C, Yellowlees LJ, Serroni S, Campagna S, Denti G, Balzani V (1999) J Am Chem Soc 121:10081
4. (a) Ruben M, Breuning E, Barboiu M, Gisselbrecht JP, Lehn JM (2003) Chem Eur J 9:291; (b) Ruben M, Breuning E, Gisselbrecht JP, Lehn JM (2000) Angew Chem Int Ed 39:4139
5. (a) Elliot CM, Hershenhart EJ (1982) J Am Chem Soc 104:7519; (b) Ohsawa Y, DeArmond MK, Hanck KW, Morris DE, Whitten DG, Neveux PE Jr (1983) J Am Chem Soc 105:6522; (c) Krejcik M, Vlcek AA (1992) Inorg Chem 31:2390; (d) Perez-Cordero, Buigas R, Brady N, Echegoyen L, Arana C, Lehn JM (1999) Helv Chim Acta 77:1222; (e) Aoki K, Chen J, Nishihara H, Hirao T (1996) J Electroanal Chem 416:151; (f) Nishihara H (2001) Bull Chem Soc Jpn 74:19; (g) Abe M, Sasaki Y, Yamada Y, Tsukahara K, Yano S, Yamaguchi T, Tominaga M, Taniguchi I, Ito T (1996) Inorg Chem 35:6724; (h) Abe M, Sasaki Y, Yamada Y, Tsukahara K, Yano S, Ito T (1995) Inorg Chem 34:4490; (i) Rulkens R, Lough AJ, Manners I, Lovelace SR, Grant C, Geiger WE (1996) J Am Chem Soc 118:12683
6. Hamaguchi T, Nagino H, Hoki K, Kido H, Yamaguchi T, Breedlove BK, Ito T (2005) Bull Chem Soc Jpn 78:591
7. Toma HE, Araki K, Alexiou ADP, Nikolaou S, Dovidauskas S (2001) Coord Chem Rev 219:187
8. (a) Spencer A, Wilkinson G (1974) J Chem Soc Dalton Trans 786; (b) Baumann JA, Salmon DJ, Wilson ST, Meyer TJ, Hatfield WE (1978) Inorg Chem 17:3342; (c) Baumann JA, Salmon DJ, Wilson ST, Meyer TJ (1979) Inorg Chem 18:2472; (d) Toma HE, Cipriano C (1989) J Electroanal Chem 263:313; (e) Toma HE, Cunha CJ, Cipriano C (1988) Inorg Chim Acta 154:63
9. (a) Yamaguchi T, Imai N, Ito T, Kubiak CP (2000) Bull Chem Soc Jpn 73:1205; (b) Baumann JA, Wilson ST, Salmon DJ, Hood PL, Meyer TJ (1979) J Am Chem Soc 101:2916; (c) Kido H, Nagino H, Ito T (1996) Chem Lett 745; (d) Toma HE, Alexiou ADP (1995) J Braz Chem Soc 6:267; e) Ota K, Sasaki H, Matsui T, Hamaguchi T, Yamaguchi T, Ito T, Kido H, Kubiak CP (1999) Inorg Chem 38:4070
10. (a) Ito T, Hamaguchi T, Nagino H, Yamaguchi T, Washington J, Kubiak CP (1997) Science 277:660; (b) Ito T, Hamaguchi T, Nagino H, Yamaguchi T, Kido H, Zavarine IS, Richmond T, Washington J, Kubiak CP (1999) J Am Chem Soc 121:4625
11. Toma HE, Alexiou ADP (1995) J Chem Res (S) 134
12. Campagna S, Denti G, Serroni S, Ciano M, Juris A, Balzani V (1992) Inorg Chem 31:2982

Molecular Architecture of Redox-Active Multilayered Metal Complexes Based on Surface Coordination Chemistry

Masa-aki Haga

Department of Applied Chemistry, Faculty of Science and Engineering, Chuo University, 1-13-27 Kasuga, Bunkyo-ku,Tokyo 112–8551, Japan

Summary Our recent research on multilayered architectures based on redox-active metal complexes and DNA nanowiring of molecular dots consisting of complexes with intercalating moiety is reviewed. The combination of self-assembly and successive metal coordination leads to novel well-defined nanostructures and functions on surfaces. To keep the vertical orientation from the surface normal, new bi- and tetrapod anchoring ligands with two and four phosphonate groups (LP and XP) have been introduced. Layer-by-layer fabrication of mono- and dinuclear metal complexes with these ligands on the surface made it possible to control the multilayering on the surface, resulting in a vectorial electron flow through potential gradients on the molecular scale. The electron transfer behaviors on the multilayer films will be discussed in detail. DNA nanowiring by point-to-point capture on a surface-immobilized metal complex on a mica surface is also described.

Abbreviations

ITO	Indium-tin oxide
SAM	Self-assembled monolayer
tpy	2,2′:6′,2″-Terpyridine
AFM	Atomic force microscope
tpy-ph-tpy	4,4′-(1,4-Phenylene)-bis-2,2′:6′,2″-terpyridine
XPS	X-ray photoelectron spectroscopy phen 1,10-phenanthroline
STEM	Scanning transmission electron microscope
vbpy	4-Methyl-4′-vinyl-2,2′-bipyridine
Mebimpy	2,6-Bis(1-methylbenzimidazol-2-yl)pyridine
BimpyH$_2$	2,6-Bis(l-benzimidazol-2-yl)pyridine

8.1
Introduction

Nanometer-sized molecular systems on surfaces are being widely explored in order to overcome the fabrication limit of silicon-based devices. The fabrication of well-defined nanometer-sized structures on a solid surface is one of the most challenging areas of research in nanoscience [1]. The top-down method based on photolithography in silicon technology allows for the downsizing of gate size to access several tens of nanometers; however, this miniaturization method is approaching its theoretical limit of ca. 50 nm. To overcome this problem, much effort has been devoted to developoing a bottom-up method based on the self-assembly of molecular units on a surface in order to bridge the gap between micrometer and nanometer in the fabrication size [2–5]. The assembly of discrete molecular units affords two- and three-dimensional topological structures through various combinations of molecular units with different geometries [6–9]. However, the inorganic–organic frameworks on surfaces have not been throughly studied. Since the coordination bond between metal ions and organic ligands has directionality and saturation, surface coordination chemistry has the potential to construct well-defined surface nanostructures from solution. As a self-assembled monolayer (SAM) on solid surfaces organic thiol is immobilized on a gold surface to form a solidlike structure on the surface. Similarly, a phosphonate group is attached as a SAM film on an indium-tin oxide (ITO) or mica surface. The SAM films on a solid surface ensure surface functionalities such as wettability, redox activity, photochemical responses, etc. However, complicated functions are not incorporated in monolayer films because of the low integration of functional molecular units [10]. On the other hand, the use of multilayer films on a surface overcomes this drawback by the incorporation of a variety of functional molecular units in each layer. For the fabrication of multilayer films on a surface, various interactions between molecular units have been employed, i.e., electrostatic interaction, hydrogen-bonding interaction, and coordination bond [11, 12]. In particular, the combination of a SAM monolayer film and the coordination bond leads to the desired extended nanoframeworks, maintaining the functionalities of each molecular unit. In this review, we focus on our approach to the layer-by-layer fabrication of redox-active metal complexes with the help of coordination bonds. Incorporation of redox-active molecular units within multilayer films makes it possible to create an appropriate potential gradient for vectorial electron transfer [13].

8.2
Fabrication of Multilayer Nanoarchitectures by Surface Coordination Chemistry

8.2.1
Layer-by-Layer Assembly on Solid Surfaces

Self-assembled monolayers have changed surface properties and additionally have added a new element of functionality to surfaces [10]. For example, surfaces become hydrophobic when a long alkyl silanol is attached on the glass slide. Even the molecules are densely packed because, in a self-assembled monolayer, the molecular density is too low to collect the photons in photoelectrochemical devices. On the other hand, multilayer films have a high molecular density, which can generate complex functionalities within layers. To fabricate a multilayer film on a surface, electrostatic interaction between anionic and cationic polyelectrolytes is commonly used as an alternating layer-by-layer method [14]. In the case of polymer electrolytes, the boundary between two layers is not well defined and each polymer component is sometimes entangled. Alternatively, the coordination bond between metal ions and anchored ligand is available for building up the layered or extended structures on the surface. As anchoring groups to solid surfaces, alkylsilanol, thiol, disulfide, and phosphonate are generally used, depending on the solid substrate, while terpyridine, carboxylate, phosphonate, and isocyanide groups are used as metal ion binding sites [15]. One of the advantages of using a coordination bond on the surface is to give directionality to the layering process, which is defined by both linker ligands and the coordination geometry around the metal ions. Therefore, the strategy for this "bottom-up" synthesis based on coordination bonding has the potential to afford versatile molecular architectures by the selection of molecular units [12, 16–20]. Starting from the metal complex unit with both anchoring group and linker group, the immobilization of the molecular unit on the solid surface creates the first layer and then the metal ion binds the terminal linker group onto the first layer. Further, the second layer can be formed simply by the immersion of the solid surface into the solution of the metal complex module. By repeating this immersion manipulation, the fabrication of a well-defined layer-by-layer film on the solid substrate can be achieved, as shown in Scheme 8.1.

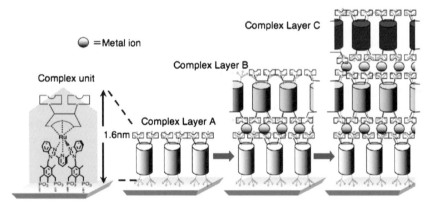

Scheme 8.1. Fabrication of layer-by-layer growth of complex units by surface coordination chemistry on solid substrate

8.2.2
Molecular Design of Anchoring Groups
for Control of Molecular Orientation on Surfaces

To make a well-defined multilayer film using a coordination bond, the molecular orientation, molecular arrangement, and molecular packing of the first layer are important factors on the solid surface. For example, the closed-packed alkyl chains of thiolates are usually tilted ~26–28° from the surface normal in the case of an alkanethiol self-assembled monolayer on an Au(111) surface [10, 21]. However, if the molecular packing is low, the molecular orientation becomes random since the simple alkyl chains are flexible. In order to control the molecular orientation on the surface, the molecular design of multipoint anchoring compounds is available, as shown in Scheme 8.2.

Rigid tripod-COOH or -SH surface-binding groups having a tetrahedral core such as tetraphenylmethane or 1,3,5,7-tetraphenyladamantane provide a stable, three-point attachment to a solid surface [11, 22–27]. This type of multipoint anchoring ligand is suitable for fixing the molecular orientation on the surface. In particular, the photochemical multicomponent system, consisting of the electron sensitizer, donor, and acceptors, plays an important role in interfacial electron transfer reactions on surfaces [28–30]. Recently, we synthesized novel bi- and tetrapod anchoring ligands (LP and XP, respectively) based on 2,6-bis(benzimidazol-2-yl)pyridine having two or four phosphonate groups and their redox-active Ru/Os complexes (Scheme 8.3) [13, 20, 31]. In the Ru complex with XP ligand, $[Ru(XP)(tpy)]^{2+}$, the X-ray crystal structure reveals that two mesityl groups are perpendicular to the 2,6-bis(benzimidazol-2-yl)pyridine plane and the rotation of the methylene group attached to the phosphonate is sterically hindered. Considering both the X-ray structure and the molecular modeling, the four legs of the phosphonate are directed to the metal oxide surface such as tin-doped indium oxide (ITO) or silicon oxide, and the auxiliary rigid linker moiety is oriented perpendicular to the surface normal. Recent advances in surface treatment allow us to obtain a relatively flat ITO surface as supplied, which makes it possible to obtain the molecular images on the ITO surfaces. Figure 8.1 shows AFM images of dinuclear $[Ru_2(XP)_2(tpy-ph-tpy)]$ complexes immobilized on a flat ITO substrate. The surface concentration of the complex on the ITO substrate was controlled by the concentration of the metal complex in the solution. At low concentrations of the metal complex, scattered dots with an average height of ~4 nm were

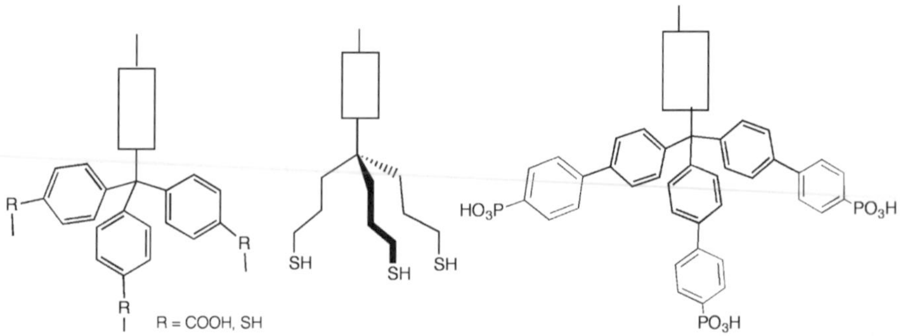

Scheme 8.2. Examples of tripod anchoring ligands

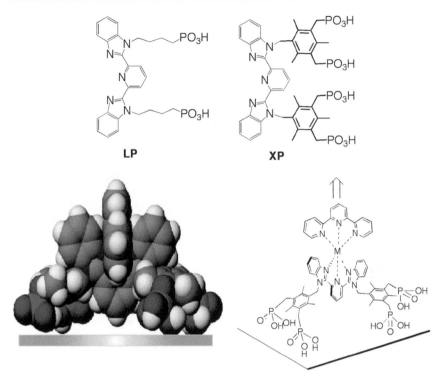

Scheme 8.3. Novel bi- and tetrapod anchoring ligands (LP and XP) and proposed molecular orientation based on both X-ray structure of Ru-XP complex and molecular modeling

observed; at higher concentrations the entire surface was covered by a closely packed monolayer. In the case of a mononuclear Ru complex, [Ru(XP)$_2$], a similar morphology was observed except that the height of the dots was ~ 1.6 nm. Molecular modeling indicates that the heights for mono- and dinuclear complexes are consistent with those expected for the vertical orientation from the surface normal.

Bis-tridentate coordination environments around an octahedral metal center ideally have a C$_{2v}$ symmetry, and two tridentate ligands are arranged in the opposite direction. Therefore, the layer-by-layer growing method can be applied to Ru/Os complexes with phosphonate groups. As shown in Scheme 8.1, the alternate immersion of the ITO electrode into each solution of a complex module ([M(L)$_2$] or [M$_2$(L)$_2$(tpy-ph-tpy)]) and metal ion (Zr^{4+} or Zn^{2+}) can form the multilayer structure on the ITO substrate: at first, the metal ion can be coordinated to the phosphonate groups on the first layer consisting of [M(L)$_2$] and [M$_2$(L)$_2$(tpy-ph-tpy)] (M = Ru(II) or Os(II), L = LP or XP), and then the second layer is formed by the binding of a metal complex module onto the metal ion (Zr^{4+} or Zn^{2+}). The multilayer formation on an ITO substrate was monitored in our experiments by the UV spectra. For the mul-

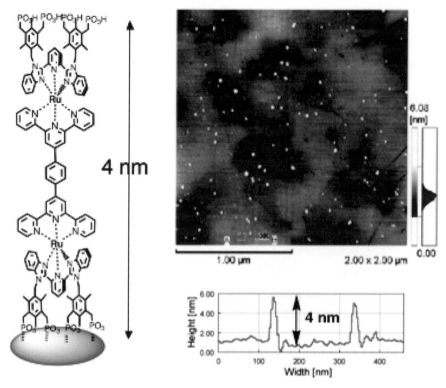

Fig. 8.1. AFM image of dinuclear $[Ru_2(XP)_2(tpy\text{-}ph\text{-}tpy)]$ complexes immobilized on flat ITO substrate and chemical structure of complex

tilayering experiments of the mono- and dinuclear complex, the plots of the absorbance of the MLCT band around 500 nm vs. number of layers reveals the linear relationship in all cases. Furthermore, the STEM images and the depth profile of XPS data were supported by the multilayer structure. Therefore, the metal-complex-based multilayers can be constructed by the combination of redox-active Ru/Os complexes and Zr^{4+} or Zn^{2+} ions. The multilayer formation between organic phosphonate and metal ion such as Zr^{4+} or Hf^{4+} was originally developed by Mallouk et al. on a variety of surfaces via sequential adsorption of metal ion and bis(phosphonic acids) from aqueous solution.

8.2.3
Molecular Design of Redox-Active Metal Complex Units for the Control of Energy Levels on Surfaces

The oxidation potentials of metal complexes are changed by the selection of heterocyclic ligands and substituents on the ligands, i.e., in some cases the oxidation potential can be predicted by Lever's ligand electrochemical parameters [32]. Further, the substitution of metal ions from Ru to Os in [M(L)$_2$], where M = Ru, Os, and L = heterocyclic ligands, induces the 0.3 ~ 0.5 V negative potential shift. The M(II/III) oxidation potentials of the [M$_2$(XP)$_2$(tpy-ph-tpy)] complexes are +0.93 V and +0.64 V as a two-electron process for M = Ru and Os, respectively. The judicious selection of ligand in complex modules can tune the redox potential, which allows us to control the direction of vectorial electron flow by arranging the redox potentials of the modular units properly. Proton-coupled electron transfer reactions also induce the potential shift of redox species in the complexes. For example, the Ru complexes having bis(benzimidazolyl)pyridine (bimpyH$_2$) show a proton-coupled electron transfer reaction [33, 34]. Scheme 8.4 shows a typical square scheme of [Ru(Mebimpy)(bimpyH$_2$)]$^{2+}$, in which the oxidation potential of

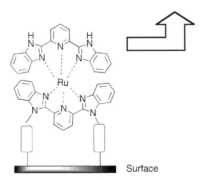

Scheme 8.4. Square scheme for proton-coupled electron transfer in [Ru(Mebimpy)(bimpyH$_2$)]$^{2+}$

[Ru(Mebimpy)(bimpyH$_2$)]$^{2+}$ in a CH$_3$CN/Britton–Robinson buffer (1:1 v/v) is shifted from +0.64 V vs. SCE to 0.14 V upon deprotonation [33]. The deprotonation of the benzimidazole N–H groups induces a large energy perturbation in the complex. Therefore, the Ru complex units having a bimpyH$_2$ ligand have the potential for molecular switch, since proton transfer reaction can act as an external stimulus. We have also found that the metal–metal interaction in dinuclear complexes can be tuned by the deprotonation/deprotonation on bridging benzimidazole ligands and that intramolecular electron transfer can be triggered by the protonation/deprotonation in tetranuclear mixed-valence Ru/Os complexes containing (2-benzimidazolyl)pyridine derivatives [35–38]. The concept of proton-induced switch can be applied to the molecular sysytem on a surface by a judicious molecular design. When a multilayer consisting of metal complex modules is immobilized on a solid surface and one of the layers can respond to the outer perturbation such as proton dissociation or photoexcitation, as shown in Scheme 8.5, this surface can possess a molecular switching property.

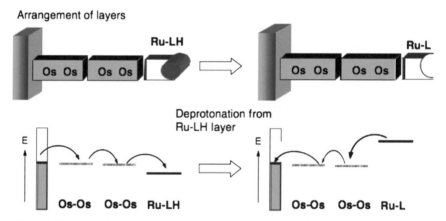

Scheme 8.5. Proton-induced molecular switch in multilayer films and energetics upon deprotonation

8.3
Chemical Functions of Redox-Active Multilayered Complexes on Surface

8.3.1
Electron Transfer Events in Multilayer Nanostructures

The layer-by-layer growth of redox-active [Ru(XP)$_2$] and [M$_2$(XP)$_2$(tpy-ph-tpy)] units (M = Ru or Os), which possess phosphonate groups at opposite ends of the unit, leads to a multilayer structure with the assistance of Zr(IV) or Zn(II)

ions on the ITO substrate. Cyclic voltammetry of the [Ru(XP)$_2$] monolayer immobilized on an ITO electrode exhibits the Ru(II/III) oxidation process at +0.65 V vs. Ag/AgCl. Similarly, the [Ru$_2$(XP)$_2$(tpy-ph-tpy)] monolayer on an ITO electrode shows one two-electron process at +0.93 V vs. Ag/AgCl. The surface coverages for [Ru(XP)$_2$] and [Ru$_2$(XP)$_2$(tpy-ph-tpy)] monolayer films are 5.60×10^{-11} and 5.54×10^{-11} mol/cm^2, respectively, which strongly suggests that the molecules are in a closely packed and self-sustained vertical orientation on the ITO surface. By increasing the number of layers with the Zr(IV) ion as a glue, the anodic peak current increases linearly up to 14 layers for [Ru(XP)$_2$] and 8 layers for dinuclear [Ru$_2$(XP)$_2$(tpy-ph-tpy)]. Above these layers, the increasing rate on the peak current is gradually decreased since the current in the multilayer film might be governed by the rate of electron hopping or electron exchange between the layers. The electron transfer reaction of [Fe(CN)$_6$]$^{4-}$ through immobilized multilayer films on an ITO electrode was examined. For the electron transfer reaction between a modified ITO electrode and [Fe(CN)$_6$]$^{4-}$ in solution, the electron mediation through the immobilized Ru units in the multilayer film was observed in the case of two or more layers.

8.3.2
Combinatorial Approach to Electrochemical Molecular Devices in a Multilayer Nanostructure on Surfaces

One of the advantages of layer-by-layer growth is that a combinatorial approach can be used for the fabrication of a multilayer nanostructure by selecting the different molecular modules as a layered unit. Since Ru and Os complexes can tune their oxidation potentials by the selection of ligands, the different ordering of Ru/Os layers from the surface will lead to a different potential distribution or gradient in a controlled manner on the nanometer scale. By changing the order of molecular units with different redox potentials on the surface, any potential sequences that might lead to a molecular rectifier or electron transfer cascade system can be constructed. The modification of electrode surfaces has been pursued for a long time in the field of electrochemistry. In particular, electrochemical polymerization of redox-active monomers such as [Ru(phen)$_2$(vbpy)]$^{2+}$ on an electrode was studied thoroughly by Murray and other researchers in the 1980s and 1990s [39, 40]. Bilayer and multilayer structures based on different combinations of redox-active Ru/Os complex polymers were also constructed in which rectification and electron hopping within these redox-active polymers were studied. The film thickness of polymers was roughly of submicron order, but it is hard to control. In contrast, the "bottom-up" approach using a self-assembled monolayer on the surface makes it possible to construct well-defined molecular films on the nanometer scale. When the coordination site or metal binding site is provided as a terminal group in the self-assembled monolayer, the metal coordination, followed by the ligation of the bridging ligand, can make the second layer. The successive

alternate binding between metal ion and bridging ligand can form the multilayered structure, and any molecular components using bridging head groups can
be bound at any position of the multilayers. Furthermore, the "complexes-as-
ligands" and/or "complexes-as-metals" strategy, which is used for the synthesis
of polynuclear or dendritic metal complexes in a homogeneous solution, can
be applied to the multilayer formation on a surface.

By using a variety of molecular units with different redox potentials such as
$[M(XP)_2]$ and $[M_2(XP)_2(tpy-ph-tpy)]$ complexes (M = Ru or Os), we can make
various types of order and combination of the modular units, which leads to
a different potential sequence in the multilayered structure. Figure 8.2 shows
two differernt structures in three-layered films and their expected potential
sequences on the surface.

In the Marcus normal region, the rate of electron transfer will be fast if
the free energy change ΔG increases, while in the inverted region the rate of
electron transfer decreases with an increase in the driving force. Therefore,
the persistent potential gradient leads to a preferential electron flow from the
outer layer to the electrode or vice versa (Fig. 8.2). Figure 8.2 also shows cyclic

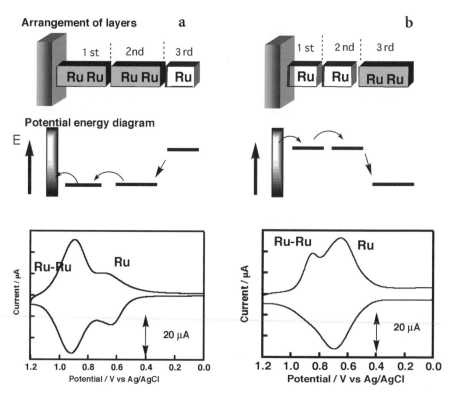

Fig. 8.2. Two different structures in three-layered films and the expected potential sequences
on the surface and their cyclic voltammograms

voltammograms of three layered films with different structures, in which the direction of current flow can be determined by the potential gradient between the inner-layer potentials and outer-layer ones. The rectification effect was obviously observed for the ratio of anodic vs. cathodic peak current in the top layer, i.e., the anodic peak current was mainly observed in the case of a [Ru(XP)$_2$] complex as the top layer, while the cathodic peak current was observed in the case of a [Ru$_2$(XP)$_2$(tpy-ph-tpy)] complex as the bottom layer, as shown in Fig. 8.2.

Consequently, the judicious selection of the molecular units makes it possible to construct the various multilayer structures with potential application for molecular electronic devices such as nonlinear optical materials, electroluminescent devices, and photovoltaic devices.

8.3.3
Surface DNA Trapping by Immobilized Metal Complexes with Intercalator Moiety Toward Nanowiring

Developing a new fabrication method for nanowiring between two molecular terminals is a challenge. For this purpose, DNA is an attractive material as it uses a wire scaffold of predetermined length [41–43]. We have prepared a series of rod-shaped Ru complexes having both naphthalene-1,4:5,8-bis(dicarboximide) (ndi) or acridine as a DNA intercalating group and phosphonate as an anchoring group [44]. These Ru complexes can interact with a double-stranded DNA line through both intercalation and electrostatic interaction between anionic DNA and a cationic Ru complex. Therefore, the molecular dots comprising the Ru complex on the electrode surface can be considered as a molecular terminal, which can capture a double-stranded DNA molecule from solution and form the nanowire (Scheme 8.6). We found that two handling methods could be used on the present Ru complexes to get the molecular dots on the mica surface: (1) a gentle rinsing of the densely covered mica surface of the Ru complex with ammonia solution (pH 11) for several minutes and (2) the immobilization on the mica surface by use of the immersing solution with low concentration of the Ru complex (< 1 µM). Once the Ru complex was immobilized on the surface as a "molecular dot", λ DNA was captured by the dots in a point-to-point manner by a meniscus transfer method. AFM measurements indicated that both clear lines of DNA and molecular dots of the Ru complex were easily discriminated, as shown in Fig. 8.3. The molecular height of each DNA line is ~ 2.0 nm, and the characteristics were connected in a point-to-point manner associated with sharp break points. Once captured by the immobilized Ru complex, these lines of DNA were not removed by simply rinsing them with water. As a controlled experiment, DNA was expanded on a bare mica surface by the meniscus transfer method, after which a clear AFM image of DNA lines was observed; however, these DNA lines were easily washed away from the bare mica surface by water, which is

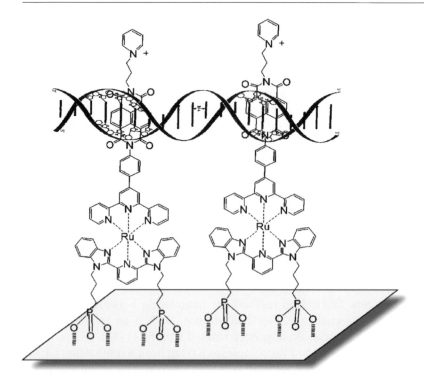

Scheme 8.6. Schematic of DNA nanowiring between two Ru complex terminals

a sharp contrast to the result above. By controlling the hooking-up points of the DNA, the nanowiring between two terminals can be achieved on an Au/SiO$_2$ prepatterned surface. This captured DNA acts as a metallization template for DNA-based nanowires. The combination of electron transfer ability and DNA intercalation for immobilized rod-shaped Ru complexes would make it possible to generate a new application for DNA sensing or for the materials science of DNA as a wire or scaffold for applications in molecular electronics [41].

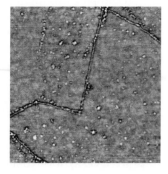

Fig. 8.3. AFM image of point-to-point DNA trapping by surface-immobilized Ru complexes on mica surface (image: 1.25 μm × 1.25 μm). The *central bending line* is hooked up by DNA to the immobilized Ru complex

8.4
Conclusion

Controlled layer-by-layer assembly by surface coordination chemistry will open new avenues in the fabrication of inorganic-organic hybrid multilayer systems on surfaces. Since self-assembly at a surface followed by surface metal coordination provides the flexibility to hook up electronically active molecules between nanometer-scale contacts, this bottom-up method will contribute to future materials science with wide-ranging applications such as sensors, molecular switches, photochemical energy conversion, and molecular devices [45].

Acknowledgement M.H. gratefully acknowledges financial support from the Institute of Science and Engineering at Chuo University and the Ministry of Education, Science, Sports and Culture for a grant-in-aid for scientific research (No. 15310076 and 16074215).

8.5
References

1. National Research Council (2003) Beyond the Molecular Frontier. National Academies Press, Washington, DC
2. Wada Y, Tsukada M, Fujihira M, Matsushige K, Ogawa T, Haga M-a, Tanaka S (2000) Jpn J Appl Phys 39:3835
3. Carroll L, Gorman CB (2002) Angew Chem Int Ed 41:4378
4. Joachim C, Gimzewski JK, Aviram A (2000) Nature 408:541
5. Jortner J, Ratner M (1997) Molecular Electronics: Chemistry for the 21st Century. Blackwell, Oxford
6. Fujita M, Umemoto K, Yoshizawa M, Fujita N, Kurukawa T, Biradha K (2001) Chem Commun p 508
7. Moulton B, Zaworotko MJ (2001) Chem Rev 101:1629
8. Michl J (1997) NATO ASI Series, vol 499. Kluwer, Dordrecht
9. Kitagawa S, Kitaura R, Noro S (2004) Angew Chem Int Ed 43:2334
10. Ulman A (1996) Chem Rev 96:1533
11. Li G, Fudickar W, Skupin M, Klyszcz A, Draeger C, Lauer M, Fuhrhop J-H (2002) Angew Chem Int Ed 2002:1828
12. Mallouk TE, Kim H-N, Ollivier PJ, Keller SW (1996) Comprehensive Supramolecular Chemistry, vol 7. Pergamon, Oxford, p 189
13. Haga M (2005) Kobunshi 54:74
14. Decher G (2003) In: Decher G, Schlenoff JB (eds) Multilayer Thin Films. Wiley-VCH, Weinheim, Germany, p 1
15. Doron-Mor I, Cohen H, Cohen SR, Popovitz-Biro R, Shanzer A, Vaskevich A, Rubinstein I (2004) Langmuir 20:10727
16. Yang JC, Aoki K, Hong H-J, Sackett DD, Arendt MF, Yau S-L, Bell CM, Mallouk TE (1993) J Am Chem Soc 115:11855
17. Liang Y, Schmehl RH (1995) J Chem Soc Chem Commun p 1007
18. Hong H-G, Mallouk TE (1991) Langmuir 7:2362
19. Cao G, Hong H-G, Mallouk TE (1992) Acc Chem Res 25:420

20. Haga M, Yutaka T (2004) In: Pombeiro AJL, Amatore C (eds) Trends in Molecular Electrochemistry. Marcel Dekker, New York, p 311
21. Nuzzo RG, Fusco FA, Allara DL (1987) J Am Chem Soc 109:2358
22. Galoppini E, Guo W, Zhang W, Hoertz PG, Qu P, Meyer GJ (2002) J Am Chem Soc 124:7801
23. Long B, Nikitin K, Fitzmaurice D (2003) J Am Chem Soc 125:5152
24. Hirayama D, Takimiya K, Aso Y, Otsubo T, Hasobe T, Yamada H, Imahori H, Fukuzumi S, Sakata Y (2002) J Am Chem Soc 124:532
25. Nikitin K, Fitmaurice D (2005) J Am Chem Soc 127:8067
26. Wei L, Padmaja K, Youngblood WJ, Lysenko AB, Lindsey JS, Bocian DF (2004) J Org Chem 69:1461
27. Muthukumaran K, Loewe RS, Ambroise A, Tamaru S, Li Q, Mathur G, Bocian DF, Misra V, Lindsey JS (2004) J Org Chem 69:1444
28. Imahori H, Arimura M, Hanada T, Nishimura Y, Yamazaki I, Sakata Y, Fukuzumi S (2001) J Am Chem Soc 123:335
29. Imahori H, Norieda H, Yamada H, Nishimura Y, Yamazaki I, Sakata Y, Fukuzumi S (2001) J Am Chem Soc 123:100
30. Imahori H, Yamada H, Nishimura Y, Yamazaki I, Sakata Y (2000) J Phys Chem B 104:2099
31. Haga M, Takasugi T, Tomie A, Ishizuya M, Yamada T, Hossain MD, Inoue M (2003) J Chem Soc Dalton Trans 2069–2079
32. Lever ABP (1990) Inorg Chem 29:1271
33. Haga M, Hong H, Shiozawa Y, Kawata Y, Monjushiro H, Fukuo T, Arakawa R (2000) Inorg Chem 39:4566
34. Haga M, Ali MM, Maegawa H, Nozaki K, Yoshimura A, Ohno T (1994) Coord Chem Rev 132:99
35. Haga M, Ano T, Kano K, Yamabe S (1991) Inorg Chem 30:3843
36. Haga M-a, Ali MM, Arakawa R (1996) Angew Chem Int Ed 35:76
37. Haga M-a, Ano T-a, Ishizaki T, Kano K, Nozaki K, Ohno T (1994) J Chem Soc Dalton Trans p 263
38. Haga M-a, Ali MM, Koseki S, Fujimoto K, Yoshimura A, Nozaki K, Ohno T, Nakajima K, Stufkens DJ (1996) Inorg Chem 35:3335
39. Abruna HD, Denisevich P, Umana M, Meyer TJ, Murray RW (1981) J Am Chem Soc 103:1
40. Murray RW (1992) In: Weissberger A, Saunders WH (eds) Techniques of Chemistry, vol 22. Wiley, New York
41. Niemeyer CM (2001) Angew Chem Int Ed 40:4128
42. Zhang J, Ma Y, Stachura S, He H (2005) Langmuir 21:4180
43. Nakao H, Gad M, Sigiyama S, Otobe K, Ohtani T (2003) J Am Chem Soc 125:7162
44. Haga M, Ohta M, Machida H, Chikira M, Tonegawa N (2006) Thin Solid Films 499:201
45. Katz E, Willner I (2004) Angew Chem Int Ed 43:6042

Chapter 9

Programmed Metal Arrays
by Means of Designable Biological Macromolecules

Kentaro Tanaka[1,2] · Tokomo Okada[1] · Mitsuhiko Shionoya[1]

[1] Department of Chemistry, Graduate School of Science, The University of Tokyo, Hongo, Bunkyo-ku, Tokyo 113-0033, Japan
[2] PRESTO, Japan Science and Technology Agency, 4-1-8 Honcho, Kawaguchi-Shi, Saitama 332-0012, Japan

Abbreviations

DNA	Deoxyribonucleic acid
EPR	Electron paramagnetic resonance
CD	Circular dichroism
Ala	Alanine
Met	Methionine
NMR	Nuclear magnetic resonance
COSY	Correlation spectroscopy
NOESY	2-D Nuclear Overhauser effect spectroscopy
ESI-TOF	Electrospray ionization-time-of-flight

9.1
Introduction

Biological macromolecules such as nucleic acids and proteins have a structural basis for programmed arrangement of the molecular building blocks into pre-designed geometries. Such bottom-up strategies in biological systems have prompted us to develop functional supermolecules from small chemical components through self-assembly processes. In this regard, metals are known as key components of self-assembled molecular architectures with discrete structures and spaces as well as metal-based chemical and physical properties. In particular, the incorporation of metals into biomolecular scaffolds is currently attracting increasing attention.

This chapter covers our recent approaches to precise programming of metal arrays templated by artificial DNA and cyclic peptides.

9.2
DNA-Directed Metal Arrays

9.2.1
Metal-Mediated Base Pairing in DNA

In natural DNA duplexes formed from two strands of poly- or oligonucleotides, hydrogen-bonded base pairs, which are attached nearly perpendicularly to the helix axis, are arranged in direct stacked contact. Among a variety of approaches to DNA-based supramolecular architectures, the strategy of replacing natural base pairs by predesigned artificial base pairs possessing a distinctive shape, size, and function is, in particular, expected to provide a general method of molecular arrangement within the DNA [1–3]. For instance, when the incorporated nucleobases have the ability to bind to metal ions, metal-mediated base pairs would be formed and aligned along the helix axis. In addition, the use of more than two kinds of metal–ligand-type nucleobases would allow heterogeneous metal arrays in a programmable manner.

Metal ions incorporated in this way would (1) affect thermal and thermodynamic stability of high-order structures of DNA such as duplex, triplex, and so on, (2) allow one-dimensional metal arrays with unique functions based on the metal sequence, and (3) assemble DNA molecules through metal coordinative linking to form two- or three-dimensional DNA networks.

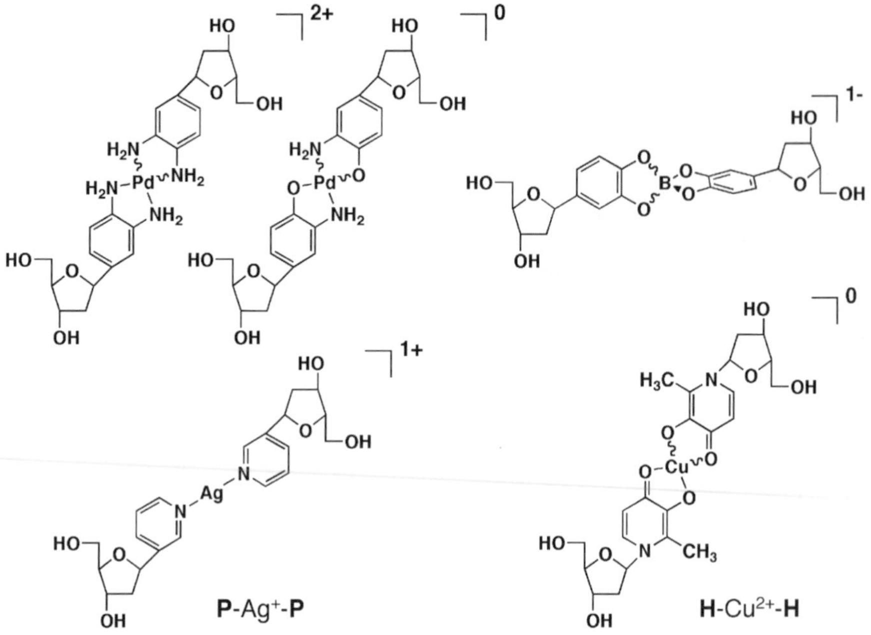

Fig. 9.1. Examples of metal-mediated base pairs using artificial nucleosides

As a novel approach to the construction of metallo-DNA, we recently reported the first Pd^{2+}-mediated base pairing by means of a metal–ligand-type nucleoside having a phenylenediamine base [4]. Since then, we have exploited alternative base pairing modes such as B^{3+}-induced base paring with catechol [5], Pd^{2+}-mediated base pairing with 2-aminophenol [6], Ag^+-assisted base pairing with pyridine [7], and Cu^{2+}-mediated base pairing with hydroxypyridone [8] (Fig. 9.1). Other groups have also reported a few examples of metal-mediated base pairing based on the same design concept [9–13]. Among these base pairs, those in a square-planar or a linear coordination geometry were expected to substitute for a flat, hydrogen-bonded natural base pair.

9.2.2
Single-Site Incorporation of a Metal-Mediated Base Pair into DNA

Incorporation of phosphoramidite derivatives of artificial nucleosides into oligonucleotides can be performed with standard protocols using a DNA synthesizer. The first example of single-site incorporation of a metal-mediated base pair into oligonucleotides was reported by Schultz and Romesberg et al. [9] using a base pair with a pyridine-2,6-dicarboxylate nucleobase as a planar tridentate ligand and a pyridine nucleobase as the complementary single donor ligand. For instance, a 15-mer DNA duplex having a base pair in the middle was significantly stabilized by Cu^{2+} ions due to the formation of a square-planar Cu^{2+} complex with the paired ligand bases inside the duplex. The stability was comparable to that of a duplex containing a natural A–T base pair instead of the Cu^{2+}-mediated base pair.

Recently, we have independently reported the single-site incorporation of an Ag^+-mediated base pair and an Ag^+-mediated base triplet into a DNA duplex and a DNA triplex, respectively, by introducing a monodentate pyridine nucleobase in the middle of each strand [7]. For example, a 21-mer DNA duplex, $d(5'-T_{10}PT_{10}-3') \cdot d(3'-A_{10}PA_{10}-5')$, containing a pyridine nucleobase (P) in the middle of the sequence was significantly stabilized in the presence of Ag^+ ions, as shown in the thermal denaturation experiments, due to the formation of positively charged P-Ag^+-P base pairing in a linear coordination geometry (Fig. 9.2). This Ag^+-dependent thermal stabilization of duplex was only slight in a reference DNA duplex, $d(5'-T_{21}-3') \cdot d(3'-A_{21}-5')$. It is to be noted that the P-Ag^+-P base pairing occurs only inside the DNA in the range of micromolar concentrations in aqueous media at neutral pH. The complexation of pyridine bases with Ag^+ in the artificial DNA should be reinforced by surrounding base pairs that are hydrogen-bonded and stacked in the hydrophobic environment within the duplex. Moreover, this Ag^+-mediated base pairing is specific because the addition of other metal ions such as Cu^{2+}, Ni^{2+}, Pd^{2+}, and Hg^{2+} showed almost no significant effects on the thermal stabilization of the duplex.

Such Ag^+-induced thermal stabilization was also observed with a DNA triplex, $d(3'-T_{10}PT_{10}-5') \cdot d(5'-A_{10}PA_{10}-3') \cdot d(5'-T_{10}PT_{10}-3')$ [7] (Fig. 9.3). In

Fig. 9.2. Ag⁺-mediated DNA duplex formation with pyridine nucleobases

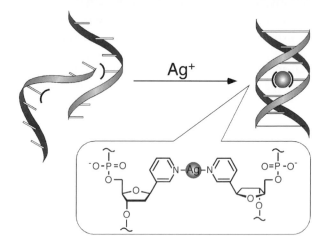

Fig. 9.3. Ag⁺-mediated DNA triplex formation with pyridine nucleobases

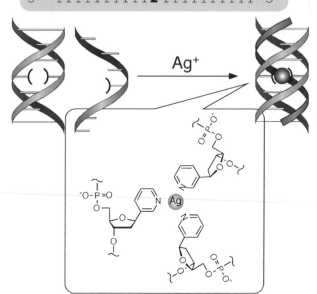

this case, this stabilization effect is believed to be due to the formation of an Ag^+-mediated base triplet in which the three pyridine nitrogen donors coordinate to an Ag^+ ion in the center of the triplex.

Along this line, a hydroxypyridone nucleobase as a flat bidentate ligand was incorporated into a 15-mer nucleotide DNA duplex, d(5'-CACATTAHTGTTG TA-3')·d(3'-GTGTAATHACAACAT-5') [8] (Fig. 9.4). In the presence of equimolar Cu^{2+} ions, a neutral Cu^{2+}-mediated base pair of hydroxypyridone nucleobases (**H**–Cu^{2+}–**H**) was quantitatively formed within the DNA, and the artificial duplex showed a higher thermal stability compared with a natural oligoduplex, d(5'-CACATTAATGTTGTA-3')·d(3'-GTGTAATTACAACAT-5'), in which the **H**–**H** base pair was replaced by an **A**–**T** base pair. In addition, EPR and CD spectra of the metallo-DNA suggested that a radical site was formed on the Cu^{2+} center within the right-handed duplex structure. This strategy was further developed for the controlled alignment of metallobase pairs along the helix axis of DNA.

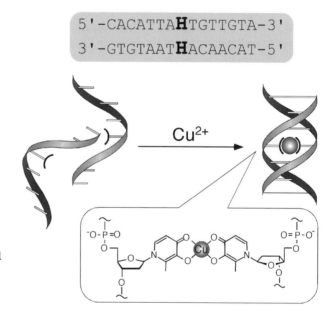

5'-CACATTA**H**TGTTGTA-3'
3'-GTGTAAT**H**ACAACAT-5'

Cu^{2+}

Fig. 9.4. Cu^{2+}-mediated DNA duplex formation with hydroxypyridone nucleobases

9.2.3
Discrete Self-Assembled Metal Arrays in DNA

DNA is a promising molecule that could act as a ligand for one-dimensional metal arrays when the nucleobases are replaced by ligand bases. That is, the multisite incorporation of metallobase pairs within DNA into direct stacked contact could lead to the construction of metal strings in DNA, in which metal ions are lined up along the helix axis in a controlled manner.

Recently, we reported the synthesis of a series of oligonucleotides, $d(5'\text{-}GH_nC\text{-}3')$ ($n = 1\text{--}5$), possessing one to five hydroxypyridone nucleobases (H) [14]. Detailed photometric titration experiments with Cu^{2+} revealed that these oligonucleotides quantitatevly produced right-handed double helices of the oligonucleotides, $nCu^{2+} \cdot d(5'\text{-}GH_nC\text{-}3')_2$ ($n = 1\text{--}5$), through Cu^{2+}-mediated base pairing ($H\text{--}Cu^{2+}\text{--}H$) (Fig. 9.5). In these metallo-DNA, the Cu^{2+} ions incorporated into each complex are aligned along the helix axes inside the duplexes with a $Cu^{2+}\text{--}Cu^{2+}$ distance of approximately 3.7 Å, as determined by EPR studies. The Cu^{2+} ions are coupled with one another through unpaired d electrons to form magnetic chains. The electron spins on adjacent Cu^{2+} centers couple in a ferromagnetic fashion with accumulating Cu^{2+} ions attaining the highest spin state, as expected from a lineup of Cu^{2+} ions. A proposed right-handed and double-stranded structure is drawn with a pentanuclear Cu^{2+} array inside the DNA (Fig. 9.6).

Our next goal is to incorporate increasing numbers of metal ions into DNA and to provide a general tool for heterogeneous metal arrays using DNA templates possessing more than two different kinds of artificial nucleobases aligned in a programmable manner.

d(5'-GHc-3')
d(5'-GHHc-3')
d(5'-GHHHc-3')
d(5'-GHHHHc-3')
d(5'-GHHHHHc-3')

$d(5'\text{-}GH_nC\text{-}3')$

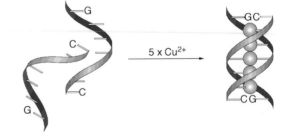

Fig. 9.5. Cu^{2+}-mediated DNA duplex formation from two artificial oligonucleotide strands bearing one to five hydroxypyridone nucleobases

$5 \times Cu^{2+}$

Fig. 9.6. A plausible structure of pentanuclear Cu^{2+} complexes within DNA

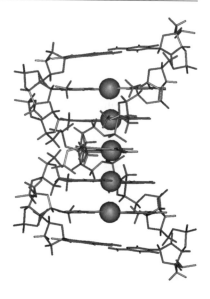

9.3
Peptide-Directed Metal Arrays

9.3.1
Design Concept

Cyclic peptides have been used extensively for ion recognition [15–18], ion channels [19, 20], metal arrays [21–23], and other applications directed towards metal-dependent functions [24–26]. These molecular functions depend largely on the number and sequence of their amino-acid constituents. Incorporation of metal binding sites at appropriate positions of the cyclic framework would provide a novel tool for supramolecular construction of discrete metal-assembled complexes in a designable manner.

The previous results with metallo-DNA prompted us to begin investigating heterogeneous metal assembly using a cyclic hexapeptide with metal coordination sites both at amino-acid side chains and the amide carbonyl groups arranged on the cyclic framework. A cyclic hexapeptide having a repeating L-Ala-L-Met sequence, *cyclo*(L-Ala-L-Met)₃ (**L**), was first designed so as to have metal binding sites both at the thioether groups for softer metals (outer circle in pale grey) and the amide carbonyl groups for harder metals (inner circle in black) (Fig. 9.7). The roundly arranged carbonyl groups of cyclic hexapeptides are known to preferentially bind to alkali or alkaline earth ions as a tridentate ligand. The cyclic hexapeptide (**L**) bearing two different types of metal binding sites was found to quantitatively and highly selectively form a tetranuclear complex with three Ag^+ ions and one Ca^{2+} ion in a capsulelike dimeric structure [27].

Fig. 9.7. Cyclic hexapeptide, *cyclo*(L-Ala-L-
Met)₃ (**L**)

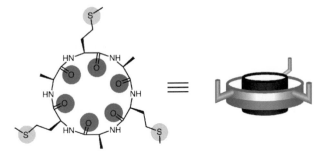

L : *cyclo*(L-Ala-L-Met)₃

9.3.2
Heterogeneous Metal Arrays Using Cyclic Peptides

The cyclic hexapeptide (**L**) with three thioether and three amide carbonyl
groups were expected to bind heterogeneously to two different metal ions possibly in a dimeric structure of the peptides. Interactions of the cyclic hexapeptide with metal ions were first examined by ^{1}H NMR titration study using Ag^+
and Ca^{2+} in combination. As a result, when the ratio of $L/Ag^+/Ca^{2+}$ reached
2:3:1, the spectrum of **L** completely changed to another set of signals, indicating
the coordination of the thioether groups to Ag^+ ions and the conformational
changes of the cyclic framework upon Ca^{2+} binding. It should be noted that the
coexistence of three Ag^+ ions and one Ca^{2+} ion is essential for the complexation.
Ultimately, the complex formed in solution was determined to be $[Ag_3CaL_2]^{5+}$
by the ESI-TOF mass spectrum of the solution ($L/Ag^+/Ca^{2+}$ = 2:3:1). These data
clearly indicate the quantitative dimeric complexation accompanying simultaneous encapsulation of three Ag^+ ions and one Ca^{2+} ion between the two cyclic
peptides (Fig. 9.8). In addition, ^{1}H–^{1}H COSY and NOESY studies successfully
determined the head-to-tail stacking of the two cyclic peptides, although there
are three possible orientations (head-to-head, tail-to-tail, or head-to-tail) in
the stacked structure since both ring faces of each cyclic peptide are different
from each other due to their asymmetric centers.

A more detailed structure of the complex was clarified by X-ray crystal
analysis. We obtained single crystals from a solution of $L/Ag^+/Ca^{2+}$ = 2:3:1,

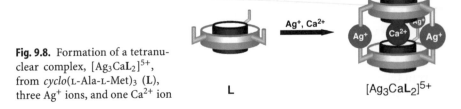

Fig. 9.8. Formation of a tetranuclear complex, $[Ag_3CaL_2]^{5+}$,
from *cyclo*(L-Ala-L-Met)₃ (**L**),
three Ag^+ ions, and one Ca^{2+} ion

which have proven to be a result of heterogeneous crystallization presumably due to its lower solubility and crystal packing. The analytical data revealed that the resulting complex is a Ca^{2+}-linked dimer of $Ag_3CaL_2^{5+}$, $[(Ag_3CaL_2)_2Ca]^{12+}$ (Fig. 9.9). In the partial structure corresponding to $Ag_3CaL_2^{5+}$ in the complex, the central Ca^{2+} ion is coordinated by the six carbonyl oxygen atoms of the two cyclic peptides in a slightly distorted octahedral geometry, and each Ag^+ ion is put between the two sulfur atoms to form a linear complex. As seen in solution, the two cyclic peptides face each other in a head-to-tail orientation with the aid of three Ag^+ ions and one Ca^{2+} ion. Without any support of Ag^+ coordination, the Ca^{2+} binding to the main chain of cyclic hexapeptides is known to be relatively weak. Thus it appears that the Ca^{2+} binding is largely reinforced by the coordination of three Ag^+ ions outside the cyclic framework.

a   b

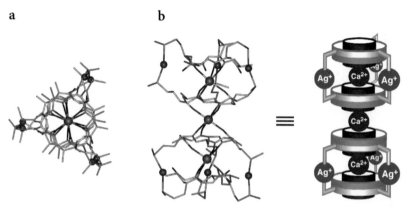

Fig. 9.9. X-ray structure of $[(Ag_3CaL_2)_2Ca]^{12+}$; **a** a *top view* and **b** a *Side view*

9.3.3
Metal Ion Selectivity in Supramolecular Complexation

The selectivity of the Ca^{2+} binding is quite high, especially compared with Mg^{2+}. The encapsulation selectivity for the Ag^+-mediated dimer formation was examined by ^1H NMR and ESI-TOF mass experiments using monovalent alkali, divalent alkaline earth, and trivalent lanthanide ions, instead of Ca^{2+} (Fig. 9.10). Only divalent Sr^{2+} and Ba^{2+} ions were found to induce the dimer formation when excess amounts of these ions were used. However, no complexation was observed under the same conditions with Mg^{2+}, alkaline metal ions, or La^{3+} ion with an ionic radius similar to Ca^{2+} and Na^+. The encapsulation of metal ions in the central cavity of the dimer was thus highly dependent on the size, charge, and solvation of metal ions.

Fig. 9.10. Effects of various ions on dimer complexation. $[L] = 2$ mM, $[Ag^+] = 3$ mM, [Metal ion] $= 1$ mM in acetone-d_6/CD$_3$OD (5:1) at 293 K. Alkaline metals include Li$^+$, Na$^+$, K$^+$, Rb$^+$, and Cs$^+$ (nd: not detected)

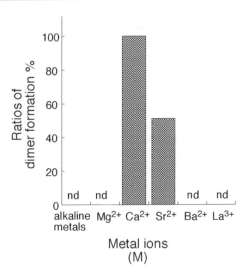

In summary, quantitative, heterogeneous metal assembly was accomplished using a predesigned cyclic hexapeptide as a template. Peptides with a predetermined number and sequence of metal coordination sites would provide template-directed arrays of metal-containing nanometer-scaled devices.

9.4
Conclusion

In this study, we demonstrate that biopolymers such as DNA and peptides can be precisely redesigned and reconstructed as templates for homogeneous or even heterogeneous assembly of metal ions in a programmable manner by chemically modiyfing their building blocks. This bottom-up strategy based on biorelated molecules would generate not only a new structural motif for metal clusters in the biomolecules but also a metal-triggered information-transfer system.

9.5
References

1. Beaucage SL, Bergstrom DE, Glick GD, Jones RA (2001) Current Protocols in Nucleic Acid Chemistry. Wiley, New York
2. Kool ET (2002) Acc Chem Res 35:936
3. Shionoya M, Tanaka K (2004) Curr Opin Chem Biol 8:592
4. Tanaka K, Shionoya M (1999) J Org Chem 64:5002
5. Cao H, Tanaka K, Shionoya M (2000) Chem Pharm Bull 48:1745
6. Tasaka M, Tanaka K, Shiro M, Shionoya M (2001) Supramol Chem 13:671
7. Tanaka K, Yamada Y, Shionoya M (2002) J Am Chem Soc 124:8802

8. Tanaka K, Tengeiji A, Kato T, Toyama N, Shiro M, Shionoya M (2002) J Am Chem Soc 124:12494
9. Meggers E, Holland PL, Tolman WB, Romesberg FE, Schultz PG (2000) J Am Chem Soc 122:10714
10. Atwell S, Meggers E, Spraggon G, Schultz PG (2001) J Am Chem Soc 123:12364
11. Zimmermann N, Meggers E, Schultz PG (2002) J Am Chem Soc 124:13684
12. Weizman H, Tor Y (2001) Chem Commun p 453
13. Weizman H, Tor Y (2001) J Am Chem Soc 123:3375
14. Tanaka K, Tengeiji A, Kato T, Toyama N, Shionoya M (2003) Science 299:1212
15. Ranganathan D, Haridas V, Karle I (1998) J Am Chem Soc 120:2695
16. Weiß T, Leipert D, Kasper M, Jung G, Göpel W (1999) Adv Mater 11:331
17. Yang D, Qu J, Li W, Zhang YH, Ren Y, Wang DP, Wu YD (2002) J Am Chem Soc 124:12410
18. Strauss J, Daub J (2002) Org Lett 4:683
19. Ghadiri MR, Granja JR, Buehler LK (1994) Nature 306:301
20. Ishida H, Qi Z, Sokabe M, Donowaki K, Inoue Y (2001) J Org Chem 66:2978
21. Kartha G, Varughese KJ, Aimoto S (1982) Proc Natl Acad Sci USA 79:4519
22. Xie P, Diem M (1995) J Am Chem Soc 117:429
23. Tanaka K, Shigemori K, Shionoya M (1999) Chem Commun p 2475
24. Bong DT, Clark TD, Granja JR, Ghadiri MR (2001) Angew Chem Int Ed 40:988
25. Lopez SF, Kim HS, Choi EC, Delgado M, Granja JR, Khasanov A, Kreahenbuehl K, Long G, Weinberger DA, Wilcoxen KM, Ghadiri MR (2001) Nature 412:452
26. Nakanishi T, Okamoto H, Nagai Y, Takeda K, Obataya I, Mihara H, Azehara H, Suzuki Y, Mizutani W, Furukawa K, Torimitsu K (2002) Phys Rev B 66:165417
27. Okada T, Tanaka K, Shiro M, Shionoya M (2005) Chem Commun p 1484

Metal-Incorporated Hosts for Cooperative and Responsive Recognition to External Stimulus

Tatsuya Nabeshima · Shigehisa Akine

Department of Chemistry, University of Tsukuba, Tsukuba, Ibaraki 305-8571, Japan

Summary Cooperative and responsive recognition of molecules and ions play very important roles to regulate molecular functions in biological as well as artificial systems. Incorporation of metal ions into artificial hosts is a very useful way to construct multi-recognition molecules with multi-functionality including characteristic properties due to their redox activity, etc. In this review we discuss mainly pseudomacrocycles and oligo-salamo (salamo: 1,2-bis(salicylideneaminooxy)ethane) compounds for heteronuclear multi-metallo-hosts as a cooperative functional system. Recognition behaviors of the pseudomacrocycles and oligo-salamo systems and regulation of the binding affinity responding to external stimuli such as metal ions and electrons are demonstrated.

Abbreviations

BAPTA Bis(o-aminophenoxy)ethane-N,N,N',N'-tetraacetic acid
K_a Association constant
salamo 1,2-Bis(salicylideneaminooxy)ethane
salbn N,N'-disalicylidene-1,4-butanediamine
salen N,N'-disalicylideneethylenediamine
saloph N,N'-disalicylidene-o-phenylenediamine
saltn N,N'-disalicylidene-1,3-propanediamine

10.1
Introduction

Cooperative and responsive recognition of molecules and ions plays very important roles in the regulation of enzymatic activity and material balances in biological systems [1]. Allostery is one of the most well-known regulation phenomena in living cells. Thus, much attention has been focused on artificial allosteric systems [2–6]. The first event of allostery is recognition of an effector (molecules and ions) to cause a conformational change in the host. This structural change eventually induces conformational change in a binding site for substrates. Consequently, artificial allosteric hosts are applicable to intelligent molecules such as logic gates, sophisticated molecular catalysts, molecular devices, etc. Incorporation of metal ions into artificial hosts should be a very useful way to construct multifunctional and multirecognition molecules. In addition, the metal center provides various coordination structures that depend on the valency and properties of ligating moieties. One of the most important characteristics of the metal centers is redox activity because redox reactions usually regulate not only the structure of metallohosts but also their functions. This is a significant advantage of metallohosts over metal-free hosts, because both chemical reactions and electrochemical methods are available to regulate functions. Since electronic and magnetic fields may well modulate properties of metallohosts, metallosystems can be used in responsive functional metalloassemblies such as metalloliquid crystals. In this review we discuss mainly pseudomacrocycles and oligosalamo (salamo: 1,2-bis(salicylideneaminooxy)ethane) compounds for heteronuclear multimetallohosts as cooperative functional systems.

10.2
Pseudomacrocycles for Cooperative Molecular Functional Systems

Pseudomacrocycles are obtained from linear precursors and metal ions [7–9]. The linear precursors contain ligating moieties at the termini of the chains. Coordinate bonding plays an important role in maintaining the cyclic structure of the pseudomacrocyclic compounds. These matallohosts have been used as allosteric hosts because the dynamic structural change from a linear form to a cyclic one is very effective for changing the recognition ability of the hosts due to a macrocyclic effect. Heavy metal ions were often chosen as effectors because various structures of heavy metal complexes are available.

Recently, we synthesized a pseudocryptand that exhibited positive and negative allosteric effects on the recognition of alkali metal ions [10]. Tripodand 1 having a 2,2′-bipyridine moiety at the three termini reacted with a Fe(II) ion to give the corresponding pseudocryptand 1·Fe(II) (Scheme 10.1). X-ray crystallographic analysis revealed that the three polyether chains were assembled in a helical fashion (Fig. 10.1). Binding constants for alkali metal ions were determined by ^{1}H NMR spectroscopy. The tripodand 1 captures Na$^+$ most

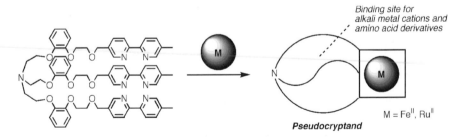

Scheme 10.1. Pseudocryptands as metalloreceptors

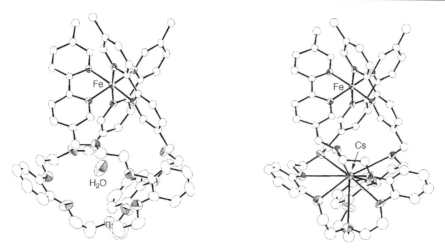

Fig. 10.1. Crystal structures of pseudocryptand $1 \cdot$Fe and its Cs^+ complex

strongly and Cs^+ most weakly among alkali metal ions. In contrast, $1 \cdot$Fe(II) showed the highest and lowest affinities to Cs^+ and Na^+, respectively. Upon complexation with Fe(II), the K_a value for Cs^+ increased by a factor of 39 (positive allostery) and for the affinity to Na^+ decreased considerably (negative allostery). A similar change in selectivity was also observed in ion transport through a liquid membrane. The molecular structure of $1 \cdot$Fe(II)Cs^+ determined by X-ray crystallography indicated that a CH-π interaction contributed to the stabilization of the Cs^+ complex (Fig. 10.1). Consequently, this interaction favors the formation of the ternary complex $1 \cdot$Fe(II)Cs^+. Furthermore, the CH-π contact disfavors shrinking of $1 \cdot$Fe(II) to prevent complexation with smaller sized guests such as Na^+ and K^+ because conformational change necessary for the effective coordination of $1 \cdot$Fe(II) to the cations is prohibited. Interestingly, $1 \cdot$Fe(II) is elongated and twisted upon complexation with Cs^+. Thus, $1 \cdot$Fe(II) is considered a molecular spring responsive to a Cs^+ ion as an external stimulus. The redox response of ion recognition of $1 \cdot$Fe(II) should be interesting. However, the addition of Cs^+ did not considerably change the redox potentials of the Fe(II) moiety. The pseudocryptand $1 \cdot$Fe(II) also works as an amino-acid binder (T. Nabeshima et al. unpubl. data). The K_a value of ammonium salts of amino-acid methyl esters decreased upon the complexation of 1 with a Fe(II) ion. Although 1 binds to phenethyl amine hydrochloride, $1 \cdot$Fe(II) shows no affinity to the ammonium guest. $1 \cdot$Fe(II), therefore, has an extremely high amino-acid selectivity over the corresponding ammonium salt. Since methyl phenylpropionate is not bound by $1 \cdot$Fe(II), cooperative interactions of the ammonium and ester moieties with the metallohost were strongly suggested. Ruthenium pseudocryptand was obtained by the reaction of 1 with $RuCl_3 \cdot 3H_2O$ (T. Nabeshima et al. unpubl. data). As seen in $1 \cdot$Fe(II), $1 \cdot$Ru(II) shows Cs^+ selectivity. Photochemical properties of $1 \cdot$Ru(II) may be applied to

fluorescent ion sensors and photoinduced electron transfer systems between
1·Ru(II) and a cationic guest.

The framework of a pseudocryptand is also useful for anion receptors
(T. Nabeshima et al. unpubl. data). Each chain of tripodand 2 bears a urea
moiety and a 2,2′-bipyridine unit at the end. Quantitative complexation of 2
with Fe(II) gives the corresponding pseudocryptand 2·Fe(II) as in the case of 1.
2·Fe(II) captures Cl⁻ ions very selectively (Fig. 10.2). The binding strength
is well regulated electrochemically. Changes in the redox potentials in the
presence of Cl⁻ clearly indicate that 2·Fe(II) has a much higher affinity to Cl⁻
than the host 2·Fe(I).

Multistep regulation of anion recognition was successfully performed using
cationic guests. We designed calix[4]arene derivative 3 according to the strategy
shown in Scheme 10.2 [11–15]. To the calixarene framework were introduced
two ester groups and two polyether chains containing a urea moiety and a 2,2′-
bipyridine unit at the ends of the chains. The polyether moieties and ester
groups provide one binding site for alkali metal ions. The bipyridine units as
another binding site for cations were expected to bind soft metal ions and to
afford a macrocyclic structure on the calix[4]arene lower rim. The urea units
should interact with anionic guests through hydrogen bonding. We imagined
that the cations bound in the two binding sites enhance the anion affinity due
to electrostatic interactions. In addition, formation of the pseudomacrocyclic
structure may be favorable for anion recognition because the two urea moieties
are assembled. As we expected, 3 recognizes Na^+ and Ag^+ simultaneously and
quantitatively and captures an anionic guest. The increase in the K_a reaches
factors of 1500 and 2000 for NO_3^- and $CF_3SO_3^-$, respectively, in the presence of
both Na^+ and Ag^+, compared to the free 3.

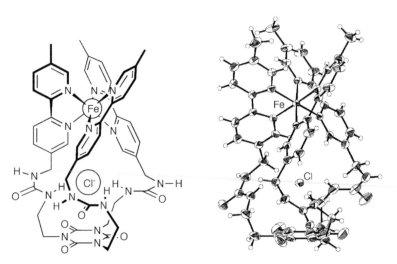

Fig. 10.2. Pseudocrypand 2·Fe(II)-Cl⁻ complex

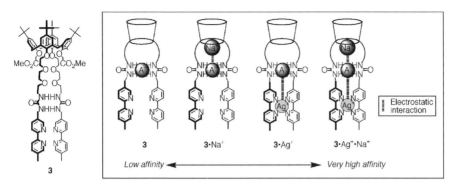

Scheme 10.2. Stepwise regulation of anion binding

Formation of coordinate bonding can be utilized to divide a recognition site into two small sites [16]. This would provide one of the most useful ways of regulating recognition ability because the size-fit principle is a very important factor in achieving large host-guest interactions. Macrocyclic hosts **4** bearing two Pd pincer complexes would be good candidates for the metal-assisted partitioning of a recognition site (Scheme 10.3). ^{1}H NMR spectroscopy indicated that **4b** quantitatively binds to a bidentate ligand such as pyrazine and 4,4′-bipyridine to give the corresponding 1:1 complexes. Pyrazine divides the cavity of **4b** into two small binding sites, but 4,4′-bipyridine would close the cavity because the long ligand stretches the cavity upon the complexation in a bridged way. The pincer moieties probably show a redox response to change the structure and functions of the pincer hosts **4**. Furthermore, the pincer moieties are also expected to exhibit catalytic activities for various organic reactions.

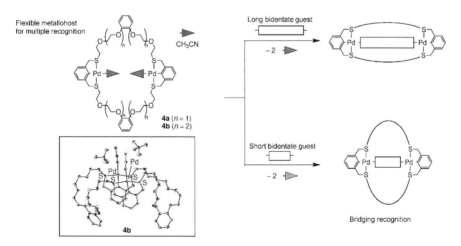

Scheme 10.3. Recognition of bidentate guest

10.3
Oligo(N$_2$O$_2$-Chelate) Macrocycles

10.3.1
Design of Macrocyclic Oligo(N$_2$O$_2$-Chelate) Ligands and Metallohosts

N$_2$O$_2$ tetradentate ligand is a useful building block for metallosupramolecules containing redox-active metal centers. Representative examples for the N$_2$O$_2$ ligands are salen and saloph, etc. These ligands (H$_2$L in Scheme 10.4) form stable chelate complexes [MLX$_n$], where M is a d-block transition metal (Mn, Fe, Co, Ni, Cu, Zn, etc.) and X is a solvent molecule or counter anion. One important property of the complex [MLX$_n$] is their coordination ability with respect to another metal M$'$ including alkali metals [17], alkaline earth [18], rare earth [19, 20], and d-block transition metals [21] due to the phenolate oxygen atoms.

Such dinuclear complexes are also obtained from a cyclic bis(N$_2$O$_2$) ligand known as a "compartmental ligand" [22]. However, larger macrocyclic oligo(N$_2$O$_2$-chelate) ligands are effective in tuning guest recognition via the metalation of N$_2$O$_2$ sites. Macrocycles consisting of salen and

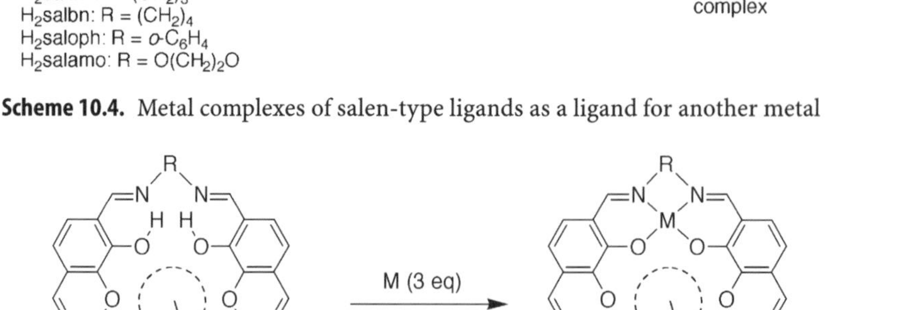

H$_2$salen: R = (CH$_2$)$_2$
H$_2$saltn: R = (CH$_2$)$_3$
H$_2$salbn: R = (CH$_2$)$_4$
H$_2$saloph: R = o-C$_6$H$_4$
H$_2$salamo: R = O(CH$_2$)$_2$O

homo- or heterodinuclear complex

Scheme 10.4. Metal complexes of salen-type ligands as a ligand for another metal

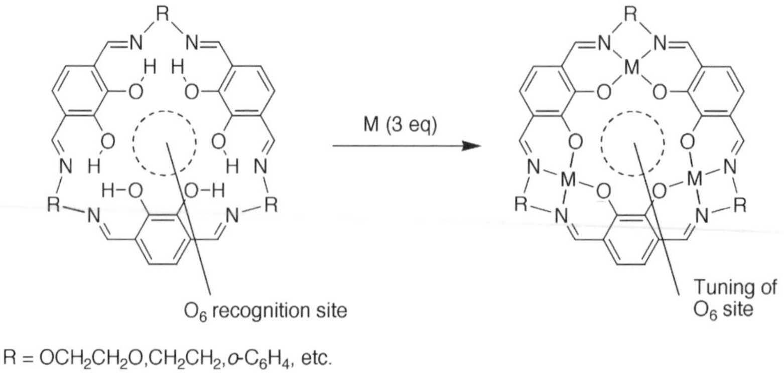

O$_6$ recognition site

Tuning of O$_6$ site

R = OCH$_2$CH$_2$O,CH$_2$CH$_2$,o-C$_6$H$_4$, etc.

Scheme 10.5. Design of tris(N$_2$O$_2$) chelates for tuning guest recognition

oligo(ethyleneoxy) chains have been reported [23, 24]. The compounds were prepared only in the presence of a template ion and isolated as a complex with the template. Thus, we design a rigid macrocyclic oligo(N_2O_2-chelate) that is expected to be synthesized without a template (Scheme 10.5). The ligands have three N_2O_2 chelate moieties as well as an O_6 site similar to that of crown ether. The guest-binding strength in the central O_6 cavity is expected to increase when the N_2O_2 chelate moieties are metalated because metal complexes of a salen-type chelate have a higher coordination ability than their free forms. Furthermore, ion recognition can be regulated stepwise if multistep redox processes modulate the oxidation state of the metals in oligometallic complexes.

10.3.2
Synthesis and Structure of Tris(N_2O_2-Chelate) Macrocycles

A 30-membered macrocyclic tris(H_2saloph) **5a** was synthesized from a dialdehyde and a diamine in 91% yield without using template ions [25]. Hydrogen bonds between the hydroxyl groups and imine nitrogens of salicylaldimine moieties were observed in solution and in the crystalline state. In addition, we obtained its water complex, in which a water molecule was incorporated in the cavity similar to 18-crown-6 via hydrogen bonds (Fig. 10.3). Alkoxy derivatives **5b–d** have higher solubility in organic solvents than **5a**. As we expected, **5** binds alkali, alkaline earth metals, or transition metals. Thus, the cavity of the macrocycles has affinity not only to hydrogen-bond donors but also to metal cations.

A 42-membered macrocycle **6** was also prepared in high yield (77%) in a similar manner [26]. In the crystal structure, six dimethyl sulfoxide molecules were bound to the macrocycle via $S{\rightarrow}O{\cdots}H{-}O$ hydrogen bonds (Fig. 10.4). The results imply that the macrocycle recognized a larger organic guest via multiple hydrogen bonds.

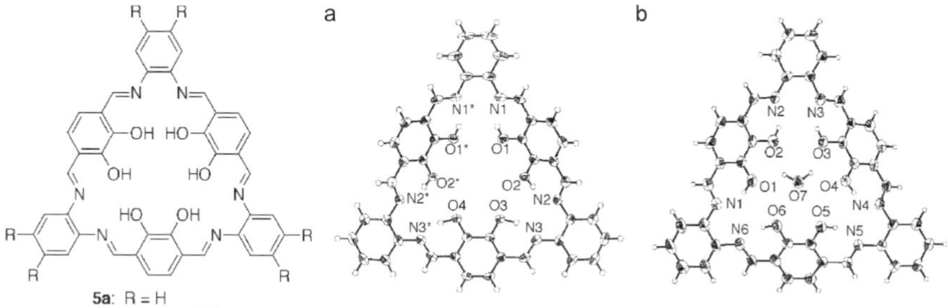

5a: R = H
5b: R = O-n-C_4H_9
5c: R = O-n-C_8H_{17}
5d: R = O-n-$C_{12}H_{25}$

Fig. 10.3. Crystal structure of tris(N_2O_2) macrocycles **5a** (**a**) and its water complex (**b**)

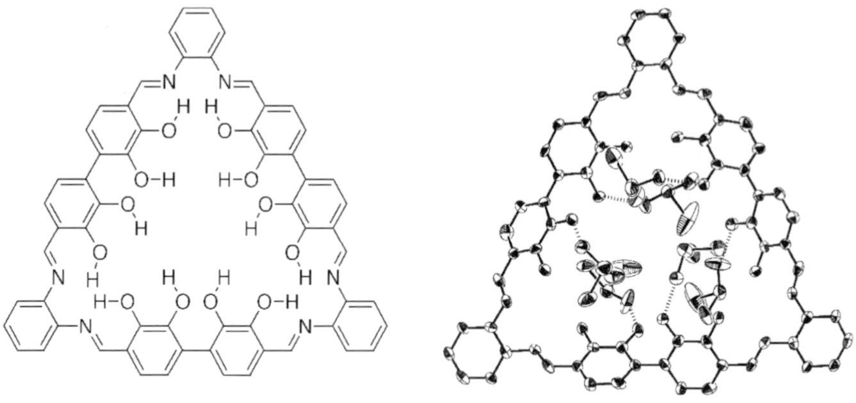

Fig. 10.4. Crystal structure of 42-membered tris(N_2O_2) macrocycle 6. Six dimethyl sulfoxide molecules bound to the macrocycle are also shown

These macrocyclic imines can be converted to metallohosts by the reaction with 3 equiv of M(OAc)$_2$ (M = Cu, Zn, Ni). Guest recognition by the free and metalated macrocyclic imines is in progress.

10.4
Acyclic Oligo(N_2O_2-Chelate) Ligands

10.4.1
Design of Acyclic Oligo(N_2O_2-Chelate) Ligands

We have already shown that conformational change from an acyclic structure to a cyclic one (pseudomacrocycle) is effective in regulating ion recognition. A rigid C-shaped structure, in which oxygen atoms are arranged in a cyclic fashion, is produced from flexible bis(N_2O_2)-chelate ligands. We expected that this C-shaped host would also provide an alternative strategy to control ion-recognition ability (Scheme 10.6). Besides the fixation effect, metalation of the chelate moiety enhances the guest-recognition ability of the central O_6 site as discussed in the previous section.

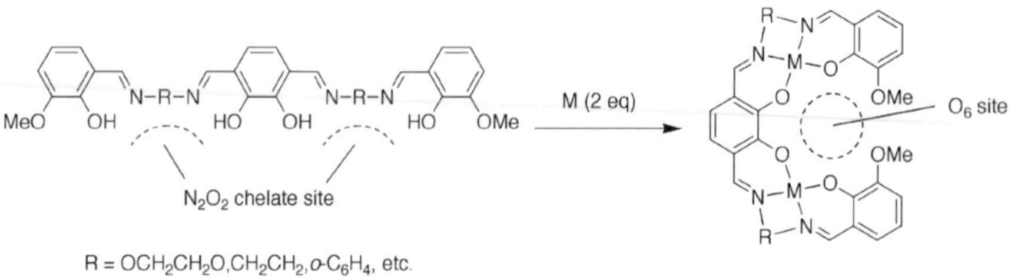

R = OCH$_2$CH$_2$O,CH$_2$CH$_2$,o-C$_6$H$_4$, etc.

Scheme 10.6. Regulation of ion recognition by metalation of bis(N_2O_2) chelate

As N_2O_2 moieties of the bis(N_2O_2)-chelate molecule, we use salamo chelate ($R = OCH_2CH_2O$), an oxime analog of salen. The salamo ligand is sufficiently more stable than imine analogs (salen, saltn, or saloph) to resist the recombination of C=N bonds [27]. Although the metal complexes of various kinds of salen-type imine ligands have been extensively investigated, there was no report on the corresponding oxime ligand, salamo, until our report on its synthesis in 2001 [28]. We also reported the first example of the complexation of the parent salamo ligand with metal ions.

10.4.2
Complexes of a New N_2O_2-Chelate Ligand, Salamo

Complexation of H_2salamo and H_2(3-MeOsalamo) with copper(II) acetate gives tetracoordinate mononuclear complexes [Cu(salamo)] and [Cu(3-MeO salamo)], respectively (Fig. 10.5) [28]. The two complexes show quasireversible redox waves, and the reduction potentials appear in a more pos-

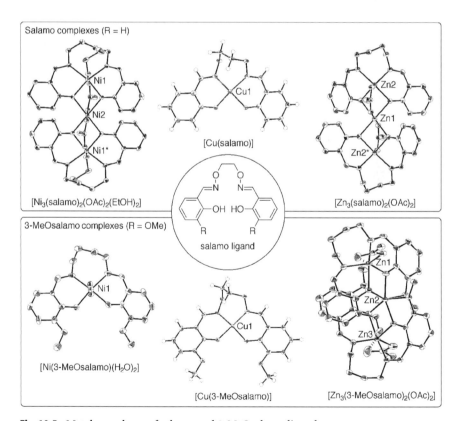

Fig. 10.5. Metal complexes of salamo and 3-MeOsalamo ligands

itive region than the corresponding salen complexes. Complexation between the salamo ligands and zinc(II) acetate affords trinuclear complexes $[Zn_3L_2(OAc)_2]$ in a highly cooperative fashion (Fig. 10.5) [29]. Such high cooperativity was not observed when salen and salbn ligands were used. Complexation of salamo and 3-MeOsalamo with cobalt(II) gave similar trinuclear complexes, whose crystal structures are very similar to those of the corresponding zinc(II) complexes. Upon addition of nickel(II) acetate, salamo ligand also gave a trinuclear nickel(II) complex, whereas a mononuclear complex was obtained from the 3-MeOsalamo ligand (Fig. 10.5) [30]. All the nickel atoms in the trinuclear and mononuclear complexes have an octahedral geometry.

10.4.3
Synthesis, Structure, and Properties of Acyclic Oligo(N_2O_2-Chelate) Ligands

We investigated the metalation of two salamo moieties of the bis(N_2O_2) ligand 7. Unexpectedly, complexation of the bis(salamo) ligand with zinc(II) acetate resulted in the quantitative and cooperative formation of $[LZn_3]^{2+}$ ($H_4L = 7$) (Scheme 10.7) instead of the dinuclear complex [31]. The central O_6 site was also metalated with zinc(II) simultaneously. In the crystal structure, only two oxygen atoms of the O_6 site coordinated to the central zinc(II) ion. Hence, a larger metal ion suitable for the O_6 cavity was expected to replace the central zinc(II) ion.

As expected, the trinuclear complex $[LZn_3]^{2+}$ strongly binds a rare earth(III) ion (Sc^{3+}, Y^{3+}, lanthanide^{3+}). During this process a free zinc(II) ion is selectively liberated and a helical heterotrinuclear complex $[LZn_2M]^{3+}$ (M = rare earth metal) is formed (Scheme 10.8) [31]. Among alkaline earth metals, only Ca^{2+} is strongly bound to the metallohost $[LZn_3]^{2+}$ [32]. The Ca^{2+}/Mg^{2+} selectivity is 10^5, which is comparable to those of excellent Ca^{2+} receptors or sensors such as BAPTA. On the other hand, the metallohost has a very weak affinity to alkali metal ions (Na^+, K^+, Rb^+, Cs^+). These results suggest that the

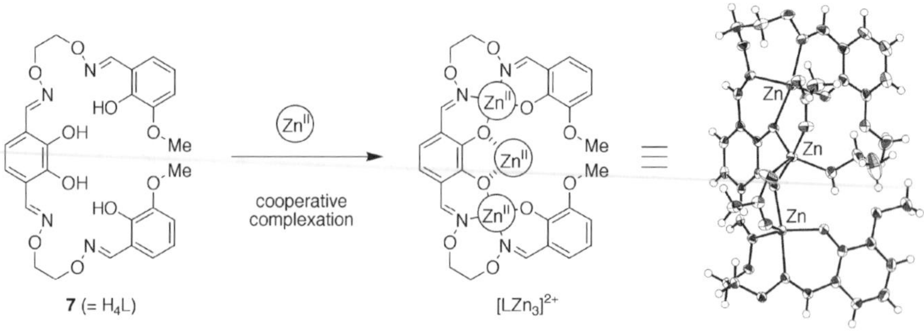

Scheme 10.7. Cooperative formation of trinuclear zinc(II) complexes $[LZn_3]^{2+}$

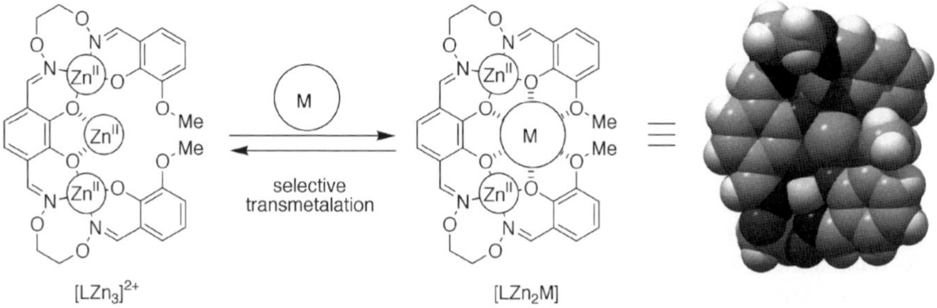

Scheme 10.8. Guest recognition by selective exchange of the central zinc(II) ion

guest charge is a significant factor for the cation-binding ability of $[LZn_3]^{2+}$. Moreover, the size-fit principle is important for this ion recognition.

Site-selective transmetalation is also applicable for a new synthetic strategy for 3d–4f heteromultimetallic complexes. A number of reports deal with the ferromagnetic interaction between Cu–Gd in [LCuGd]-type complexes where L is a salen-type N_2O_2 chelate [33–35]. A Cu_2Gd complex is also obtained from the bis(salamo) ligand 7, and the magnetic interaction between Cu–Gd is ferromagnetic [36]. This strongly suggests that the copper ions in the N_2O_2 sites electronically affect magnetic properties of metal ions in the central O_6 site. This is also supported by the difference in reduction potentials of Cu^{II}/Cu^{I} in the N_2O_2 sites depending on the guest ion bound in the central O_6 cavity.

10.5
Conclusion

Conformational change and transformation of host frameworks by using coordinate bonding work very efficiently to regulate molecular recognition. Since these metal centers accept some oxidation states, a variety of redox activities such as redox response of molecular recognition and catalytic activity are expected. Thus, we believe that these structural transformations by metal coordination are applicable to many important and interesting developments, e.g., modulation of structure and functions of large bioactive molecules such as DNA and proteins because these molecules contain ligating moieties, anionic, and/or cationic groups. The combination of self-assembly and the redox control of functions will also open a new way to the construction of functional nanosystems responding multiply to external factors, i.e., various ions and molecules, photons, strength of electronic and magnetic fields, etc.

10.6
References

1. Perutz MF (1989) Mechanisms of Cooperativity and Allosteric Regulation in Proteins. Cambridge University Press, Cambridge

2. Nabeshima T (1996) Coord Chem Rev 148:151
3. Nabeshima T, Akine S, Saiki T (2000) Rev Heteroat Chem 22:219
4. Rebek Jr J (1984) Acc Chem Res 17:258
5. Tabushi I (1988) Pure Appl Chem 60:581
6. Shinkai S, Ikeda M, Sugasaki A, Takeuchi M (2001) Acc Chem Res 34:494
7. Nabeshima T, Inaba T, Furukawa N (1987) Tetrahedron Lett 28:6211
8. Nabeshima T, Inaba T, Furukawa N, Hosoya T, Yano Y (1993) Inorg Chem 32:1407
9. Nabeshima T, Inaba T, Sagae T, Furukawa N (1990) Tetrahedron Lett 31:3919
10. Nabeshima T, Yoshihira Y, Saiki T, Akine S, Horn E (2003) J Am Chem Soc 125:28
11. Nabeshima T, Saiki T, Iwabuchi J, Akine S (2005) J Am Chem Soc 127:5507
12. Saiki T, Iwabuchi J, Akine S, Nabeshima T (2004) Tetrahedron Lett 45:7007
13. Nabeshima T, Saiki T, Sumitomo K (2002) Org Lett 4:3207
14. Nabeshima T, Saiki T, Sumitomo K, Akine S (2004) Tetrahedron Lett 45:4719
15. Nabeshima T, Saiki T, Sumitomo K, Akine S (2004) Tetrahedron Lett 45:6761
16. Nabeshima T, Nishida D, Akine S, Saiki T (2004) Eur J Inorg Chem, p 3779
17. Cunningham D, McArdle P, Mitchell M, Chonchubhair NN, O'Gara M, Franceschi F, Floriani C (2000) Inorg Chem 39:1639 and references therein
18. Carbonaro L, Isola M, La Pegna P, Senatore L, Marchetti F (1999) Inorg Chem 38:5519 and references therein
19. Condorelli G, Fragalà I, Giuffrida S, Cassol A (1975) Z Anorg Allg Chem 412:251
20. Casellato U, Guerriero P, Tamburini S, Vigato PA, Benelli C (1993) Inorg Chim Acta 207:39
21. Gruber SJ, Harris CM, Sinn E (1968) J Inorg Nucl Chem 30:1805
22. Guerriero P, Tamburini S, Vigato PA (1995) Coord Chem Rev 139:17
23. van Staveren CJ, van Eerden J, van Veggel FCJM, Harkema S, Reinhoudt DN (1988) J Am Chem Soc 110:4994
24. van Doorn AR, Schaafstra R, Bos M, Harkema S, van Eerden J, Verboom W, Reinhoudt DN (1991) J Org Chem 56:6083
25. Akine S, Taniguchi T, Nabeshima T (2001) Tetrahedron Lett 42:8861
26. Akine S, Hashimoto D, Saiki T, Nabeshima T (2004) Tetrahedron Lett 45:4225
27. Akine S, Taniguchi T, Dong W, Masubuchi S, Nabeshima T (2005) J Org Chem 70:1704
28. Akine S, Taniguchi T, Nabeshima T (2001) Chem Lett, p 682
29. Akine S, Taniguchi T, Nabeshima T (2004) Inorg Chem 43:6142
30. Akine S, Nabeshima T (2005) Inorg Chem 44:1205
31. Akine S, Taniguchi T, Nabeshima T (2002) Angew Chem Int Ed 41:4670
32. Akine S, Taniguchi T, Saiki T, Nabeshima T (2005) J Am Chem Soc 127:540
33. Winpenny REP (1998) Chem Soc Rev 27:447
34. Sakamoto M, Manseki K, Okawa H (2001) Coord Chem Rev 219–221:379
35. Benelli C, Gatteschi D (2002) Chem Rev 102:2369
36. Akine S, Matsumoto T, Taniguchi T, Nabeshima T (2005) Inorg Chem 44:3270

Chapter 11

Synthesis of Poly(binaphthol) via Controlled Oxidative Coupling

Shigeki Habaue · Bunpei Hatano

Department of Chemistry and Chemical Engineering, Faculty of Engineering, Yamagata University, Yonezawa 992-8510, Japan

Summary Dinuclear metal complexe is one of the best candidates as a catalyst of the oxidative coupling with high stereo-, regio-, and activity controls, because two redox sites are fixed to simultaneously activate two phenolic substrates and smoothly promote a homolytic coupling. In this chapter, the controlled oxidative coupling of 2-naphthols and polymerizations with various dinuclear-type copper and oxovanadium complexes are reviewed.

Abbreviations

OCP	Oxidative coupling polymerization
DHN	Dihydroxynaphthalene
PPOH	4-Phenoxyphenol
M_n	Number average molecular weight
M_w	Weight average molecular weight
TMEDA	N,N,N',N'-tetramethylethylenediamine
HMBN	3,3'-Dihydroxy-2,2'-dimethoxy-1,1'-binaphthalene
(+)PMP	(+)-1-(2-Pyrrolidinylmethyl)pyrrolidine
(−)Sp	(−)-Sparteine
(S)Phbox	(S)-2,2'-Isopropylidenebis(4-phenyl-2-oxazoline)
BMX	N,N'-bis(2-morpholinoethyl)-p-xylylenediamine
SEC	Size exclusion chromatography

11.1
Introduction

1,1'-Bi-2-naphthol is one of the most important artificial chiral auxiliaries and has been significantly used in asymmetric synthetic reactions and chiral discriminations. Therefore, polymers including the binaphthol moieties are very attractive as a functional chiral material, and numerous reports can be found on their syntheses and applications [1, 2]. In contrast, some attention is now being paid to the poly(binaphthol)s, such as poly(2,6-dihydroxy-1,5-naphthylene) [3, 4] and poly(2,3-dihydroxy-1,4-naphthylene) [5, 6] (Scheme 11.1). These polymers can be prepared by the catalytic oxidative coupling polymerization (OCP) of the commercially available 2,6- or 2,3-dihydroxynaphthalene (2,6- or 2,3-DHN), respectively, and are

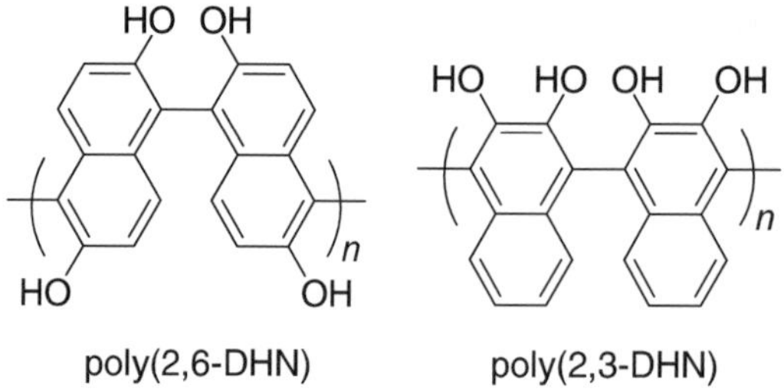

poly(2,6-DHN)   poly(2,3-DHN)

Scheme 11.1.

unique because of their rigid main chain with continuous axial dissymmetry. The structure of the ···*RRRR*···/···*SSSS*··· and ···*RSRS*··· isomers are quite different from each other. The former possesses a stable one-handed helical conformation, whereas in the latter, all the hydroxyl groups exist on one side of a plane through the main chain shown in Scheme 11.2. Accordingly, the synthesis of these poly(binaphthol)s is interesting from the viewpoint of developing novel functional polymer materials.

The catalytic OCP of phenolic compounds is a very important industrial process affording poly(2,6-dimethyl-1,4-phenylene oxide) (PPO), an amorphous engineering plastic [7, 8]. However, it is generally very difficult to control the coupling reaction between the generated phenoxy radical intermediates, except for the 2,6-disubstituted phenols. On the other hand, asymmetric oxidative coupling reactions of 2-naphthol derivatives with metal complexes, such as Cu(I) [9–11], Ru(II) [12], and V(IV) [13–16], affording the 1,1'-bi-2-naphthol skeleton, have been reported. For example, the copper(I) complex with a chiral diamine gave several promising results for oxidative coupling, especially for 3-alkoxycarbonyl-2-naphthols (up to 94% ee) [11].

Further design of the catalyst system will realize a well-controlled oxidative coupling, which is widely applicable to the polymerization as well as the coupling reaction of various 2-naphthol derivatives. Oxidative coupling takes place between two in situ generated metal complexes with 2-naphthol derivatives [9]. Accordingly, a significant acceleration of the reaction rate and enhancement of the stereoselectivity should be expected when two redox sites are fixed to simultaneously activate two substrates and smoothly promote a homolytic

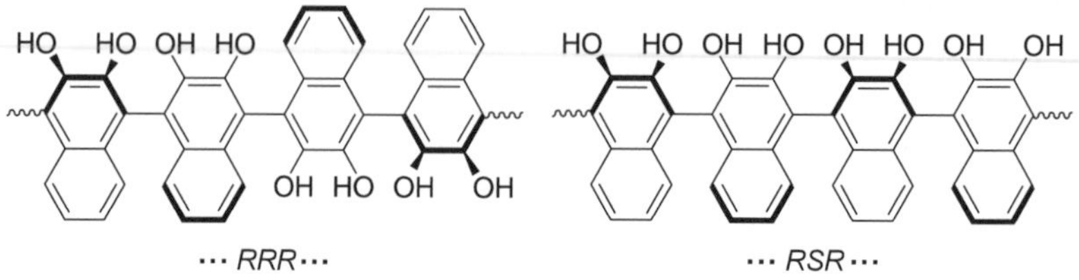

··· *RRR* ···   ··· *RSR* ···

Scheme 11.2.

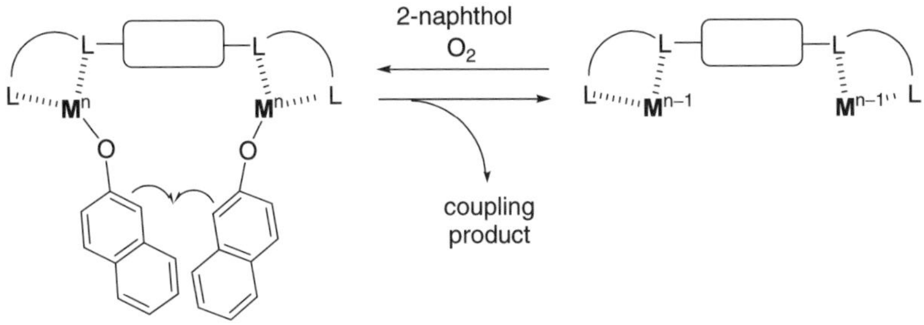

Scheme 11.3.

coupling between these two intermediates, as shown in Scheme 11.3, that is, the dinuclear metal complex is one of the best candidates as a catalyst of oxidative coupling with higher stereo-, regio-, and activity controls.

11.2
Asymmetric Oxidative Coupling with Dinuclear Metal Complexes

Various optically active oxovanadium(IV) complexes, prepared from vanadyl sulfate, aldehydes, and amino acids, for the asymmetric oxidative coupling of 2-naphthols have been reported [13–15]. These catalysts showed moderate to good enantioselectivities while generally showing low activities often taking more than several days to complete a reaction.

Dinuclear vanadium complexes **1** and **2** were developed by Gong et al. (Scheme 11.4), and the oxidative coupling of 2-naphthol in CCl_4 at $0\,^\circ C$ under an O_2 atmosphere was carried out (Scheme 11.5) [17, 18]. Reaction with complex **1** (10 mol%) for 6 d gave a coupling product in 93% yield with an enantioselectivity of 83% ee (R). A higher selectivity (90% ee (R)) was observed for the 7-d reaction with catalyst **2** (5 mol%), a structural analog of **1** having an achiral linkage, while the coupling with C_1-symmetric model complex **3** (10 mol%) resulted in a much lower yield and stereoselectivity (40% yield, 57% ee (R)).

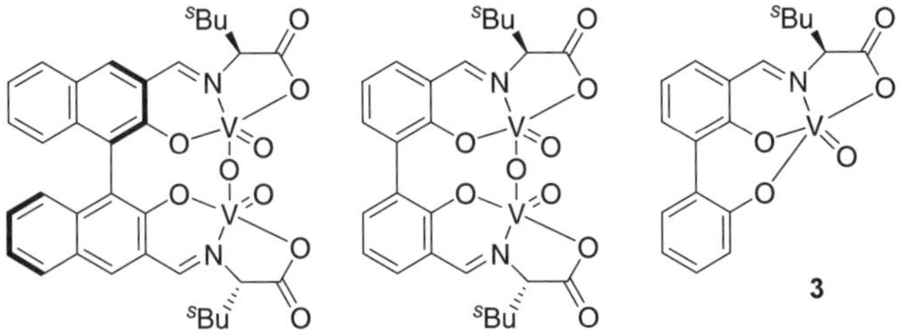

Scheme 11.4.

A structurally very similar oxovanadium catalyst, **4**, derived from (*S*)-*tert*-leucine, was also reported by Sasai and coworkers (Scheme 11.6) [19]. The catalyst **4** (5 mol %) promoted the coupling reaction of 2-naphthol in CH_2Cl_2 at 30 °C for 24 h under an O_2 atmosphere to give binaphthol in 83% yield with 83% ee (*S*), whereas reaction with the mononuclear model catalyst **5**, prepared from an equimolar amount of $VOSO_4$ with the corresponding ligand, under the same conditions afforded (*R*)-isomer with 13% ee in 15% yield. In addition, the rate constants for **4** and model **6** catalyzed coupling reaction (temp. = 20 °C, under air) were evaluated to be $k_4 = 0.1738 \, M^{-1} \, h^{-1}$ and $k_6 = 0.0036 \, M^{-1} \, h^{-1}$, respectively. These results indicate that the dinuclear-type catalyst is quite effective for both stereocontrol and rate enhancement during asymmetric oxidative coupling reactions.

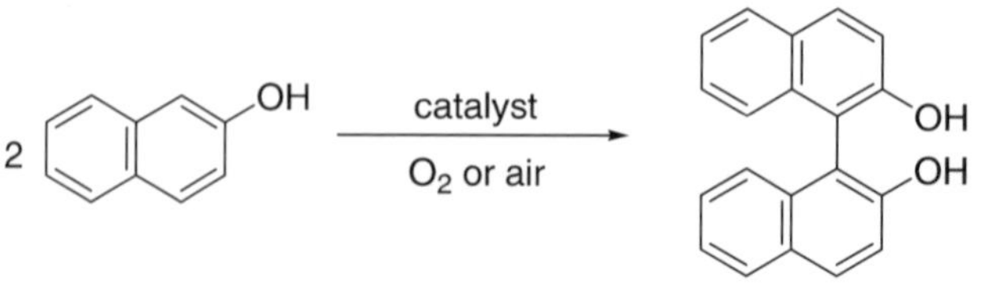

Scheme 11.5.

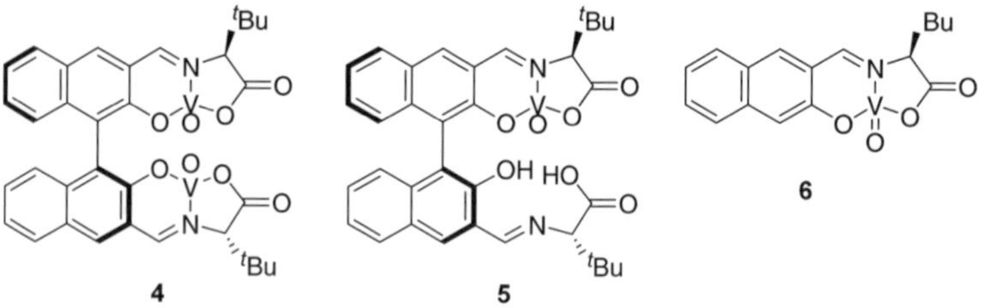

Scheme 11.6.

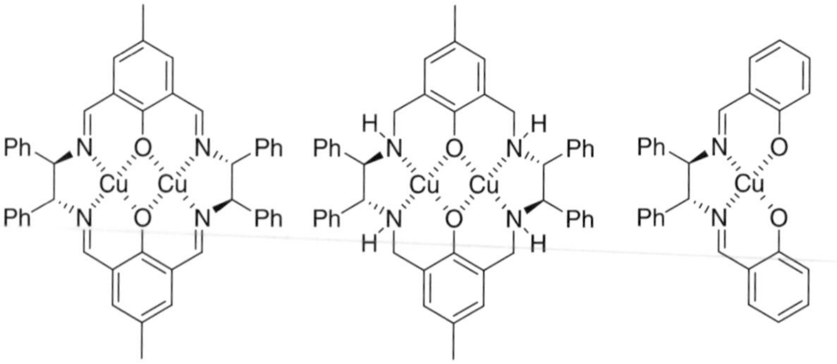

Scheme 11.7.

Dinuclear copper catalysts for asymmetric oxidative coupling were also reported by Gao and coworkers [20]. Optically active Schiff-base and tetramino dicopper complexes with a macrocyclic structure, such as **7** and **8**, were prepared (Scheme 11.7), and the coupling reaction of 2-naphthol was conducted in the presence of a catalyst (10 mol %) in CCl_4 at 0 °C for 7 d with molecular oxygen as the oxidant. Coupling with the Schiff-base catalyst **7** resulted in a yield of 80% and an enantioselectivity of 84% ee (S). A slight increase in the ee value (88%, 85% yield) was observed when complex **8** was used as a catalyst. In contrast, the control coupling reaction in the presence of complex **9** with one metal center afforded a coupling product with a much lower enantioselectivity, 19% ee, suggesting that a rigid dinuclear copper structure is essential and plays an important role in controlling the coupling stereochemistry.

11.3
Oxidative Coupling Polymerization of Phenols

The oxidative polymerization of 2,6-disubstituted phenols, where the o-positions are protected, provides a linear polymer consisting of 1,4-oxyphenylene units. However, the OCP of the 2,6-unsubstituted phenols affording poly(1,4-phenylene oxide)s was unsuccessful because of the difficult regiocontrol during the phenoxy radical coupling reaction. Recently, a highly regioselective OCP of phenol derivatives using tyrosinase model complexes, such as **10** and **11**, was attained by Higashimura et al. (Scheme 11.8) [21–23].

The OCP of 4-phenoxyphenol (PPOH) in the presence of a copper catalyst (5 mol %) at 40 °C under a dioxygen atmosphere was performed. For instance, polymerization with **11** in toluene for 19 h gave a methanol-insoluble polymer in 89% yield [98% conv., number average molecular weight (M_n) = 1.2 × 10^3, weight average molecular weight (M_w = 4.7 × 10^3)]. The regioselectivity was estimated by analysis of the dimers generated during the initial polymerization stage. The dimeric products **12** and **13** were formed based on the C–O coupling, while the C–C coupling led to **14** and **15** (Scheme 11.9). Following 0.2 h of polymerization, the dimers were obtained in 8% yield (9% conv.) with a ratio of **12/13/14/15** = 93/7/0/0. None of the C–C coupling products was detected,

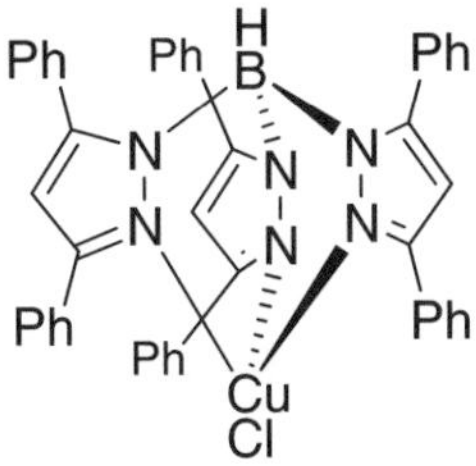

Scheme 11.8.   **10**

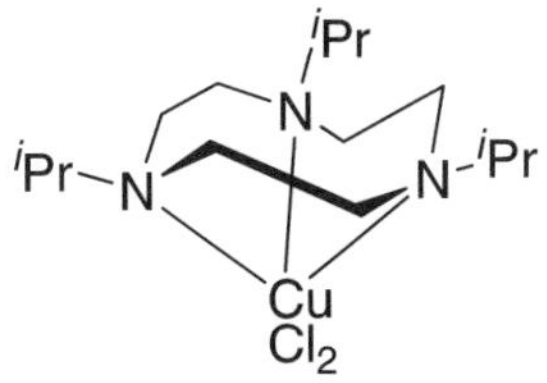

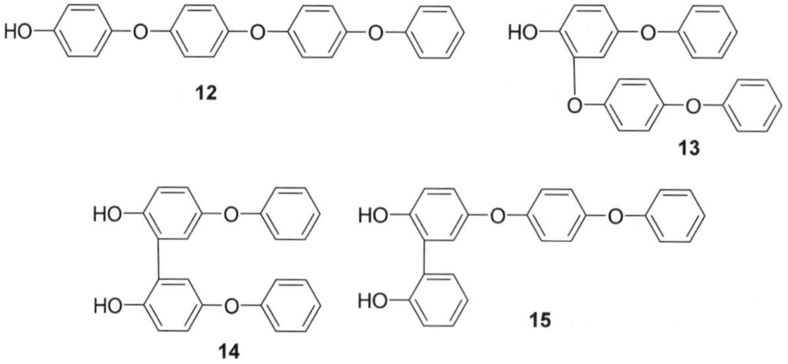

Scheme 11.9.

indicating that the polymerization should proceed in a highly regioselective manner.

The OCP of PPOH with dicopper complex **16** in toluene/acetonitrile (4/1 (v/v)) at 40 °C was also conducted (Scheme 11.10) [24]. The PPOH conversion reached 96% after 98 h of polymerization, and the methanol-insoluble polymer was obtained in 73% yield ($M_n = 2.0 \times 10^3$, $M_w = 3.5 \times 10^3$). The dimer ratio was determined to be 89/10/0/1 after polymerization for 0.2 h (16% conv., 11% yield); therefore, the C–O coupling reaction should be selectively promoted during polymerization. Although no significant difference in the regioselectivity of the dimer formation (12/13/14/15 = 87/11/1/1) was observed for the polymerization with the mononuclear model copper complex **17**, the methanol-insoluble polymer was not obtained in this polymerization system (PPOH conv. = 16%).

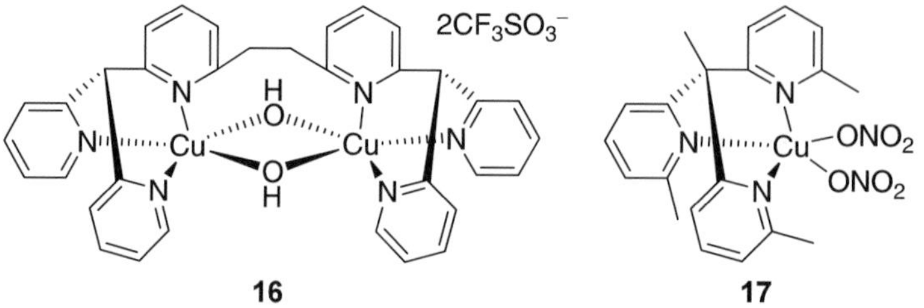

Scheme 11.10.

11.4
Oxidative Coupling Polymerization of 2,3-Dihydroxynaphthalene

Recently, poly(2,6-DHN) was synthesized by the solid-state OCP with an excess amount of $FeCl_3 \cdot 6H_2O$ [3] or by the OCP with CuCl(OH)-N,N,N',N'-

tetramethylethylenediamine (TMEDA) in 2-methoxyethanol using 2,6-DHN as the monomer (Scheme 11.11) [4]. For example, the latter polymerization system at 25 °C for 3 h in air (catalyst: 10 mol %) gave a polymer in 98% yield with an M_n value of 3.1×10^4 (M_w/M_n = 2.3).

On the other hand, the asymmetric OCP of 2,3-DHN with the CuCl-(+)-1-(2-pyrrolidinylmethyl)pyrrolidine [(+)PMP] and CuCl$_2$-(−)-sparteine [(−)Sp] complexes (Scheme 11.12), which are well known as effective reagents for the oxidative coupling reaction that produces the 1,1′-bi-2-naphthol derivatives [9,25], could not produce a polymer [5,26], although the poly(2,3-DHN) derivative was obtainable by the OCP with these conventional reagents using 3,3′-dihydroxy-2,2′-dimethoxy-1,1′-binaphthalene (HMBN) as the monomer (Scheme 11.13) [26–29].

The asymmetric OCP of 2,3-DHN was successfully accomplished using a novel catalyst system for the oxidative coupling, the complex of CuCl with the bisoxazolines, such as (S)-2,2′-isopropylidenebis(4-phenyl-2-oxazoline) [(S)Phbox] [5]. For instance, polymerization in THF at room temperature for 24 h under an O$_2$ atmosphere afforded a methanol-insoluble polymer in

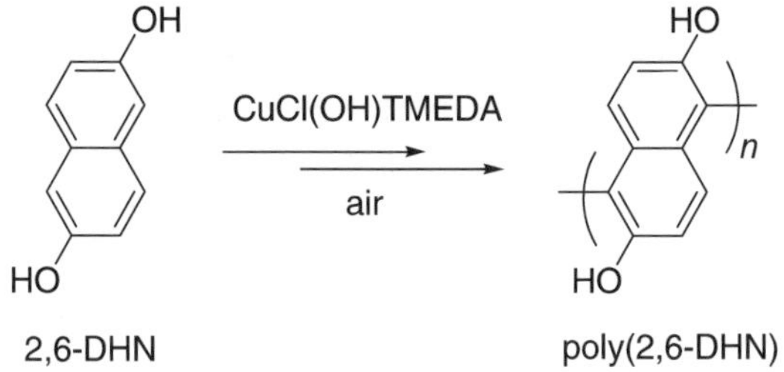

Scheme 11.11.

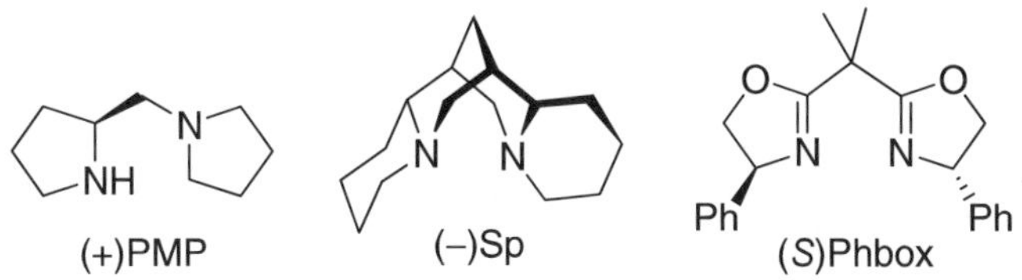

Scheme 11.12.

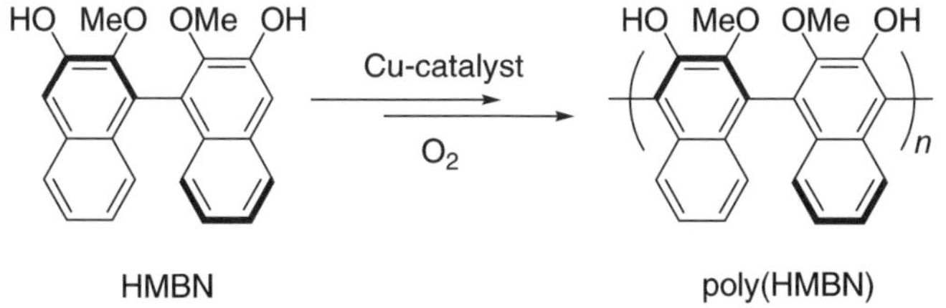

Scheme 11.13.

30% yield ($M_n = 1.1 \times 10^4$, $M_w/M_n = 2.6$) after acetylation of the hydroxyl groups, and the polymer was quantitatively obtained after 48 h of polymerization (Scheme 11.14). The obtained polymer was rich in the S-configuration, and the enantioselectivity was estimated to be 43% ee from the model reaction, the coupling of 3-benzyloxy-2-naphthol, although the actual selectivity is not clear at present.

Although typical diamines were not effective for producing a polymer, a tetraamine ligand, consisting of two ethylenediamine moieties attached to an aromatic ring, N,N'-bis(2-morpholinoethyl)-p-xylylenediamine (BMX), was synthesized (Scheme 11.15) and used as a dinuclear-type catalyst in combination with copper salts for the OCP of 2,3-DHN [6]. An unsuccessful result was obtained for polymerization with the 2CuCl-BMX complex (Table 11.1, entry 1). In marked contrast, a methanol-insoluble polymer was obtained in good yields from the OCP with a 2CuCl$_2$-BMX catalyst, where the mixed solvent CH$_2$Cl$_2$-methanol was used to completely dissolve the complex (entries 2 and 3). The polymer yield and M_n value for the polymerization performed using CuCl$_2$ and BMX (1:1) were significantly reduced (entry 5). These results indicate that the dinuclear-type copper(II) complex is effective for the OCP of 2,3-DHN (Scheme 11.16).

The complex comprised of CuCl$_2$ and m-BMX afforded a polymer with an $M_n = 3.7 \times 10^3$ in 72% yield (entry 6) whose values were almost comparable to those for polymerization with BMX. The tetraamine ligand bearing the bisphenol A linkage 18 [30], however, showed a lower polymer productivity (entry 7). In addition, the OCP with a typical diamine ligand, such as (−)Sp

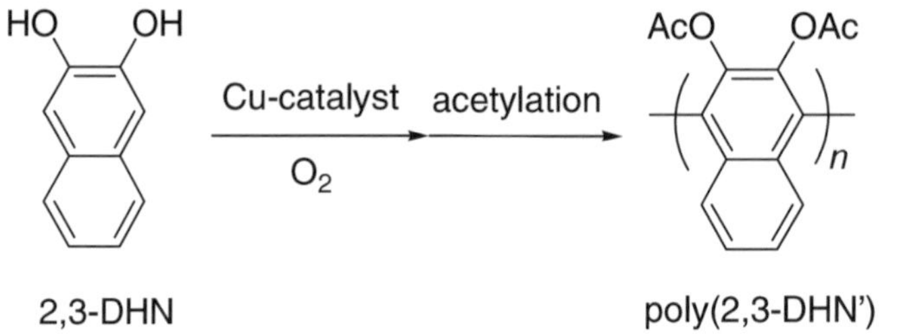

2,3-DHN   poly(2,3-DHN')

Scheme 11.14.

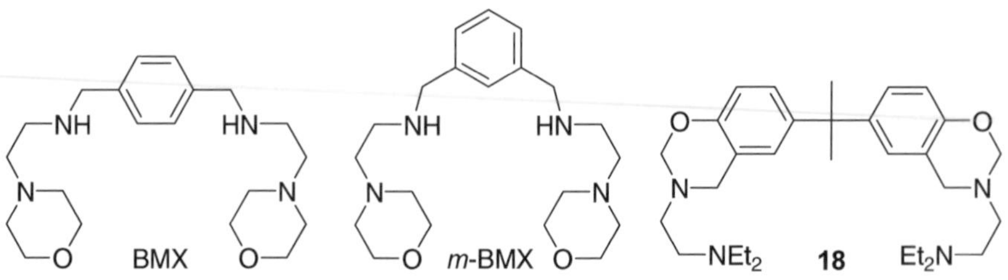

BMX m-BMX   18

Scheme 11.15.

Table 11.1. OCP of 2,3-DHN with BMX based copper catalyst[a]

Entry	Catalyst	Solvent	Yield (%)[b]	$M_n \times 10^{-3}$ (M_w/M_n)[c]
1	2CuCl-BMX	CH_2Cl_2	6	1.7 (–)
2	2CuCl$_2$-BMX	CH_2Cl_2-MeOH	74	3.8 (2.4)
3	2CuCl$_2$-BMX	CH_2Cl_2-MeOH[d]	63	4.4 (3.0)
4	2CuCl$_2$-BMX	MeOH	23	2.1 (–)
5	CuCl$_2$-BMX[e]	CH_2Cl_2-MeOH	29	2.6 (1.3)
6	2CuCl$_2$-m-BMX	CH_2Cl_2-MeOH	72	3.7 (2.3)
7	2CuCl$_2$-**18**	CH_2Cl_2-MeOH	52	3.0 (1.9)
8	CuCl$_2$-(–)Sp	CH_2Cl_2-MeOH	28	2.7 (1.5)
9	CuCl$_2$-(+)PMP	CH_2Cl_2-MeOH	0	–

[a]Conditions: [Cu]/[2,3-DHN] = 0.1, solvent: CH_2Cl_2/MeOH = 7/1 (v/v), time: 48 h, temp.: room temperature, O_2 atmosphere
[b]MeOH-insoluble part of poly(2,3-DHN′)
[c]Determined by SEC (polystyrene standard)
[d]CH_2Cl_2/MeOH = 1/1 (v/v), [Cu]/[2,3-DHN] = 0.2
[e][CuCl$_2$]/[BMX]/[2,3-DHN] = 0.1/0.1/1

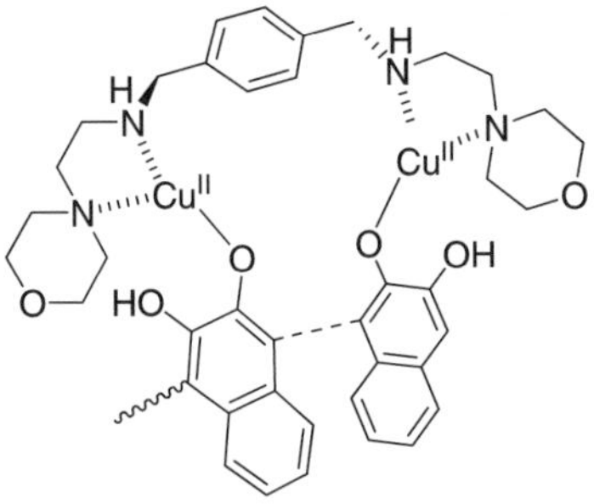

Scheme 11.16.

or (+)PMP, resulted in a much lower or no yield (entries 8 and 9). The ligand structure significantly affected the catalyst activity.

The OCP of *rac*-HMBN with a dinuclear-type copper catalyst in CH_2Cl_2-methanol (7/1 (v/v)) at room temperature was also examined (Scheme 11.13). The 2CuCl-BMX complex produced a polymer with an M_n of 3.3×10^3 $(M_w/M_n = 5.1)$ in 65% yield after acetylation of the hydroxyl groups, similar to the results reported for the conventional Cu(I)-diamine reagents [26–29]. Polymerization with a copper(II) catalyst (2CuCl$_2$-BMX), however, resulted in a much lower yield (28%, $M_n = 2.3 \times 10^3$, $M_w/M_n = 1.4$). Accordingly, the monomer structure significantly affected the catalyst activity for polymerization using a dinuclear-type Cu(II) catalyst system, and this may be due to the stable 1:1 complex formation between 2CuCl$_2$-BMX and HMBN during polymerization.

11.5
Conclusion

Dinuclear copper and oxovanadium complexes, in which the two redox sites are precisely arranged, simultaneously caused the one-electron oxidation of the redox active ligands, such as phenoxides and naphthoxides, and effectively promoted a coupling reaction along with elimination of the reduced metals. Further design of the dinuclear-type complex should provide a catalyst system able to significantly control the stereochemistry during the oxidative coupling reactions and produce a polymer with novel functions.

11.6
References

1. Pu L (1998) Chem Rev 98:2405
2. Pu L (2000) Macromol Rapid Commun 21:795
3. Suzuki M, Yatsugi Y (2002) Chem Commun, p 162
4. Sasada Y, Shibasaki Y, Suzuki M, Ueda M (2003) Polymer 44:355
5. Habaue S, Seko T, Okamoto Y (2003) Macromolecules 36:2604
6. Habaue S, Muraoka R, Aikawa A, Murakami S, Higashimura H (2005) J Polym Sci Part A Polym Chem 43:1635
7. Hay AS (1998) J Polym Sci Part A Polym Chem 36:505
8. Kobayashi S, Higashimura H (2003) Prog Polym Sci 28:1015
9. Nakajima M, Miyoshi I, Kanayama K, Hashimoto S, Noji M, Koga K (1999) J Org Chem 64:2264
10. Li X, Yang J, Kozlowski MC (2001) Org Lett 3:1137
11. Kim KH, Lee D-W, Lee Y-S, Ko D-H, Ha D-C (2004) Tetrahedron 60:9037
12. Irie R, Masutani K, Katsuki T (2000) Synlett, p 1433
13. Hon S-W, Li C-H, Kuo J-H, Barhate NB, Liu Y-H, Wang Y, Chen C-T (2001) Org Lett 3:869
14. Barhate NB, Chen C-T (2002) Org Lett 4:2529
15. Chu C-Y, Hwang D-R, Wang S-K, Uang B-J (2001) Chem Commun, p 980
16. Hirao T (1997) Chem Rev 97:2707
17. Luo Z, Liu Q, Gong L, Cui X, Mi A, Jiang Y (2002) Chem Commun, p 914
18. Luo Z, Liu Q, Gong L, Cui X, Mi A, Jiang Y (2002) Angew Chem Int Ed 41:4532
19. Somei H, Asano Y, Yoshida T, Takizawa S, Yamataka H, Sasai H (2004) Tetrahedron Lett 45:1841
20. Gao J, Reibenspies JH, Martell AE (2003) Angew Chem Int Ed 42:6008
21. Higashimura H, Fujisawa K, Moro-oka Y, Kubota M, Shiga A, Terahara A, Uyama H, Kobayashi S (1998) J Am Chem Soc 120:8529
22. Higashimura H, Kubota M, Shiga A, Fujisawa K, Moro-oka Y, Uyama H, Kobayashi S (2000) Macromolecules 33:1986
23. Higashimura H, Fujisawa K, Namekawa S, Kubota M, Shiga A, Moro-oka Y, Uyama H, Kobayashi S (2000) J Polym Sci Part A Polym Chem 38:4792
24. Higashimura H, Kubota M, Shiga A, Kodera M, Uyama H, Kobayashi S (2000) J Mol Cat A: Chem 161:233
25. Smrcina M, Polakova J, Vyskocil S, Kocovsky P (1993) J Org Chem 58:4534

26. Habaue S, Seko T, Okamoto Y (2002) Macromolecules 35:2437
27. Habaue S, Seko T, Isonaga M, Ajiro H, Okamoto Y (2003) Polym J 35:592
28. Habaue S, Seko T, Okamoto Y (2003) Polymer 44:7377
29. Habaue S, Ajiro H, Yoshii Y, Hirasa T (2004) J Polym Sci Part A Polym Chem 42:4528
30. Agag T, Takeichi T (2003) Macromolecules 36:6010

Part III

Redox Systems via Molecular Chain Control

Chapter 12

Nano Meccano

Yi Liu · Amar H. Flood · J. Fraser Stoddart

California NanoSystems Institute and Department of Chemistry and Biochemistry,
University of California, Los Angeles, 405 Hilgard Avenue, Los Angeles, CA 90095, USA

Summary Switchable catenanes and rotaxanes, as well as a class of self-complexing donor–acceptor cyclophanes, can generate molecular motions when appropriate redox conditions are applied, thus turning them into functional nanoscale machines for a range of different applications. The key to switching lies in the fact that the redox states of the molecular components determine the vast preference of one (co-)conformation over all others. The switching of these molecular machines has been investigated in considerable detail in solution using ^{1}H NMR and UV-visible spectroscopic methods, as well as by cyclic voltammetry and spectroelectrochemistry. Such machines have been demonstrated to switch in electronic devices, in the context of which molecular switch tunnel junctions and an 8×8 molecular memory have been constructed. The machines' abilities to generate induced molecular motions in closely packed Langmuir–Blodgett films have been established under redox conditions. Mechanical devices have been constructed from these molecular machines by attachment to the surfaces of different substrates. In one example, switchable [2]rotaxanes have been shown to function as molecular valves for regulating the release of guest molecules embedded in porous silica materials. In another example, specially designed molecular machines function as biomimetic molecular muscles that are capable of bending an array of microscopic cantilever beams up and down under redox control. Such examples suggest that their cooperative ability to generate nanoscale mechanical motions when married with different substrates means that molecular machines can be utilized to move objects around across length scales that reach up to, and possibly into, the world of their macroscopic counterparts.

Abbreviations and Symbols

CBPQT^{4+}	Cyclobis(paraquat-p-phenylene)
CV	Cyclic voltammetry
DNP	1,5-Dioxynaphthalene
GSCC	Ground state co-conformation
HQ	Hydroquinone
LB	Langmuir–Blodgett
MSCC	Metastable state co-conformation
NEMS	Nanoelectromechanical systems
NP	Naphthalene
RAM	Random access memory
SAM	Self-assembled monolayer
SC	Self-complexing

SCE   Saturated calomel electrode
TTF   Tetrathiafulvalene
UC   Uncomplexed
XPS   X-Ray photoelectron spectroscopy

12.1
Introduction

Molecular machine refers to [1, 2] a molecule that contains multiple movable components whose locations and motions can be controlled by external stimuli. The nanoscale size of such molecular machines renders them small and efficient building blocks in the search for innovative applications in biological systems and materials science. An efficient molecular machine requires reversible actions with speeds that are tunable in response to a given stimulus and are associated with the subsequent generation of distinctive signals. Recent examples of such systems (Fig. 12.1) include switchable bistable [2]catenanes [3–14], bistable [2]rotaxanes [15–31], and self-complexing bistable compounds [32–36]. The first two examples are molecular systems with two ring components or one ring and one dumbbell component, respectively, that are interlocked with each other by mechanical bonds. The relative movements of the two rings in [2]catenanes give rise to circumrotary motions, while, in the case of [2]rotaxanes, linear sliding motions can be generated from the relative movements between the ring and dumbbell components. In non-interlocked self-complexes, a moveable arm is covalently attached to a ring component, from which motions can be generated by controlling the arm to be in or out of the cavity of the ring. In all cases, the relative locations (ring-to-ring, ring-

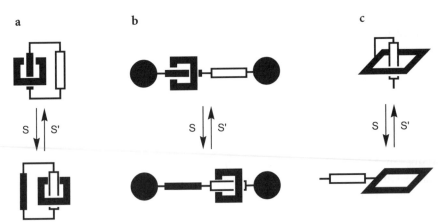

Fig. 12.1. Switchable molecular machines. **a** Bistable [2]catenane. **b** Bistable [2]rotaxane. **c** Covalently linked donor–acceptor macrocyle. S and S′ indicate external stimuli

to-dumbbell, or ring-to-arm) are dictated by molecular recognition based on a variety of noncovalent bonding interactions controlled by redox-active π systems.

It is obvious from inspection of the graphical representations (Fig. 12.1) that, in order to direct the ring movement in a controllable and reversible fashion, one has to modulate the ring's binding to the two recognition sites by swapping the relative strengths of the sites. Among the different reported approaches to controlling the molecular recognition of the two sites, redox processes have been the most extensively used [37–42]. Such processes take account of not only the input stimuli, but also the readout signals (e.g., relative mechanical motions) that are necessary for the efficient operation of molecular machines. By changing the redox state of the components of the molecular machine using chemical [43–50], electrochemical [51–60] or photochemical [61–73] means, the relative locations of the components can be changed accordingly and so generate machinelike molecular motions.

During the past two decades, our group has developed [74] convenient template-directed syntheses of redox-switchable [2]catenanes and [2]rotaxanes. In such interlocking systems, the key component is usually a π-accepting tetracationic cyclophane, cyclobis(paraquat-p-phenylene) (CBPQT^{4+}). A collection of π-donors, such as 1,5-dioxynaphthalene (DNP), 1,4-dihydroquinone (HQ), and tetrathiafulvalene (TTF), has been incorporated into the other component to serve as the recognition units to hold the CBPQT^{4+} ring in place. Building on early successes, truly switchable and therefore bistable [2]catenanes and [2]rotaxanes with two different π-donors in their molecular structures have been prepared. Some of them display an "all-or-nothing" preference for the CBPQT^{4+} ring, it being located on only one (e.g., TTF) of the π-donors for all intents and purposes. The vastly different recognition affinities of two different π-donors toward the CBPQT^{4+} ring are the basis for such a preference. The excellent redox properties of the TTF unit provide a means of switching off its binding when it is oxidized to its mono- or dicationic forms, thus enabling the movement of the CBPQT^{4+} ring to an alternative recognition unit (e.g., DNP or HQ). On the other hand, molecular recognition can be universally shut down by reduction of the two π-electron-deficient bipyridinium units in the CBPQT^{4+} rings. The combination of the redox control – oxidation of the TTF unit and reduction of the CBPQT^{4+} ring – gives rise to the efficient operation of nanoscale molecular machines.

In this chapter, we begin with examples of redox-switchable molecular machines in solution. Ultimately, it will be shown that they form the basis for the successful application of these switchable molecular machines in electronic devices. Appropriately designed systems have been demonstrated to do real mechanical work by harnessing of the redox-activated movements in such nanosized molecules. In all of these contexts the redox sites within the molecular machines lead to controllable relative movements of components on the nanoscale, and for electronic devices they also provide orbital energy levels that tune the conductance between high and low states.

12.2
Redox-Controllable Molecular Switches in Solution

12.2.1
Bistable [2]Catenanes

The [2]catenane $1 \cdot 4PF_6$ (Scheme 12.1), which has been prepared [7, 8] by template-directed synthesis, is comprised of a $CBPQT^{4+}$ ring and a macro-cyclic polyether with two different π-electron-rich recognition sites, namely, a DNP ring system and a TTF unit. The vastly different recognition interactions between the $CBPQT^{4+}$ ring and the two π-donors gives rise to a nearly "all-or-nothing distribution" of two possible translational isomers

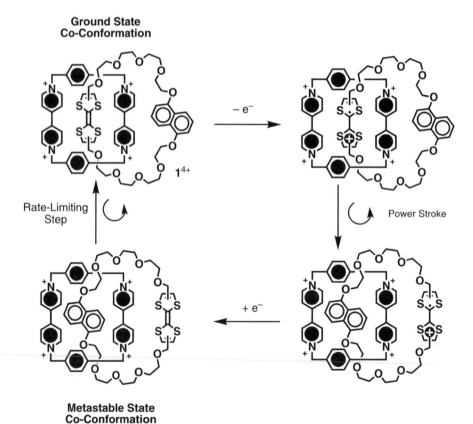

Scheme 12.1. A [2]catenane 1^{4+} that can be switched by both chemical and electrochemical means

with the CBPQT^{4+} ring selectively encircling the TTF unit. The X-ray crystal structure of $1 \cdot 4PF_6$ reveals that the TTF unit resides inside the cavity of the tetracationic cyclophane, an observation that is consistent with its solution-state behavior as characterized by ^{1}H NMR and UV-visible spectroscopy. The desirable all-on/off switch of the [2]catenane $1 \cdot 4PF_6$ in solution has been demonstrated by both chemical and electrochemical means. When a positive potential is exerted on a MeCN solution of $1 \cdot 4PF_6$, the TTF unit can be oxidized to the TTF$^+$ radical cation (at $E_{1/2}$ 650 mV vs. saturated calomel electrode, SCE) or indeed to the TTF^{2+} dication (at $E_{1/2}$ 720 mV vs. SCE), while the DNP unit remains strictly neutral. A charge–charge repulsive force between the generated positive charge and the four positive charges on the CBPQT^{4+} ring causes (Scheme 12.1) the macrocyclic polyether to circumrotate and present its alternative recognition unit, viz., the DNP ring system, to the inner reaches of the cavity of the CBPQT^{4+} ring. Upon reduction of the oxidized tetrathiafulvalene back to its neutral state, the macrocyclic polyether circumrotates through the cavity of the tetracationic cyclophane to position itself once again around the neutral TTF unit, thus returning the [2]catenane to its original state. The ability to generate reversible, controllable redox-driven circumrotation motions in [2]catenanes renders them the smallest nanomachines to date. They have many potential applications.

12.2.2
Bistable [2]Rotaxanes

The preference of the CBPQT^{4+} ring to encircle the TTF unit rather than the DNP ring system in the bistable [2]catenane led to the successful construction of a collection of bistable [2]rotaxanes. A [2]rotaxane $2 \cdot 4PF_6$ has been synthesized [30, 48] with a CBPQT^{4+} ring encircling its dumbbell component incorporating TTF and DNP recognition sites (Scheme 12.2). Similar to the bistable [2]catenane, an "all-or-nothing" situation is observed such that the ring is located exclusively around the TTF unit in the ground state. The movement of the CBPQT^{4+} ring from the TTF unit to the DNP ring system can be controlled by changing the redox states of the TTF unit, as demonstrated by both UV-visible and ^{1}H NMR spectroscopies. When the TTF unit is oxidized chemically to the TTF^{2+} form by Fe(ClO$_4$)$_3$, the CBPQT^{4+} ring moves a remarkable 1.9 nm along the rigid p-terphenyl spacer from the TTF to the DNP recognition site. The reduction of the TTF^{2+} dication back to its neutral state (TTF) switches the unit back to being a strong π-donor-based recognition site, thus allowing the CBPQT^{4+} ring to move back to its starting position – the ground state of the [2]rotaxane. Complementary to the circumrotary motion observed in bistable [2]catenanes, the linear sliding motion of the moveable CBPQT^{4+} ring along the long axis enables us to generate nanoscale one-dimensional mechanical movements.

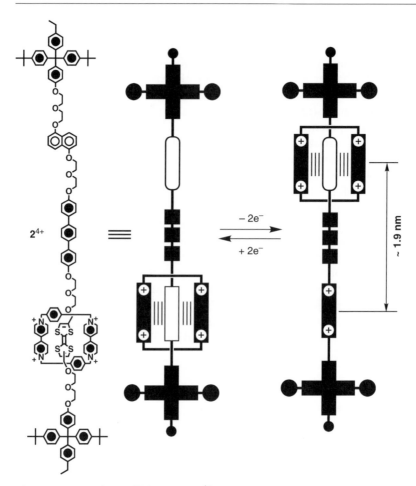

Scheme 12.2. Switching of [2]rotaxane 2^{4+}

12.2.3
Self-Complexing Molecular Switches

Self-complexing molecular switches [35, 36] usually contains an arm component, covalently linked to a macrocycle, with sufficient flexibility such that
the arm can be included inside the macrocycle's cavity by virtue of stabilizing
noncovalent bonding interactions. In contrast with switchable [2]catenanes
and [2]rotaxanes, such compounds do not require a secondary recognition
unit for switching. Rather, the switching behavior is based on reversible movement of the arm into and out of the macrocycle's cavity in response to a redox
stimulus.

The self-complexing compound 3^{4+} (Scheme 12.3) was prepared with a TTF-
bearing arm covalently linked to a π-accepting $CBPQT^{4+}$ ring. Just as in the

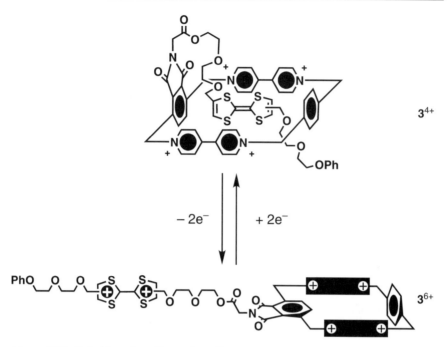

3^{4+}

$-2e^-$ $+2e^-$

3^{6+}

Scheme 12.3. Switching of a self-complex 3^{4+}

case of the bistable [2]catenanes and [2]rotaxanes, molecular recognition between a TTF donor and a CBPQT^{4+} acceptor in the self-complexing compound 3^{4+} can be turned "off" by the oxidation of the TTF unit to its TTF$^{+\cdot}$ cation radical or TTF^{2+} dication and "on" by their reduction back to the neutral (TTF) form. The switching process has been investigated by both chemical and electrochemical means. For example, ^{1}H NMR spectroscopy has been employed (Fig. 12.2) to monitor the chemical redox cycle at 253 K in CD$_3$COCD$_3$ solution. The ^{1}H NMR spectrum of 3^{4+} (Fig. 12.2a) is characteristic of a self-complexing (SC) conformation with averaged C_s symmetry. A mixture of *cis/trans* isomers, which are able to interconvert when irradiated by light, is observed in a 2:3 ratio as a result of the substitution mode of the TTF unit. No resonances are detected for uncomplexed (UC) 3^{4+}. Upon addition of 2 equiv. of tris(*p*-bromophenyl)aminium hexachloroantimonate into the CD$_3$COCD$_3$ solution of 3^{4+}, a much simpler spectrum (Fig. 12.2b) is observed, indicating that, upon oxidation, the TTF^{2+} dication no longer resides inside the CBPQT^{4+} cavity and its protons resonate at $\delta = 9.83$ ppm. The remainder of the spectrum is commensurate with the UC-3^{4+} conformation having averaged C_{2v} symmetry. When Zn dust is added to the NMR tube, the original spectrum (Fig. 12.2c) is regenerated, indicating a return to the SC-3^{4+} conformation with the neutral TTF unit back inside the cavity of the CBPQT^{4+} ring.

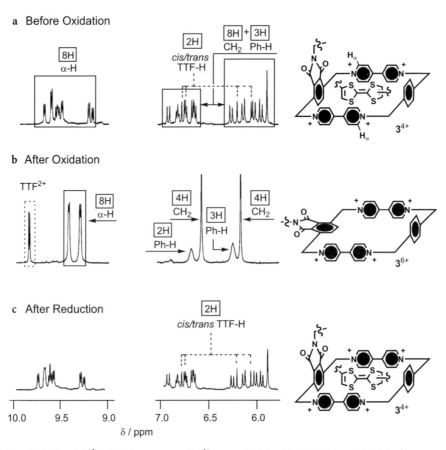

Fig. 12.2. Partial ^{1}H NMR spectra of 3^{4+} recorded in CD$_3$COCD$_3$ at 253 K before oxidation (**a**), after addition of 2 equiv. of tris(*p*-bromophenyl)aminium hexachloroantimonate (**b**), and after addition of Zn dust as a reductant (**c**)

The electrochemical switching behavior of 3^{4+} in MeCN is revealed by cyclic voltammetry (CV). At a scan rate of 1 000 mV s^{-1}, the CV displays (Fig. 12.3) only one two-electron anodic peak at +0.91 V vs. SCE, together with two single-electron cathodic peaks on the return sweep. The behavior differs from that (Fig. 12.3) of the control **4**, which displays two reversible, well-separated one-electron oxidation processes at +0.36 and +0.70 V vs. SCE. The CV indicates that the TTF arm in SC-3^{4+} undergoes oxidative dethreading in concert with the direct production of the TTF^{2+} dicationic form – and that it rethreads following formation of the charge-neutral TTF unit. Both chemical and electrochemical studies testify to the redox-controllable switching ability of such self-complexing donor–acceptor pairs, rendering them viable candidates for constructing nanoscale molecular machinery.

Fig. 12.3. CV of 1.0×10^{-3} M
MeCN solutions of $\mathbf{3^{4+}}$ and $\mathbf{4}$ at
scan rate of $1\,000$ mV s^{-1}

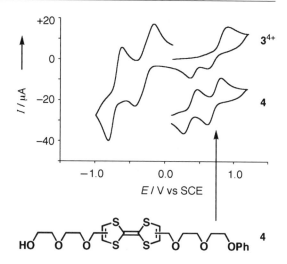

12.3
Application of Redox-Controllable Molecular Machines in Electronic Devices

Although early efforts were directed mainly at switchable catenanes and rotaxanes in the context of solution-phase mechanical processes, it has also been demonstrated [10, 25] that the redox-controllable mechanical movements in interlocked molecules can be stimulated electrically within the setting of solid-state devices. Molecular switch tunnel junctions employing a crossbar architecture have been fabricated utilizing redox-switchable mechanically interlocked molecules as the key active components.

Early fabrication of such a crossbar device was based (Fig. 12.4) on the switchable [2]catenane $\mathbf{1^{4+}}$. The switchable molecules were self-assembled as a Langmuir monolayer at the air–water interface and then deposited on a polysilicon electrode. A top electrode of Ti, followed by Al, was subsequently vapor-deposited on top of the molecular monolayer. High-quality, closely packed films were produced from switchable [2]catenanes employing a cosurfactant in the form of dimyristoylphosphatidyl anions (DMPA^{-}) as the counterions [10, 12, 24], which are essential for preventing the top Ti layer from penetrating through the monolayer and thus causing a short circuit.

The molecular devices display switching between high and low conductance states at different applied voltages. Each device is characterized by applying a "write" voltage, V, and recording the "read" current, I. The remnant molecular signature (Fig. 12.4b) tracks the read current (y axis) as the writing voltage (x axis) is cycled around a loop in 40-mV steps, from 0.0 V to +2.0 V, down to −2.0 V, and back to 0.0 V. In order to record the read current, a read voltage of +100 mV is applied in between each of the 40-mV pulses. For the [2]catenane-based crossbar, a low current is recorded, corresponding to the OFF state of the device, until the threshold voltage of +2 V is reached, whereupon a higher

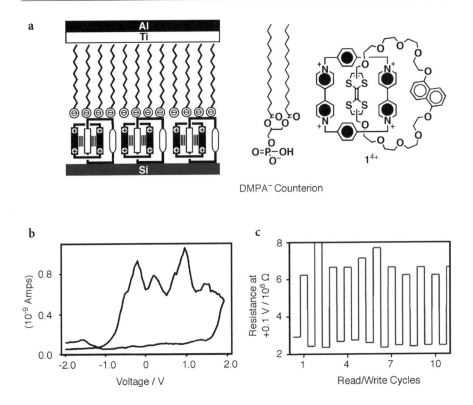

Fig. 12.4. An electronic device based on the bistable [2]catenane 1^{4+} (**a**), remnant molecular signature (**b**), and binary switching behavior (**c**) of the device

current is read, effectively switching the device into an ON state. The ON state is maintained until the threshold voltage for switching the device back into a low current or OFF state at −1 V is reached. The binary behavior is recorded (Fig. 12.4c) in a separate experiment by reading the current after the device is alternately cycled between ON and OFF states by writing at the threshold voltages. These data reveal that, in addition to the reversible voltage-gated switching of the device ON and OFF more than ten times, the ON state is metastable, displaying a temperature-dependent resetting of the device back to the OFF state.

From the reversible electronically driven switching observed in devices incorporating the bistable [2]catenane 1^{4+} has emerged a simple logic circuit and a random access memory (RAM) circuit [25] that were fabricated recently by using a related amphiphilic bistable [2]rotaxane 4^{4+} (Scheme 12.4). Crossbar devices built around this [2]rotaxane have been investigated [25] for their ability to switch. It was found that the devices displayed very similar behavior to that of the bistable catenane-based ones in terms of remnant molecular

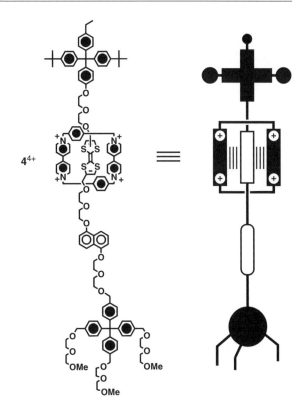

Scheme 12.4. Structural formula and graphical representation of amphiphilic [2]rotaxane 4^{4+}

4^{4+}

signatures and the binary switching behavior of the devices. By contrast, control devices based on dumbbell components and non-redox-active molecules did not display any switching behavior. First-principles computation has been employed [75–77] to evaluate the physical basis of the measured conductance difference between the ON and OFF states. The computational results confirm that the ON and OFF states correspond to the co-conformations with the DNP ring system and TTF unit encircled by the $CBPQT^{4+}$ ring, respectively.

A 64-bit memory was constructed, comprised in total of 64 crossbars containing a monolayer of bistable [2]rotaxane molecules. Fifty-six bits out of the total operated, allowing the acronyms such as DARPA to be written successfully into and read out of the chip in ASCII code. A feature of such a circuit is that the metastabilities of the ON state in these molecular analogs display half-lives of 15 to 60 min, which are quite different from standard RAM where the memory addresses need to be continually rewritten every tens of milliseconds.

For the redox-activated switching of the bistable [2]rotaxanes, a universal nanoelectromechanical mechanism (Fig. 12.5) has been proposed [78–81]. An electrochemically driven translation of the $CBPQT^{4+}$ ring from the TTF unit to the DNP ring systems forms a metastable state co-conformation (MSCC) that is able to relax thermally back to the ground state co-conformation (GSCC).

Ground State Co-Conformation

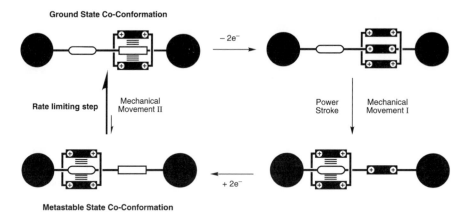

Metastable State Co-Conformation

Fig. 12.5. Proposed nanoelectromechanical switching mechanism of bistable [2]rotaxanes

The free energy barriers for relaxation of the MSCC to the GSCC have been unravelled in solution [80], in self-assembled monolayers [78], and in polymer matrices [79] to be 16, 18, and 18 kcal mol^{-1}, respectively. These values correlate well with a barrier of 21 kcal mol^{-1} observed in molecular switch tunnel junction devices.

In an attempt to miniaturize molecular electronic devices, a semiconducting, single-walled carbon nanotube (SWNT) was incorporated [14] as the bottom electrode in two-terminal molecular switch tunnel junctions. They contain a molecular monolayer of a related switchable [2]catenane coated noncovalently along the wall of SWNT. Such SWNT-based molecular devices operate very much like the reported silicon-based ones.

12.4
Application of Redox-Controllable Molecular Machines in Mechanical Devices

The redox-switchable properties of bistable catenanes and bistable rotaxanes in both solution and in electronic devices offer the potential for those molecular switches to be incorporated into wholly mechanical devices, such as actuators [82, 83] and molecular valves [84–86], as the active components. Nature provides some insight into how simple molecular components can be organized to perform complex mechanical tasks. Taking a lesson from biomolecular motors [87, 88], such as myosin and actin in muscle fiber, mechanical motions generated from functional molecular machines can be amplified and harnessed in nanoelectromechanical systems (NEMS) cooperatively through self-organization of the bistable molecules on surfaces. The first verification of the redox-stimulated mechanical movements was demonstrated in *closely packed* condensed phases of bistable rotaxanes on solid substrates.

12.4.1
Switching in Langmuir–Blodgett Film

A bistable rotaxane 6^{4+} (Fig. 12.6a) possessing regular redox-switching properties was synthesized with hydrophobic and hydrophilic stoppers installed at both ends to introduce amphiphilicity. Such amphiphilicity in solution enables the molecules to readily self-assemble [89] into Langmuir monolayers at the air–water interface. These molecular monolayers were subsequently transferred to SiO_2 substrates for further investigation.

Langmuir isotherms of both the amphiphilic bistable [2]rotaxane 6^{4+} and its dumbbell 7 were recorded in order to establish their ability to form stable monolayers, as well as their capacity to switch in closely packed condensed phases. A comparison of Langmuir layers prepared on neutral and oxidizing subphases provides insightful information on the switching behavior. Two different isotherms for the [2]rotaxane 6^{4+} were obtained (Fig. 12.6b) when water or water premixed with the oxidant $Fe(ClO_4)_3$ was used as the subphase. In general, the unoxidized [2]rotaxane 6^{4+} molecules in the monolayer appear to occupy a smaller mean molecular area compared to the oxidized ones at almost every pressure. By comparison, the dumbbell 7 was subjected to the same set of experimental conditions. Only a negligible difference in the Langumir isotherms was displayed before and after oxidation for this control. Such behavior is in contrast to that of the amphiphilic, bistable rotaxane 6^{4+} and supports the interpretation that the redox-driven mechanical movement of the $CBPQT^{4+}$ ring is the cause of the change observed in the rotaxane's isotherm.

X-Ray photoelectron spectroscopic (XPS) studies provided further evidence for the movement of the $CBPQT^{4+}$ ring in the LB monolayers [89]. The intensity of the photoemission from the 1s orbital on the nitrogen atoms in the $CBPQT^{4+}$ ring in the bistable [2]rotaxane 6^{4+} was recorded for both the unswitched and switched LB monolayers transferred onto a SiO_2 substrate. An increase in the relative intensity of the nitrogen signal was observed (Fig. 12.6c) when the oxidant was added to the subphase. The intensity increase was calculated to be equal to a 44% change in height with respect to the monolayers' thickness. The 3.7-nm movement of the $CBPQT^{4+}$ ring upwards from the TTF unit to the DNP ring system is believed to be the cause of the intensity change. By contrast, the dumbbell 7 displayed no signal. Furthermore, an amphiphilic bistable [2]rotaxane 8^{4+}, in which the order of the TTF unit and DNP ring system were reversed with respect to the SiO_2 substrate, displayed consistent spectroscopic changes.

In situ switching was performed in a closely packed LB monolayer transferred onto a SiO_2 substrate as an ultimate test of the mechanical movement of the $CBPQT^{4+}$ ring. A LB double layer of 8^{4+} was prepared such that a hydrophilic outer surface was accessible to the aqueous solution of the oxidant $Fe(ClO_4)_3$, into which the LB double layer was immersed. An intensity in-

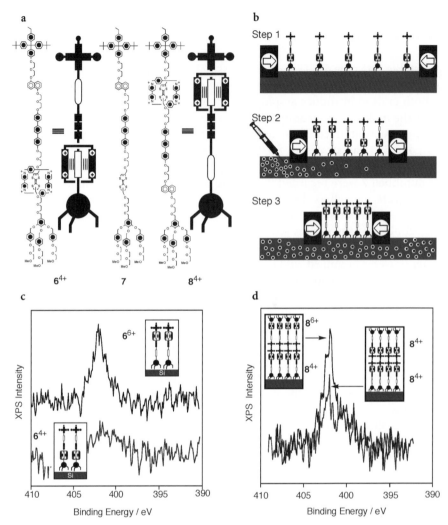

Fig. 12.6. a Structural formulas and graphical representations of amphiphilic rotaxanes 6^{4+} and 8^{4+}, and a dumbbell 7. **b** In situ Langmuir switching of the monolayers displays changes in **c** XPS spectra of LB monolayer of SiO_2 substrate that correlate with movement of tetracationic cyclophane. **d** Mechanical movement of cyclophane in LB monolayer was correlated with changes in XPS spectra

crease of the nitrogen signal was observed (Fig. 12.6d) following exposure to the oxidizing conditions, an observation consistent with the $CBPQT^{4+}$ ring's movement. This experiment indicates that, even when it is located in the closely packed condensed phase of a monolayer, the $CBPQT^{4+}$ ring in bistable [2]rotaxanes is still able to undergo mechanical movement following oxidation of the TTF unit. The verification of relatively linear mechanical movements within switchable rotaxanes in condensed phases opens up a window for developing switchable rotaxane-based NEMS devices.

12.4.2
Molecular Machines Functioning as Nanovalves

A supramolecular nanomachine has been demonstrated [86] to function as a nanovalve that opens and closes the orifices to molecular-sized pores and releases a small number of molecules on demand. The supramolecular valve that has been used to open and close the nanocontainer is a pseudorotaxane composed (Fig. 12.7a) of two components – a DNP-containing thread **9**, which is attached to the silica surface, and the moving part, the CBPQT^{4+} ring **10**$^{4+}$, which controls access to the interior of the nanopore. Operating the nanovalve involves three steps: (1) filling the container, (2) closing the valve, and (3) opening the valve to release the contents of the container on demand. The tubular pores, which are approximately 2 nm wide, are filled (Fig. 12.7b) with stable luminescent Ir(ppy)$_3$ molecules by allowing them to diffuse from solution into the open pores. The orifices are then "closed" by pseudorotaxane formation.

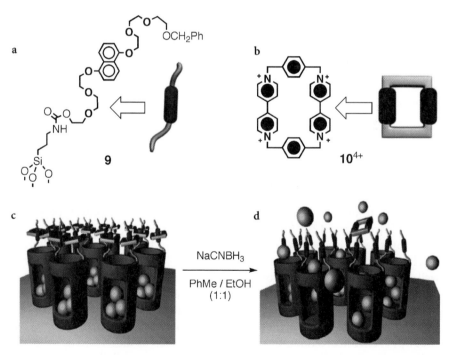

Fig. 12.7. **a** A DNP-containing thread **9**. **b** A cyclophane **10**$^{4+}$. **c** Nanopores whose orifices are covered with pseudorotaxanes formed between **9** and **10**$^{4+}$. **d** Upon reduction, the **10**$^{2+}$ bisradical dications are released and so allow the Ir(ppy)$_3$ to escape

An external reducing reagent ($NaCNBH_3$) is used to reduce the $CBPQT^{4+}$ ring, effecting dethreading of the pseudorotaxane so as to unlock the nanopores and allowing the release of the dye as measured by following its luminescent signal in the surrounding solution.

A switchable [2]rotaxane 11^{4+} (Scheme 12.5) was employed [84] in order to obtain a nanovalve that is reversibly operational. The rotaxane was tethered around the 2-nm pores of a silica framework (MCM-41) and was demonstrated subsequently to be fully functional under redox conditions. Translocation of the mobile $CBPQT^{4+}$ ring brings it close to or away from the orifice of the nanopores by redox reagents, thus controlling the closing and opening of the nanovalve. The controlled positioning of the $CBPQT^{4+}$ ring thus regulates the capturing and subsequent release of guest dye molecules trapped in the nanopores. Compared to the supramolecular system, activation of the nanovalve is achieved by chemical oxidation–reduction cycles of the TTF unit instead of destructive reduction of the $CBPQT^{4+}$ ring and its diffusion away into solution. The nanovalve with tethered bistable [2]rotaxanes, by contrast, constitutes a reusable molecular valve.

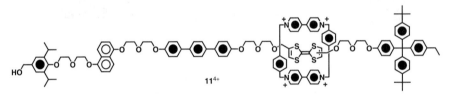

11^{4+}

Scheme 12.5. Structural formula of a [2]rotaxane 11^{4+} that functions as a molecular nanovalve

12.4.3
Artificial Molecular Muscles

Expanding upon the reported switchable [2]rotaxanes, a prototype palindromic [3]rotaxane 12^{8+} (Fig. 12.8) was created [82–84] according to a design that is reminiscent of a "molecular muscle" for the purpose of amplifying and harnessing molecular mechanical motions. The design takes advantage of well-established donor–acceptor recognition chemistry to control the localization of its two $CBPQT^{4+}$ rings on its two TTF units, as opposed to on its two naphthalene (NP) ring systems. The location and thus the switching of the two $CBPQT^{4+}$ rings can be controlled to be on either the TTF or NP recognition sites by (1) chemical oxidation and reduction cycles, as illustrated by ^{1}H NMR spectroscopy, and (2) electrochemical redox processes, as illustrated by CV and UV-visible spectroelectrochemistry. The switching of interring distances from 4.2 to 1.4 nm mimics the contraction and extension of skeletal muscle.

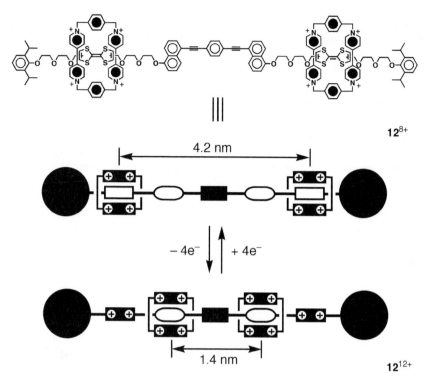

Fig. 12.8. Structural formula of prototypical molecular muscle 12^{8+} and graphical representation of its switching that generates contraction and extension motion

The reversible switching of 12^{8+} in solution provides a model for its mechanical motion when it is attached to a surface. In order for a rotaxane-based mechanical device to become an engineering reality, however, it is imperative to self-organize these nanoscale machines at interfaces and to demonstrate that the mechanical switching behavior exhibited in solution is preserved when they are mounted on solid supports. In order to achieve this objective, [3]rotaxane 13^{8+} (Fig. 12.9a) with disulfide tethers covalently attached to the two mobile CBPQT^{4+} rings was prepared for the purpose of self-assembling it onto a gold surface. A gold-coated silicon cantilever array was then coated with a SAM of 13^{8+} and placed in a transparent fluid cell (Fig. 12.9b). The position of each cantilever beam was monitored by an optical lever method, while aqueous Fe(ClO$_4$)$_3$ (oxidant) and ascorbic acid (reductant) solutions were sequentially introduced into the fluid cell. Addition of the oxidant caused the cantilever beams to bend upward by ca. 35 nm to an apparent saturation point (Fig. 12.9c, top series of traces). Entry of the reductant caused the beams to bend back downward to their starting positions. This behavior was observed for all four cantilever beams for 25 complete cycles (only 3 cycles were shown here) with a noticeable decrease in the bending magnitude.

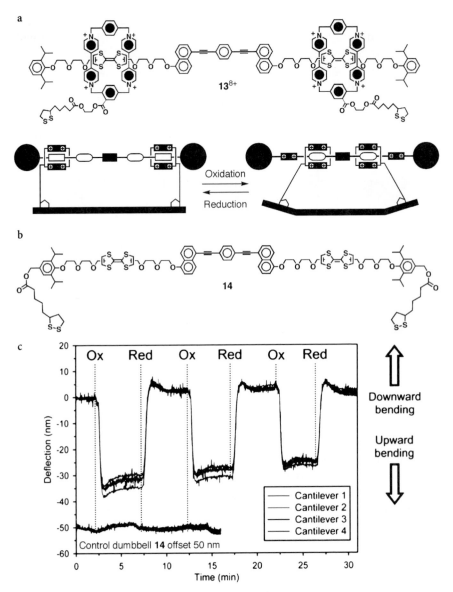

Fig. 12.9. **a** Structural formula of molecular muscle 13^{8+} in a SAM on gold surface. **b** Control dumbbell. **c** Experimental data showing three cycles of upward and downward bending of four cantilever beams coated with 13^{8+} (*top traces*) and control dumbbell 14 (*bottom traces*) upon injection of oxidant and reductant solutions

To establish that such a bending is not the result of mundane conformational changes and/or electrostatic changes associated with the oxidation of the TTF unit in the dumbbell, a disulfide-containing control compound 14

(Fig. 12.9c), which lacks the mobile $CBPQT^{4+}$ rings but contains recognition sites with the same relative geometries as are present in the [3]rotaxane, was synthesized and subjected to redox conditions identical to those described for the bis-bistable [3]rotaxane-based device. Only the slightest of movements of the control-coated cantilever array were observed following sequential injections of the same oxidant and reductant (Fig. 12.9d, bottom series of traces). The different device behavior suggests that the presence of the movable disulfide-tethered $CBPQT^{4+}$ ring in the bis-bistable [3]rotaxane are essential for the redox-controlled bending of the cantilever beams.

These results correlate directly with the collective effect of molecular-scale contractions and extensions of the interring distances in the disulfide-tethered palindromic [3]rotaxane molecules. They support the hypothesis that the cumulative nanoscale movements within surface-bound "molecular muscles" can be harnessed to perform larger-scale mechanical work.

12.5
Conclusions

The construction of nanoscale molecular machines with controllable mechanical movements has been achieved by incorporation of recognition sites with distinctive yet different binding affinities to the central moveable part – a tetracationic $CBPQT^{4+}$ ring – in bistable catenanes and rotaxanes. It is the redox properties of the recognition sites in those functional machines that enable the precise control of the nanoscale movements. By changing the redox state of the recognition sites, thermodynamic factors have been identified that make it possible for the mechanical motions of molecular components to be controlled precisely. Firstly, we have demonstrated that the stimulation of the mechanical movements can be realized conveniently by chemical, electrochemical, or photochemical means in solution. Secondly, voltage-gated movements of such machines in more confined media, from polymer electrolyte matrixes [79] and, ultimately, to self-assembled monolyers [78] to solid-state electronic devices [10, 14], lead to the vindication of their practical utility. A universal nanomechanical mechanism has been unraveled for their redox-activated performance. Finally, the potential of those nanoscale machines in mechanical devices has been demonstrated by the actuation of microscale cantilevers with a SAM of a specially designed molecular machine. In addition, the mechanical movements of bistable rotaxanes can be harnessed to regulate the release of guest molecules from a nanoscale container. Based on these early successes, future efforts will be directed from operating machines at the nanoscale to move other objects around across scales that reach into the world of their macroscopic counterparts, thus bringing the realization of molecular machines back to one of the original sources of their inspiration.

12.6
References

1. Balzani V, Credi A, Raymo FM, Stoddart JF (2000) Angew Chem Int Ed 39:3348
2. Balzani V, Venturi M, Credi A (2003) Molecular devices and machines – a journey into the nanoworld. Wiley-VCH, Weinheim
3. Livoreil A, Dietrich-Buchecker C, Sauvage J-P (1994) J Am Chem Soc 116:9399
4. Cárdenas DJ, Livoreil A, Sauvage J-P (1996) J Am Chem Soc 118:11980
5. Baumann F, Livoreil A, Kaim W, Sauvage J-P (1997) Chem Commun, p 35
6. Livoreil A, Sauvage J-P, Armaroli N, Balzani V, Flamigni L, Ventura B (1997) J Am Chem Soc 119:12114
7. Asakawa M, Ashton PR, Balzani V, Credi A, Hamers C, Mattersteig G, Montalti M, Shipway AN, Spencer N, Stoddart JF, Tolley MS, Venturi M, White AJP, Williams DJ (1998) Angew Chem Int Ed 37:333
8. Balzani V, Credi A, Mattersteig G, Matthews OA, Raymo FM, Stoddart JF, Venturi M, White AJP, Williams DJ (2000) J Org Chem 65:1924
9. Asakawa M, Higuchi M, Mattersteig G, Nakamura T, Pease AR, Raymo FM, Shimizu T, Stoddart JF (2000) Adv Mater 12:1099
10. Collier CP, Mattersteig G, Wong EW, Luo Y, Beverly K, Sampaio J, Raymo FM, Stoddart JF, Heath JR (2000) Science 289:1172
11. Hamilton DG, Montalti M, Prodi L, Fontani M, Zanello P, Sanders JKM (2000) Chem Eur J 6:608
12. Pease AR, Jeppesen JO, Stoddart JF, Luo Y, Collier CP, Heath JR (2001) Acc Chem Res 34:433
13. Pease AR, Stoddart JF (2001) Struct Bond 99:189
14. Diehl MR, Steuerman DW, Tseng HR, Vignon SA, Star A, Celestre PC, Stoddart JF, Heath JR (2003) Chem Phys Chem 4:1335
15. Bissell RA, Córdova E, Kaifer AE, Stoddart JF (1994) Nature 369:133
16. Benniston AC (1996) Chem Soc Rev 25:427
17. Gong C, Gibson HW (1997) Angew Chem Int Ed Engl 36:2331
18. Collin J-P, Gaviña P, Sauvage J-P (1997) New J Chem 21:525
19. Murakami H, Kawabuchi A, Kotoo K, Kunitake M, Nakashima N (1997) J Am Chem Soc 119:7605
20. Collin JP, Gaviña P, Heitz V, Sauvage JP (1998) Eur J Inorg Chem, p 1
21. Armaroli N, Balzani V, Collin J-P, Gaviña P, Sauvage J-P, Ventura B (1999) J Am Chem Soc 121:4397
22. Blanco M-J, Jimenez MC, Chambron J-C, Heitz V, Linke M, Sauvage J-P (1999) Chem Soc Rev 28:293
23. Jeppesen JO, Perkins J, Becher J, Stoddart JF (2001) Angew Chem Int Ed 40:1216
24. Collier CP, Jeppesen JO, Luo Y, Perkins J, Wong EW, Heath JR, Stoddart JF (2001) J Am Chem Soc 123:12632
25. Luo Y, Collier CP, Jeppesen JO, Nielsen KA, De Ionno E, Ho G, Perkins J, Tseng H-R, Yamamoto T, Stoddart JF, Heath JR (2002) Chem Phys Chem 3:519
26. Jimenez-Molero MC, Dietrich-Buchecker C, Sauvage J-P (2002) Chem Eur J 8:1456
27. Da Ross T, Guldi DM, Morales AF, Leigh DA, Prato M, Turco R (2003) Org Lett 5:689
28. Jeppesen JO, Nielsen KA, Perkins J, Vignon SA, Di Fabio A, Ballardini R, Gandolfi MT, Venturi M, Balzani V, Becher J, Stoddart JF (2003) Chem Eur J 9:2982
29. Yamamoto T, Tseng H-R, Stoddart JF, Balzani V, Credi A, Marchioni F, Venturi M (2003) Collect Czech Chem Commun 68:1488

30. Tseng T-H, Vignon SA, Celestre PC, Perkins J, Jeppesen JO, Di Fabio A, Ballardini R, Gandolfi MT, Venturi M, Balzani V, Stoddart JF (2004) Chem Eur J 10:155
31. Kang S, Vignon SA, Tseng H-R, Stoddart JF (2004) Chem Eur J 10:2555
32. Ashton PR, Ballardini R, Balzani V, Boyd SE, Credi A, Gandolfi MT, Gómez-López M, Iqbal S, Philp D, Preece JA, Prodi L, Ricketts HG, Stoddart JF, Tolley MS, Venturi M, White AJP, Williams DJ (1997) Chem Eur J 3:152
33. Amendola V, Fabbrizzi L, Mangano C, Pallavicini P (2001) Acc Chem Res 34:488
34. Fabbrizzi L, Foti F, Licchelli M, Maccarini PM, Sacchi D, Zema M (2002) Chem Eur J 8:4965
35. Liu Y, Flood AH, Stoddart JF (2004) J Am Chem Soc 126:9150
36. Liu Y, Flood AH, Moskowitz RM, Stoddart JF (2005) Chem Eur J 11:369
37. Shinkai S (1990) In: Inoue Y, Gokel GW (eds) Cation Binding by Macrocycles. Marcel Dekker, New York, p 397
38. Kaifer AE (1999) Supramolecular Electrochemistry. Wiley-VCH, Weinheim
39. Niemz A, Rotello VM (1999) Acc Chem Res 32:42
40. Kaifer AE (1999) Acc Chem Res 32:62
41. Boulas PL, Gomez-Kaifer M, Echegoyen L (1998) Angew Chem Int Ed 110:226
42. Venturi M, Credi A, Balzani V (1999) Coord Chem Rev 185/186:233
43. Lane AS, Leigh DA, Murphy A (1997) J Am Chem Soc 119:11092
44. Ashton PR, Ballardini R, Balzani V, Baxter I, Credi A, Fyfe MCT, Gandolfi MT, Gómez-López M, Martínez-Díaz M-V, Piersanti A, Spencer N, Stoddart JF, Venturi M, White AJP, Williams DJ (1998) J Am Chem Soc 120:11932
45. Jiménez MC, Dietrich-Buchecker C, Sauvage J-P (2000) Angew Chem Int Ed 39:3284
46. Lee JW, Kim KP, Kim K (2001) Chem Commun, p 1042
47. Elizarov AM, Chiu HS, Stoddart JF (2002) J Org Chem 67:9175
48. Tseng H-R, Vignon SA, Stoddart JF (2003) Angew Chem Int Ed 41:1491
49. Badjic JD, Balzani V, Credi A, Silvi S, Stoddart JF (2004) Science 303:1845
50. Kaiser G, Jarrosson T, Otto S, Ng Y-F, Bond AD, Sanders JKM (2004) Angew Chem Int Ed 43:1959
51. Raehm L, Kern J-M, Sauvage J-P (1999) Chem Eur J 5:3310
52. Bermudez V, Capron N, Gase T, Gatti FG, Kajzar F, Leigh DA, Zerbetto F, Zhang S (2000) Nature 406:608
53. Kern J-M, Raehm L, Sauvage J-P, Divisia-Blohorn B, Vidal P-L (2000) Inorg Chem 39:1555
54. Ballardini R, Balzani V, Dehaen W, Dell'Erba AE, Raymo FM, Stoddart JF, Venturi M (2000) Eur J Org Chem, p 591
55. Collin J-P, Kern J-M, Raehm L, Sauvage J-P (2000) In: Feringa BL (ed) Molecular Switches. Wiley-VCH, Weinheim, p 249
56. Colasson BX, Dietrich-Buchecker C, Jimenez-Molero MC, Sauvage J-P (2002) J Phys Org Chem 15:476
57. Altieri A, Gatti FG, Kay ER, Leigh DA, Paolucci F, Slawin AMZ, Wong JKY (2003) J Am Chem Soc 125:8644
58. Poleschak I, Kern J-M, Sauvage J-P (2004) Chem Commun, p 474
59. Katz E, Lioubashevsky O, Willner I (2004) J Am Chem Soc 126:15520
60. Katz E, Sheeney-Haj-Ichia, Willner I (2004) Angew Chem Int Ed 43:3292
61. Ballardini R, Balzani V, Gandolfi MT, Prodi L, Venturi M, Philp D, Ricketts HG, Stoddart JF (1993) Angew Chem Int Ed Engl 32:1301
62. Ashton PR, Ballardini R, Balzani V, Credi A, Dress R, Ishow E, Kocian O, Preece JA, Spencer N, Stoddart JF, Venturi M, Wenger S (2000) Chem Eur J 6:3558

63. Brouwer AM, Frochot C, Gatti FG, Leigh DA, Mottier L, Paolucci F, Roffia S, Wurpel GWH (2001) Science 291:2124
64. Collin J-P, Laemmel A-C, Sauvage J-P (2001) New J Chem 25:22
65. Leigh DA, Wong JKY, Dehez F, Zerbetto F (2003) Nature 424:174
66. Bottari G, Leigh DA, Pérez EM (2003) J Am Chem Soc 125:13360
67. Gatti FG, Len S, Wong JKY, Bottari G, Altieri A, Morales MAF, Teat SJ, Frochot C, Leigh DA, Brouwer AM, Zerbetto F (2003) Proc Natl Acad Sci USA 100:10
68. Altieri A, Bottari G, Dehez F, Leigh DA, Wong JKY, Zerbetto F (2003) Angew Chem Int Ed 42:2296
69. Brouwer AM, Fazio SM, Frochot C, Gatti FG, Leigh DA, Wong JKY, Wurpel GWH (2003) Pure Appl Chem 75:1055
70. Mobian P, Kern J-M, Sauvage J-P (2004) Angew Chem Int Ed 43:2392
71. Flamigni L, Talarico AM, Chambron J-C, Heitz V, Linke M, Fujita N, Sauvage J-P (2004) Chem Eur J 10:2689
72. Hernandez JV, Kay ER, Leigh DA (2004) Science 306:1532
73. Perez EM, Dryden DTF, Leigh DA, Teobaldi G, Zerbetto F (2004) J Am Chem Soc 126:12210
74. Flood AH, Liu Y, Stoddart JF (2004) In: Gleiter R, Hopf H (eds) Modern Cyclophane Chemistry. Wiley-VCH, Weinheim, p 485
75. Deng W-Q, Muller RP, Goddard WA (2004) J Am Chem Soc 126:13562
76. Flood AH, Ramirez RJA, Deng W-Q, Muller RP, Goddard WA, Stoddart JF (2004) Aust J Chem 57:301
77. Jang SS, Jang YH, Kim YH, Goddard WA, Flood AH, Laursen BW, Tseng HR, Stoddart JF, Jeppesen JO, Choi JW, Steuerman DW, DeIonno E, Heath JR (2005) J Am Chem Soc 127:1563
78. Tseng H-R, Wu D, Fang N, Zhang X, Stoddart JF (2003) Chem Phys Chem 5:111
79. Steuerman D, Tseng H-R, Peters AJ, Flood AH, Jeppesen JO, Nielsen KA, Stoddart JF, Heath JR (2004) Angew Chem Int Ed 43:6486
80. Flood AH, Peters AJ, Vignon SA, Steuerman DW, Tseng H-R, Kang SS, Heath JR, Stoddart JF (2004) Chem Eur J 10:6558
81. Flood AH, Steuerman D, Heath JR, Stoddart JF (2004) Science 306:2055
82. Huang TJ, Brough B, Ho C-M, Liu Y, Flood AH, Bonvallet PA, Tseng H-R, Stoddart JF, Baller M, Magonov S (2004) Appl Phys Lett 85:5391
83. Liu Y, Flood AH, Bonvallet PA, Vignon SA, Northrop BH, Tseng H-R, Jeppesen JO, Huang TJ, Brough B, Baller M, Magonov S, Solares SD, Goddard WA, Ho C-M, Stoddart JF (2005) J Am Chem Soc 127:9745
84. Nguyen T, Tseng H-R, Celestre PC, Flood AH, Liu Y, Zink AI, Stoddart JF (2005) Proc Natl Acad Sci USA 102:10029
85. Chia S, Cao J, Stoddart JF, Zink JI (2001) Angew Chem Int Ed 40:2447
86. Hernandez R, Tseng H-R, Wong JW, Stoddart JF, Zink JI (2004) J Am Chem Soc 126:3370
87. Goodsell DS (1996) "Our Molecular Nature: The Body's Motors, Machines, and Message". Springer, Berlin Heidelberg New York
88. Boyer PD (2002) J Biol Chem 277:39045
89. Huang TJ, Tseng H-R, Sha L, Lu W, Brough B, Flood AH, Yu B-D, Celestre PC, Chang JP, Stoddart JF, Ho C-M (2004) Nano Lett 4:2065

Through-Space Control of Redox Reactions Using Interlocked Structure of Rotaxanes

Nobuhiro Kihara[1] · Toshikazu Takata[2]

[1] Department of Chemistry, Faculty of Science, Kanagawa University, Tsuchiya, Hiratsuka 259-1293, Japan
[2] Department of Organic and Polymeric Materials, Tokyo Institute of Technology, Ookayama, Meguro, Tokyo 152-8552, Japan

Summary The redox potential of the ferrocene moiety decreased when rotaxane had a crown ether wheel that was free of the intercomponent interaction. It was assumed that the redox reaction of the ferrocene moiety accompanied the transposition of the wheel component on the axle. The reduction of ketone by the action of rotaxane bearing the dihydronicotinamide group was also investigated. The activity of the dihydronicotinamide group was considerably enhanced by the presence of the crown ether wheel. Furthermore, enantioselective reduction was achieved using an optically active wheel component. The effect of rotaxane structure and intercomponent interaction on the redox behavior are discussed.

Abbreviations

DB24C8 Dibenzo-24-crown-8 ether
TTF   Tetrathiafulvalene
NMR   Nuclear magnetic resonance
NADH   Nicotinamide adenine dinucleotide

13.1
Introduction

Rotaxane is a typical interlocked compound consisting of wheel and axle components arranged spatially by mechanical bonding [1]. The axle component threading into the macrocyclic wheel component has two bulky substituents at the two termini that prevent dethreading from the wheel component.

The preparation of the rotaxanes requires special strategies to construct the interlocked structure. The simplest technique is end-capping. The inclusion complex of the wheel component with the axle component without the bulky terminal substituent(s) is prepared before the termini of the axle component are introduced to fix the interlocked structure. To prepare the inclusion complex, denoted as pseudorotaxane, various attractive intermolecular interactions have been utilized. One of the most versatile complex systems is the combination of crown ether and secondary ammonium salt [2]. Using this combination, a variety of components can be designed for the preparation of rotaxanes, while the modification of the resulting rotaxanes is also an easy path to new rotaxanes.

The intermolecular interaction that is necessary to produce the pseudoro-taxane complex often remains in rotaxane as a strong intercomponent inter-action. Because of the unique structure of rotaxane, special features favorable for its application to molecular devices and supramolecular materials are ex-pected [3]. In the presence of the strong intercomponent interaction, however, the structural features of the rotaxanes would be limited because the good com-ponent mobility, which is the most noteworthy feature, is restricted. The control of the intercomponent interaction has been investigated based on the prop-erties of rotaxanes. In the case of the crown ether-secondary ammonium salt system, the intercomponent interaction can be removed by acylative neutral-ization of the ammonium moiety [4]. In the intercomponent-interaction-free rotaxane, the wheel component can change its position on the axle component without any change in the conformation, when the axle component is fixed.

Since all rotaxane components are situated very closely together, they strongly affect each other. When a redox-active functional group is placed on the axle of rotaxane, the redox behavior depends on the components' rel-ative positions, or how the wheel component surrounds the axle. Even in intercomponent-interaction-free rotaxanes, the wheel component affects the redox behavior, whereas the redox reaction can control the position of the wheel component on the axle component in rotaxane.

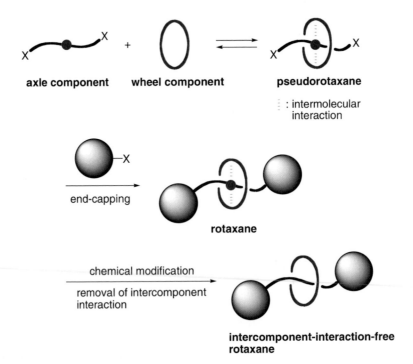

Fig. 13.1. Preparation of rotaxane by end-capping

13.2
Redox Behavior and Conformation of Ferrocene-End-Capped Rotaxane

Since in rotaxane the relative movements of a component against its counter-part are similar to basic mechanical motions, much attention has been paid to the construction and motion of rotaxane-based molecular machines [3]. The redox reaction is one of the most promising methods of controlling molecular motion via the control of the intercomponent interaction [5]. Stoddart et al. pioneered the redox-reaction-induced transposition of a wheel component on a counterpart axle [6]. In the reduced form, a viologen-based macrocycle is placed on the benzidine or tetrathiafulvalene (TTF) moiety by the strong in-tercomponent charge transfer (CT) interaction. By the one-electron oxidation of the benzidine or TTF moiety, the CT interaction is lost, and the electrostatic repulsion between the cations enforces the transposition. We designed a novel

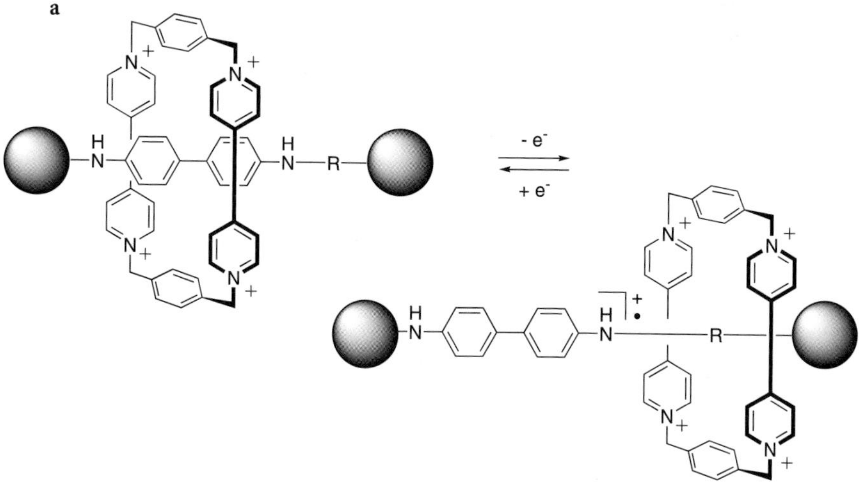

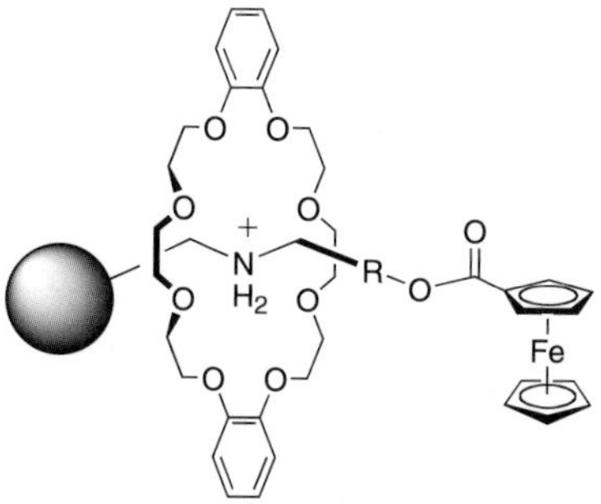

Fig. 13.2. (a) Stoddart's redox-induced transposition systems based on CT interaction. (b) Our ferrocene-based system

ferrocene-containing rotaxane **1** with a crown ether wheel as the electron-rich component that may interact with the ferrocenium cation [7].

Rotaxane **1a** was prepared as shown in Scheme 13.1. The secondary ammonium salt **2** with a hydroxy group at the terminus was treated with ferrocenecarboxylic anhydride and a catalytic amount of tributylphosphane in the presence of DB24C8 to afford **1a** in 87% yield [8]. The X-ray crystallographic structure of **1a** as a benzene adduct clearly confirmed its rotaxane structure. Since there is a strong intercomponent hydrogen-bonding interaction between the DB24C8 component and the ammonium group, the transposition of the crown ether component on the axle was not expected. Rotaxane **1a** was treated with excess triethylamine and acetic anhydride to convert it to the corresponding nonionic rotaxane **3a** [4] in which both components are free of the hydrogen-bonding interaction. The X-ray crystallographic structure of **3a** reveals that the weak intercomponent interaction became working after

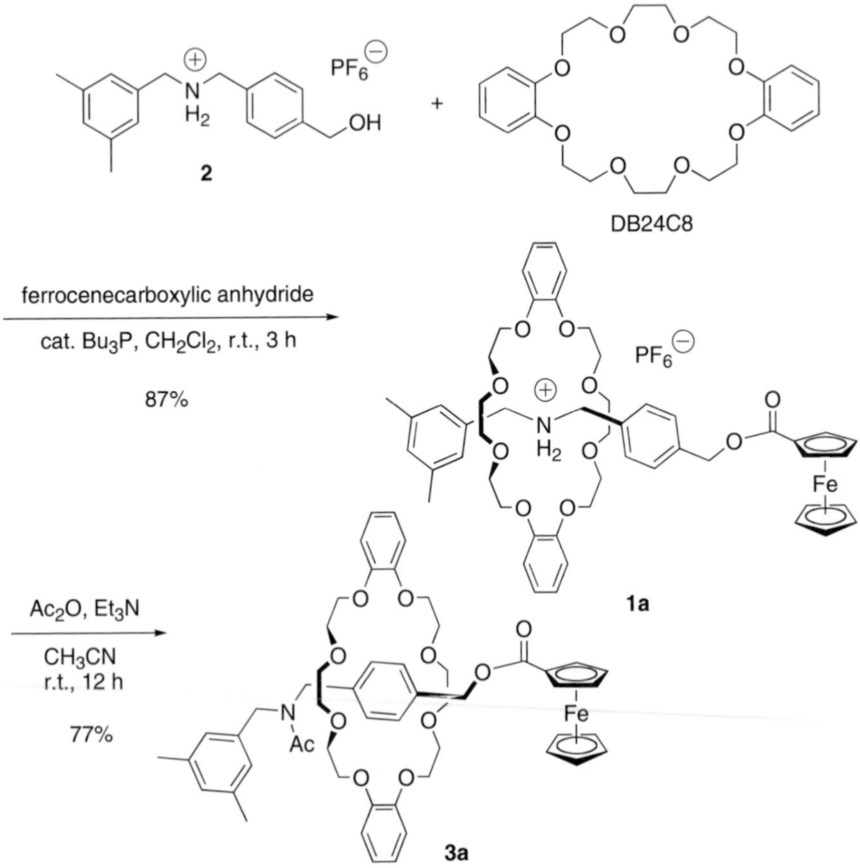

Scheme 13.1.

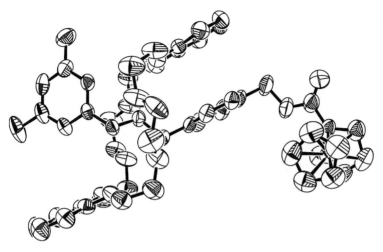

Fig. 13.3. ORTEP drawing of **1a**·C_6H_6. Hydrogens, benzene, and PF_6 omitted for clarity

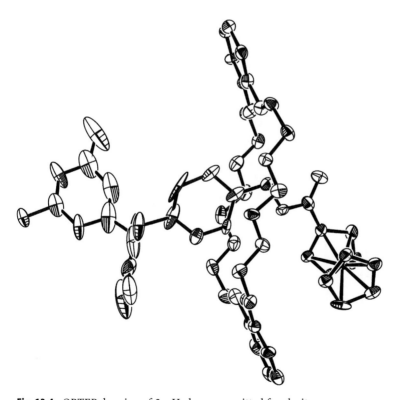

Fig. 13.4. ORTEP drawing of **3a**. Hydrogens omitted for clarity

the elimination of the strong hydrogen-bonding interaction. The crown ether component was moved to the *p*-phenylene group, and a CH/π interaction between the γ-methylene groups of the DB24C8 component and the *p*-phenylene group occurred. The ¹H NMR spectrum of **3a** is shown in Fig. 13.5. The considerable down-field shift of the *p*-phenylene group and the split of the γ-

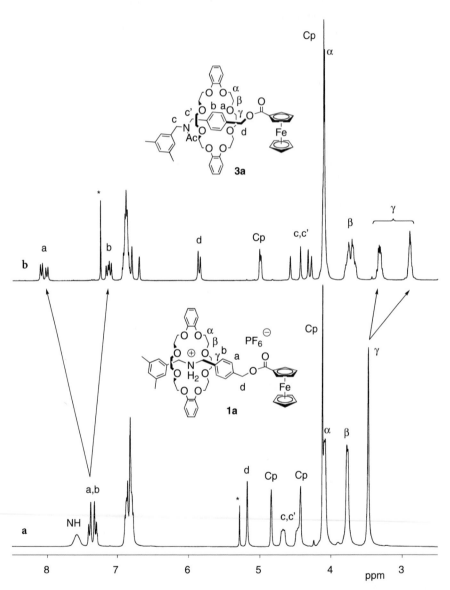

Fig. 13.5. ¹H NMR spectra of (a) **1a** (270 MHz, CDCl₃) and (b) **3a** (270 MHz, CDCl₃). Asterisk (*) denotes residual CHCl₃ or CH₂Cl₂

methylene protons in the DB24C8 component clearly indicate the occurrence of a CH/π interaction even in the solution state. Since the CH/π interaction is rather weak, the DB24C8 component is loosely bound to the p-phenylene group.

Cyclic voltammetry analyses (vs. Ag/AgCl) of 1a and related compounds were carried out in acetonitrile. The results are summarized in Table 13.1. 1a showed a slightly higher $E_{1/2}$ than benzyl ferrocenecarboxylate and model compound 4 because of its cationic character. Due to the strong hydrogen-bonding interaction between the DB24C8 component and the ammonium group, the wheel component did not affect the ferrocene moiety of 1a. On the other hand, nonionic rotaxane 3a showed an $E_{1/2}$ approximately 80 mV lower than that of 1a. The DB24C8 component in 3a stabilized the oxidized state and reduced its $E_{1/2}$. Since the $E_{1/2}$ of 4 did not change after the addition of DB24C8, it was concluded that the stabilization was induced by the crown ether-ferrocenium cation interaction. Since this interaction is very weak, it occurs only when the approach of the crown ether component to the ferrocene moiety is structurally allowed, as in the rotaxane system driven by the very weak attractive interaction.

To demonstrate the interaction between the wheel component and the ferrocene group, the electronic spectra of 3a, 4, and their oxidized states was measured. One-electron oxidation was carried out chemically with silver tetrafluo-

Table 13.1. Redox potentials of ferrocene-containing compounds[a]

Ferrocene derivative	$E_{1/2}$/mV
Benzyl ferrocenecarboxylate	331
4	335[b]
4 + DB24C8	335[b,c]
1a	353
3a	271

[a]vs. Ag/Ag$^+$ with glassy carbon electrode at 298 K in 0.1 mol/L tetrabutylammonium perchlorate in acetonitrile (0.5 mmol/L).
[b]With 0.1 mol/L tetraethylammonium perchlorate in acetonitrile (5 mmol/L).
[c]1 equivalent of DB24C8 was added to solution of 4.

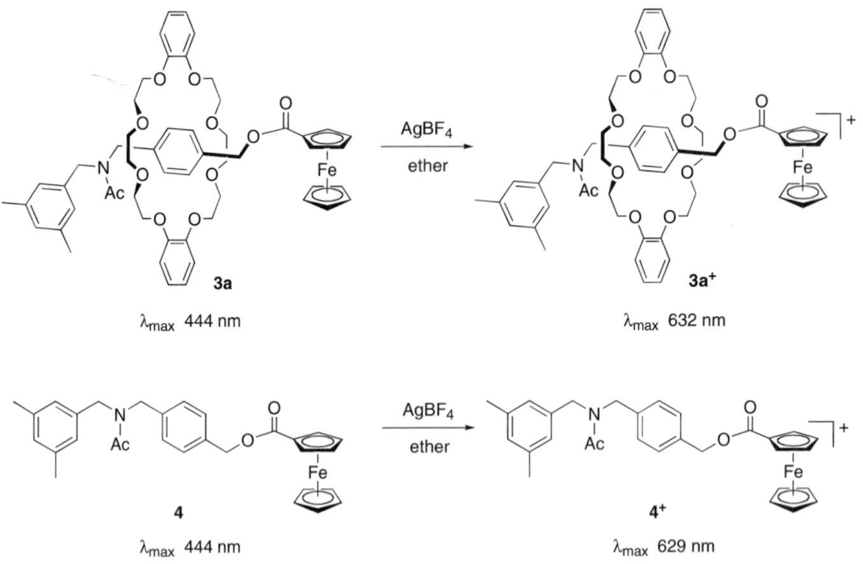

Scheme 13.2.

roborate in ether under an argon atmosphere. The absorption of **3a** at 444 nm, corresponding to the ferrocene moiety, disappeared after the oxidation, while the absorption at 632 nm, which was assigned to the ferrocenium cation moiety, appeared [9]. The reversible redox behavior of **3a** was easily confirmed; the electronic spectrum of the solution of **3a**$^+$ readily returned to that of **3a** upon exposure to air or water. The electronic spectrum of **4** was identical to that of **3a** in the visible region. However, λ_{max} of the ferrocenium cation of **4**$^+$ was observed at 629 nm. The redshift of the ferrocenium cation of **3a**$^+$ as compared to that of **4**$^+$ suggests a stabilizing factor like the attractive interaction between the crown ether and the ferrocenium cation. Since there is no perturbation in the reduced state, it can be deduced that the ferrocene group attracted the wheel component by the interaction induced by the one-electron oxidation.

To ensure the nature of the crown ether-ferrocenium cation interaction and the transposition of the wheel component on the axle component, some ferrocene derivatives with rotaxane structures were prepared and their $E_{1/2}$ values measured. Rotaxanes **1b** and **3b**, containing dicyclohexano-24-crown-8 as the wheel component, were prepared to examine the effect of the electron-rich benzene rings of DB24C8. Since the $\Delta E_{1/2}$ value between **3b** and **1b** was similar to that between **3a** and **1a**, the ferrocenium cation interacted with the ether oxygen, but not with the electron-rich benzene ring. Rotaxanes **1c**, **1d**, **3c**, and **3d** having axles with longer spacers were prepared. The ^{1}H NMR spectra of **1d** and **3d** are shown in Fig. 13.6. The occurrence of a CH/π interaction between the wheel and the axle in **3d** is evident from the down-field shift of the *p*-phenylene group and the split of the γ-methylene groups in the DB24C8 component and indicates that the DB24C8 wheel is located at the p-phenylene group. The $E_{1/2}$ of **3c** was approximately 65 mV lower than that of **1c**, and the $E_{1/2}$ of **3d** was approximately 45 mV lower than that of **1d**. A considerable decrease in $E_{1/2}$ was observed even in **3d**, which has a dodecamethylene spacer between the p-phenylene, which is the "station" of the wheel component in reduced form, and the ferrocene group. The large $\Delta E_{1/2}$ cannot be explained without the transposition of the wheel component on the axle that leads it

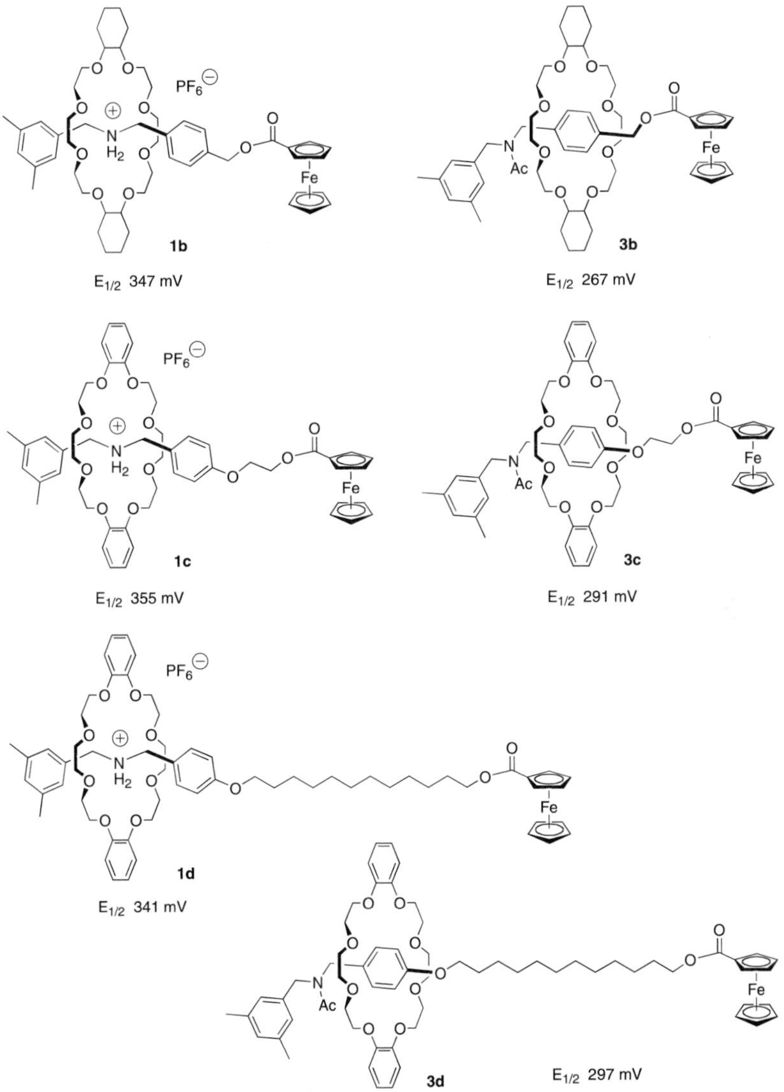

Fig. 13.6. Redox potentials of ferrocene derivatives with rotaxane structure

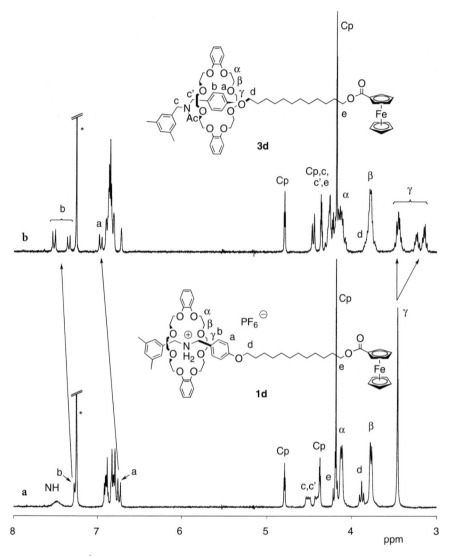

Fig. 13.7. Partial ^{1}H NMR spectra of (a) **1d** (270 MHz, CDCl$_3$) and (b) **3d** (270 MHz, CDCl$_3$). Asterisk (*) denotes the residual CHCl$_3$

to interact with the ferrocenium cation. The rather small $\Delta E_{1/2}$ for **3c** and **3d** presumably result from the difference in the structures of the *p*-phenylene moieties of the axles – dialkylbenzene for **3a** and **3b**, and alkylalkoxybenzene for **3c** and **3d**.

At the present time, it is not clear whether the attractive interaction between the DB24C8 component and the ferrocenium cation moiety is a simple ion-dipole interaction or the electrostatic interaction that accompanies electron donation from the crown ether oxygen to the ferrocenium cation. However, it is clear that the weak crown ether-ferrocenium cation interaction did not induce the transposition when there was a strong intercomponent hydrogen-bonding interaction between the wheel and the axle components. However, when the wheel component was labile on the axle, the transposition occurred even by the weak attractive intercomponent interaction induced by the redox reaction. As a consequence, rotaxanes with a controllable weak intercomponent interaction, such as **3**, constitute an excellent scaffold for the construction of a sensitive molecular device.

13.3
Reduction of Ketone by Rotaxane Bearing a Dihydronicotinamide Group

In many biological reduction reactions, NADH acts as the hydride source. Since the reaction center of NADH is dihydronicotinamide, various biomimetic reduction systems using dihydronicotinamide derivatives have been developed [10]. The general reaction mechanism of dihydronicotinamide reduction of a carbonyl compound is illustrated in Scheme 13.3.

The reduction is carried out in the presence of a Lewis acid, which promotes the reduction reaction through the coordination of the carbonyl groups of both dihydronicotinamide and the substrate [11]. The electron density of the carbonyl group of ketone decreases upon the coordination to the Lewis acid to activate the ketone. In the Lewis acid-dihydronicotinamide-ketone three-component complex, hydride is transferred from dihydronicotinamide to ketone with the assistance of the aromatization energy of dihydronicotinamide. Since the aromatization energy of dihydronicotinamide is not very large, the activation of the ketone by the Lewis acid is the key to effective reduction. Therefore, much attention has been directed at the efficient formation of the three-component complex. Although various metal salts have been examined for use as the Lewis acid, magnesium ion is known to be the most effective metal cation.

Kellogg's asymmetric reduction of ketones is one of the most impressive examples of dihydronicotinamide reduction [12] (Scheme 13.4). The magnesium ion is coordinated by chiral macrocycle **5** in which dihydronicotinamide is included. Since the complex firmly fixes the carbonyl group of ketone in the chiral environment, effective asymmetric reduction occurs.

When the dihydronicotinamide group is introduced onto the axle component of rotaxane, the dihydronicotinamide group is placed under the strong influence of the wheel component. When a crown ether is used as the wheel component, the coordination of the Lewis acid to the crown ether may enhance the reduction reaction.

Rotaxane **6** bearing dihydronicotinamide was prepared as shown in Scheme 13.5. As in the preparation of **1**, ammonium salt **2** was acylated by 3,5-dimethylbenzoic anhydride in the presence of DB24C8 to afford rotaxane **7**

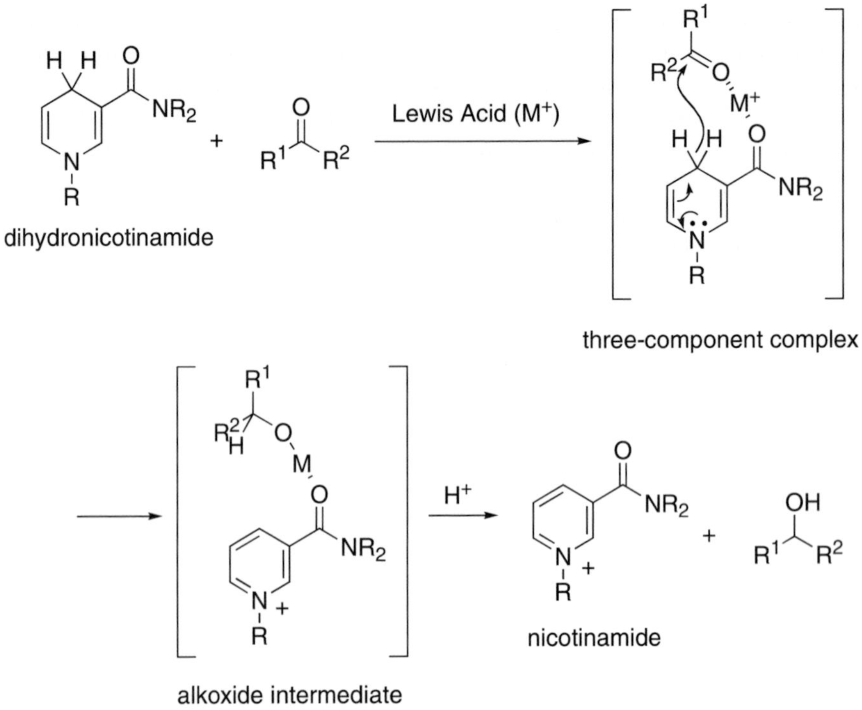
dihydronicotinamide
Lewis Acid (M+)
three-component complex
alkoxide intermediate
H+
nicotinamide

Scheme 13.3.

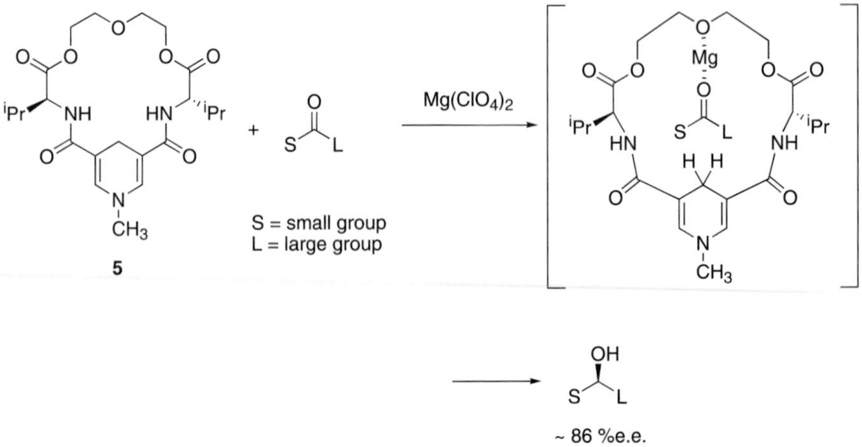
5
S = small group
L = large group
Mg(ClO4)2
~ 86 %e.e.

Scheme 13.4.

in 92% yield. The acylative neutralization of **7** with nicotinoyl chloride followed by treatment with benzyl chloride afforded intercomponent-interaction-free rotaxane **8** (oxidized form) bearing the nicotinamide group in good yield [13]. The reduction of **8** was carried out using sodium hydrosulfite in a biphase reaction system to give rotaxane **6** (reduced form) bearing the dihydronicotinamide group [14]. In the same manner, dihydronicotinamide derivative **9**, which has no wheel component, was prepared.

The reduction of methyl benzoylformate by **6** and **9** was investigated. The results are summarized in Table 13.2. Magnesium perchlorate and scandium triflate could promote the reaction when used as the Lewis acid. Rotaxane **6**

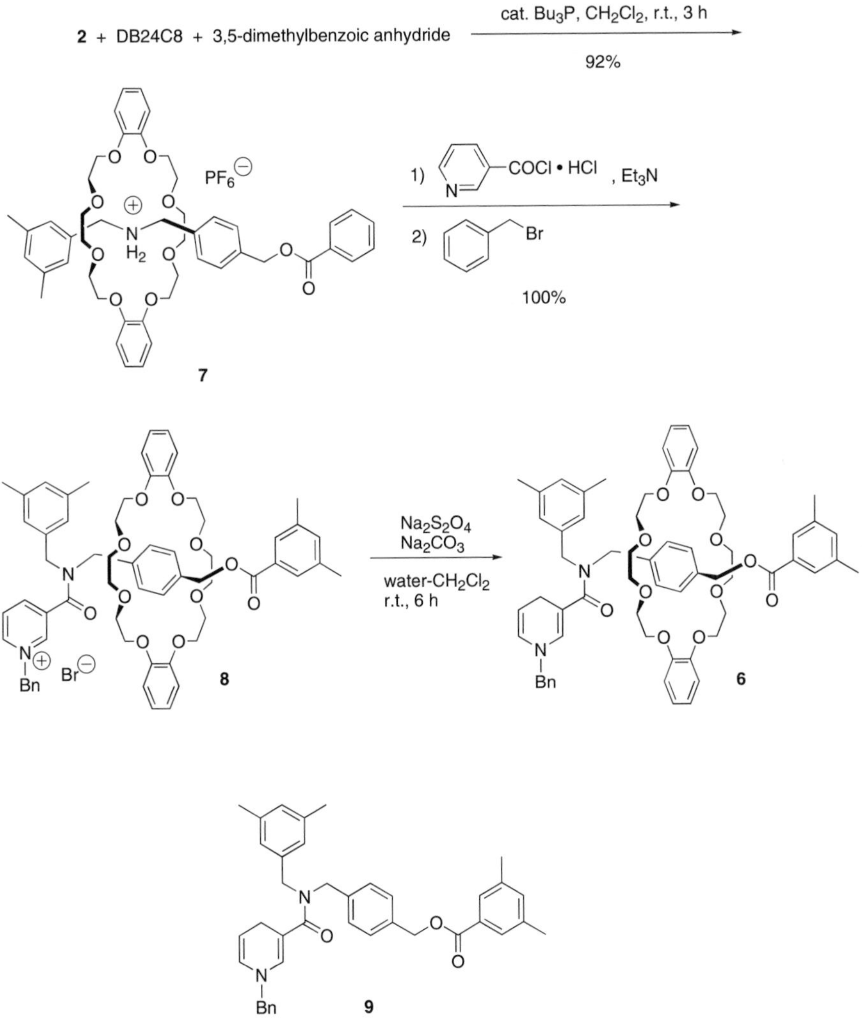

Scheme 13.5.

gave methyl mandelate in better yield than did **9**; in contrast, the addition of DB24C8 to **9** decreased the yield. The rotaxane structure enhanced the reactivity of the dihydronicotinamide moiety. It is inferred that the crown ether wheel coordinated the magnesium ion to form the effective reaction field for the reduction. This is supported by the solvent effect: when the reactions were carried out in donative solvent such as acetonitrile, rotaxane **6** was more effective than nonrotaxane **9**. Meanwhile, when the reactions were carried out in noncoordinative solvent such as dichloromethane, the difference in the yields was very small. Therefore, the wheel component of **6** provided excellent coordination circumstances for magnesium ion, and the reduction occurred even in the highly coordinative solvent.

For the formation of chelate-type magnesium complex, it seems important that the wheel crown ether is close to the dihydronicotinamide group but is free from the axle component. Since the wheel component is captured on the axle component, the wheel component can move to the appropriate position for forming the ternary complex with low activation energy without any change of the conformation. The change of the configuration during molecular recognition, which is called "induced fit," has been thought to be the key feature of enzymes that are highly effective and selective natural catalysts. However, effective molecular recognition with induced fit in an artificial receptor has been unsuccessful because of the great loss of entropy. Rotaxane can provide

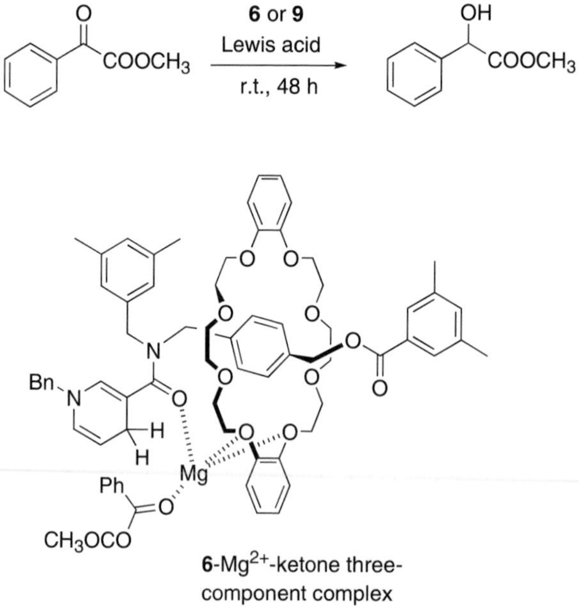

6-Mg^{2+}-ketone three-
component complex

Scheme 13.6.

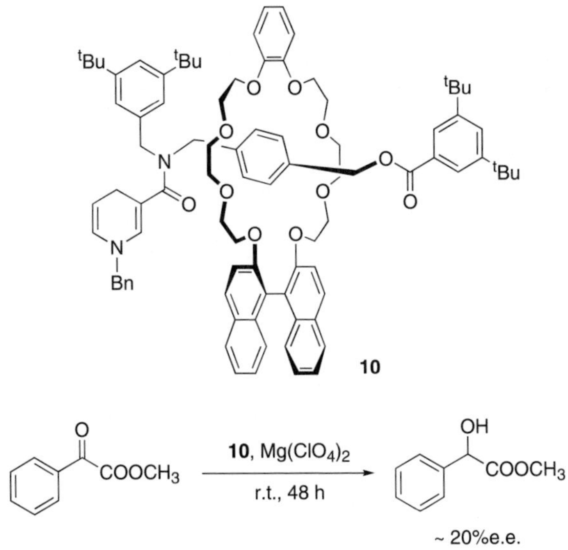

10

~ 20%e.e.

Scheme 13.7.

a novel type of reaction field where induced fit occurs with the minimum loss of entropy.

The main problem with this reaction is the low yield of the alcohol. The long reaction time decreased the yield. The mass spectra analysis of the reaction mixture revealed that the reaction of magnesium alkoxide, formed during the reduction, with **6** or **9** occurred simultaneously with the reduction. To suppress

Table 13.2. Reduction of methyl benzoylformate by **6** or **9**[a]

Dihydronicotinamide	Solvent	Lewis	Acid yield/%
6	Acetonitrile	None	4
6	Acetonitrile	$Mg(ClO_4)_2$	19
6	Acetonitrile	$Sc(OTf)_3$	16
6	Dichloromethane	$Mg(ClO_4)_2$	12
6	Acetonitrile[b]	$Mg(ClO_4)_2$	13
9	Acetonitrile	$Mg(ClO_4)_2$	11
9	Acetonitrile	$Sc(OTf)_3$	9
9	Dichloromethane	$Mg(ClO_4)_2$	10
9	Acetonitrile[c]	$Mg(ClO_4)_2$	8

[a] 0.04 mmol (calculated) of dihydronicotinamide was reacted with 0.4 mmol of methyl benzoylformate in the presence of 0.04 mmol of Lewis acid in 1.0 mL of solvent at room temperature for 48 h.
[b] Reaction was carried out for 173 h.
[c] 0.04 mmol of DB24C8 was added.

Table 13.3. Effect of proton source[a]

Dihydronicotinamide	Proton source	Yield/%
6	None	19
6	H_2O	24
6	NH_4Cl	19
9	None	11
9	H_2O	9
9	NH_4Cl	9

[a] 0.04 mmol (calculated) of dihydronicotinamide was reacted with 0.4 mmol of methyl benzoylformate in the presence of 0.04 mmol of magnesium perchlorate and 0.04 mmol of proton source in 1.0 mL of acetonitrile at room temperature for 48 h

the reaction of the alkoxide intermediate, the reduction reaction was carried out in the presence of a proton source. The results are summarized in Table 13.3.

When **6** was used as the reductant, the addition of water increased the yield, since the water hydrolyzed the alkoxide intermediate. However, when **9** was used as the reductant, the addition of water decreased the yield. The formation of the ternary complex was suppressed in the case of **9**, as reported previously [15], because the coordination of water decreased the Lewis acidity of the magnesium ion. In the case of **6**, the crown ether wheel strongly coordinated to the magnesium ion to avoid the effect of the Lewis basicity of the water. This result clearly showed the effect of the rotaxane structure.

Furthermore, the asymmetric version of dihydronicotinamide reduction in the rotaxane system was also investigated. Novel rotaxane **10**, containing a chiral wheel crown ether, was prepared. When the reduction of methyl benzoylformate was carried out, methyl madelate of 20% ee was obtained. This result clearly indicates that the reaction occurs under the strong influence of the wheel component, and the rotaxane provides an effective asymmetric reaction field through the cooperation of the wheel and the axle components.

13.4
Conclusion

The authors have introduced two types of redox reactions based on the structural characteristics of rotaxanes in the present chapter. One is the redox reaction of the ferrocene-end-capped rotaxane, where the redox potential of the ferrocene moiety decreased when the rotaxane had a crown ether wheel that was free of the intercomponent interaction. From the results obtained, it is suggested that the redox reaction of the ferrocene moiety accompanied the transposition of the wheel component on the axle.

The other redox system involves the reduction of ketone by the action of the rotaxane bearing the dihydronicotinamide group. It was concluded that the activity of the dihydronicotinamide group was considerably enhanced by the presence of the crown ether wheel. Furthermore, the enantioselective reduction was achieved using the rotaxane possessing an optically active wheel component.

These two examples reveal the significance of the nanometer-sized reaction field on rotaxane of which components are linked by flexible mechanical connection allowing to interact mutually. Since the components of rotaxane are located closely, the change of redox state of a component sensitively affects the other components. The relative configuration of components, intercomponent interaction, coordination ability, and the shape of the reaction field are strongly correlated to the redox behavior in the rotaxane system.

Acknowledgement   We are grateful for the financial support through a Grant-in-Aid for Scientific Research (Priority Area Research) from the Ministry of Education, Culture, Sports, Science, and Technology, Iketani Science and Technology Foundation, and Yazaki Memorial Foundation for Science and Technology.

13.5
References

1. Selected books: (a) In: Molecular Catenanes, Rotaxanes and Knots, Sauvage J-P, Dietrich-Buchecker C (eds) Wiley-VCH: Weinheim, 1999; (b) Raymo FM, Stoddart JF (2000) In: Templated Organic Synthesis, Diederich F, Stang PJ (eds) Wiley-VCH: Weinheim 75; (c) Steed JW, Atwood JL (2000) Supramolecular Chemistry. Wiley: Chichester, 511
2. (a) Kolchinski AG, Busch DH, Alcock NW (1995) J Chem Soc, Chem Commun 1289; (b) Ashton PR, Glink PT, Stoddart JF, Tasker PA, White AJP, Williams DJ (1996) Chem Eur J 2, 729; (c) Amabilino DB, Stoddart JF (1995) Chem Rev 95, 2725; (d) Hubin TJ, Busch DH (2000) Coord Chem Rev 200–205, 5; (e) Takata T, Kihara N (2000) Rev Heteroat Chem, 22 197; (f) Kihara N, Takata T (2001) J Synth Org Chem Jpn 59, 206
3. (a) Balzani V, Venturi M, Credi A (2003) In: Molecular Devices and Machines. Wiley-VCH: Weinheim; (b) Feringa BK (ed) (2001) In: Molecular Switches. Wiley-VCH, Weinheim
4. Kihara N, Tachibana Y, Kawasaki H, Takata T (2000) Chem Lett 506
5. Pease AR, Jeppesen JO, Stoddart JF, Luo Y, Collier CP, Heath JR (2001) Acc Chem Res 34, 433
6. (a) Ballardini R, Balzani V, Credi A, Gandolfi MT, Venturi M (2001) Acc Chem Res 34, 445; (b) Segura JL, Martín N (2001) Angew Chem Int Ed 40, 1372; (c) Jeppesen JO, Nielsen KA, Perkins J, Vignon SA, Di Fabio A, Ballardini R, Gandolfi MT, Venturi M, Balzani V, Becher J, Stoddart JF (2003) Chem Eur J 9, 2982
7. Kihara N, Hashimoto M, Takata T (2004) Org Lett 6, 1693. Other examples of ferrocene-containing interlocked compounds: (a) Kaifer AE (1999) Acc Chem Res, 32, 62; (b) Isnin R, Kaifer AE (1993) Pure Appl Chem 65, 495; (c) Benniston AC, Harriman A (1993) Angew Chem Int Ed Engl 32, 1459; (d) Isnin R, Kaifer AE (1991) J Am Chem Soc 113, 8188; (e) Liu J, Gomez-Kaifer M, Kaifer AE (2001) Mol Machines Motors 99, 141

8. (a) Kawasaki H, Kihara N, Takata T (1999) Chem Lett, 1015; (b) Kihara N, Nakakoji N, Takata T (2002) Chem Lett 924; (c) Watanabe N, Yagi T, Kihara N, Takata T (2002) Chem Commun 2720

9. (a) Connelly NG, Geiger WE (1996) Chem Rev 96, 877; (b) Guillon C, Vierling P (1994) J Organomet Chem 464, C42

10. (a) Burgess VA, Davies SG, Skerlj RT (1991) Tetrahedron: Asymm 2, 299; (b) Eisner U, Kuthan J (1972) Chem Rev 72, 1

11. (a) Ohnishi Y, Kagami M, Ohno A (1975) J Am Chem Soc 97, 4766; (b) Shinkai S, Ikeda T, Hamada H, Manabe O, Kunitake T (1977) J Chem Soc, Chem Commun 848; (c) Gase RA, Pandit UK (1979) J Am Chem Soc 101, 7059; (d) Hood RA, Prince RH (1979) J Chem Soc Chem Commun 163; (e) Ohno A, Yasui S, Oka S (1980) Bull Chem Soc Jpn 53, 2651; (f) Ohno A, Shio T, Yamamoto H, Oka S (1981) J Am Chem Soc 103, 2045; (g) Fukuzumi S, Fujii Y, Suenobu T (2001) J Am Chem Soc 123, 10191

12. (a) Kellogg RM (1984) Angew Chem Int Ed Engl 23, 782; (b) Talma R, Jouin AG, Vries JGD, Troostwijk CB, Buning GHW, Waninge JK, Visscher J, Kellogg RM (1985) J Am Chem Soc 107, 3981

13. Wallenfels von K, Gellrich M (1959) Liebigs Ann Chem 621, 149

14. (a) Kosower EM, Bauer SW (1690) J Am Chem Soc 82, 2191; (b) Ohno A, Kimura T, Kim SG, Yamamoto H, Oka S (1977) Bioorg Chem 6, 21; (c) Ohno A, Ikeguchi M, Kimura T, Oka S (1979) J Am Chem Soc 101, 7036; (d) Ohno A, Kimura T, Yamamoto H, Kim S-G, Oka S, Ohnishi Y (1977) Bull Chem Soc Jpn 50, 1535; (e) Burgess VA, Davies SG, Skerlj RT, Whittaker M (1992) Tetrahedron Asymm 3, 871; (f) Kanomata N, Nakata T (1997) Angew Chem Int Ed 36, 1207

15. Ohno A, Kobayashi H, Nakamura K, Oka S (1985) Tetrahedron Lett 24, 1263

Chapter 14

Metal-Containing Star and Hyperbranched Polymers

Masami Kamigaito

Department of Applied Chemistry, Graduate School of Engineering, Nagoya University, Nagoya 464-8603, Japan

Abbreviations

CL ε-caprolactone
LA   Lactide
MMA   Methyl methacrylate
tBMA *Tert*-butyl methacrylate
T_m Melting temperature
T_g Glass transition temperature

14.1
Introduction

One of the best ways to construct excellent redox systems with synthetic polymers is to put a metal into a well-defined nanospace built by the polymer chains, as suggested by the outstanding functions of natural macromolecules such as metalloproteins and enzymes. Synthetic metal-containing polymers [1–3] have already been prepared by attaching metals to various polymers via covalent linkage or coordination, although the nanoenvironments around the metals have not been well arranged in comparison to the natural ones. These synthetic metal-containing polymers, which are called polymer-metal complexes, polymer-supported metal complexes, macromolecular metal complexes, organometallic polymers, metallopolymers, etc. depending on the situation, can be employed for many applications like catalysts, conductive polymers, photochemical molecular devices, sensors, etc., whose functions are influenced by the redox nature of the metals therein. To construct more efficient systems, more precise arrangements of the polymer chains around the redox-active metal sites are needed in some cases. Therefore, the use of an architecturally controlled polymer along with an appropriate metal would lead to novel functional materials based on well-constructed redox-active sites.

Although the controlled structure of synthetic polymers is far inferior to that of natural macromolecules, recent advances in polymer synthesis have enabled the preparation of various architecturally or three-dimensionally controlled polymers. Typical examples of these polymers include globular macromolecules like dendrimers, hyperbranched polymers, and star or star-shaped polymers (Fig. 14.1).

These branched polymers basically consist of a number of arm chains radially attached to a central core moiety. Thus, they are sometimes called "star-branched polymers"; star polymers consist of linear arm chains and branched cores, whereas dendrimers and hyperbranched polymers have "tree-like" branched chains. The degree of branching in these globular polymers increases in the following order: star polymers, hyperbranched polymers, and dendrimer. In addition to the difference in the degree of branching, they possess different spatial and segment densities as well. For example, on going from the inside (core) to the outside (surface or periphery), the spatial and segmental densities decrease in star polymers but increase in dendrimers, which means that star polymers are denser inside with more free volumes for arm chains around the periphery, whereas dendrimers contain some voids near the core and a highly congested outer surface. Furthermore, star and hyperbranched polymers are synthesized by polymerization reaction in one-pot systems, whereas dendrimers require multistep condensation and other reactions where separation and purification of intermediates are needed. The former two polymers are much more accessible and more suitable for applications than the latter. However, the former are the mixtures of a series of analogous macromolecules with similar but different molecular weights, while the latter has basically a single molecular weight and the most regulated structure among them.

Irrespective of such differences, all these have almost globular shapes that have special nanospaces constructed by a number of arm chains. Therefore, putting a metal into the inner or the outer nanospaces would give some excellent redox functions to these polymers. There have been a relatively large number of original papers and excellent reviews concerning the metal-containing dendrimers [4–9] in comparison to star and hyperbranched polymers. This chapter is thus devoted to a review of the synthesis and the functions of the metal-containing star and hyperbranched polymers.

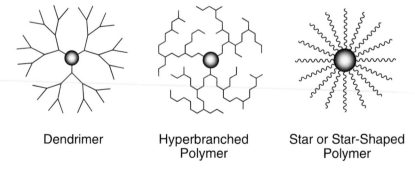

Dendrimer   Hyperbranched   Star or Star-Shaped
                    Polymer   Polymer

Fig. 14.1. Dendrimer, hyperbranched polymer, and star or star-shaped polymer

14.2
Metal-Containing Star Polymers

The synthetic strategy of the metal-containing star polymers is based on the combination of star polymer synthesis and the coordination chemistry of metals. Among a variety of synthetic methods for star polymers, living polymerization is one of the best in terms of well-defined architectures of the resulting star polymers (i.e., overall molecular weights, number and length of arm chains per molecule). There are three general methods for the synthesis, as illustrated in Scheme 14.1: (a) a living polymerization with a multifunctional initiator, (b) a coupling reaction of linear living polymers with a multifunctional coupling agent (terminator), and (c) a linking reaction of linear living polymers with a divinyl compound (polymer linking method).

In principle, the star polymers prepared by methods (a) and (b) have predetermined numbers (f) of arms per molecule, but the f is relatively small (mostly below 10) due to synthetic difficulties of the multifunctional initiators and coupling agents. In contrast, the linking method (c) leads to star polymers with a crosslinked microgel core and a large number of arms (f up to several hundred), though f certainly involves a statistical distribution. All three methods are also applicable for the synthesis of metal-containing star polymers.

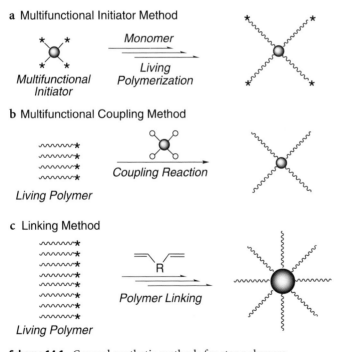

a Multifunctional Initiator Method

Multifunctional
Initiator

Monomer

Living
Polymerization

b Multifunctional Coupling Method

Living Polymer

Coupling Reaction

c Linking Method

Living Polymer

R

Polymer Linking

Scheme 14.1. General synthetic methods for star polymers

14.2.1
Metal-Containing Star Polymers with a Small and Well-Defined Number of Arms

Metal-containing star polymers with a small and well-defined number of arms have been obtained by a metal-complex-based multifunctional initiator possessing plural initiating sites in its ligands [method (a) in Scheme 14.2]. Another method for a similar star polymer is to prepare the polymers with the end-functional group that can subsequently become the ligand for the metal complex [method (b)]. Both methods give metal-centered star polymers, where the arm polymers are attached to the central metals via coordination. Fraser et al. coined the former as the metalloinitiator or divergent synthesis while the latter as the macrolignad chelation or convergent synthesis [10].

The first example of a metal-centered star polymer was reported by Chujo and coworkers. They prepared the star polymers (1) based on the complexation of bipyridyl-terminated poly(oxylethylene)s with ruthenium(II) ion (Fig. 14.2) [11]. The end-functionalized linear polymer was derived from commercially available poly(oxyethylene) monomethyl ether and 2,2′-bypyridine and coordinated to Ru(II) to afford the ruthenium-centered three-arm star polymers. They also prepared similar star polymers with three arms (2) from diketone-terminal poly(oxyethylene) and Cr(III) [12] and with four arms (3) from pyridine-terminal poly(oxyethylene) and Ru(II) [13]. The star polymers with a Ru-core exhibited metal-to-ligand charge transfer absorption in the visible region.

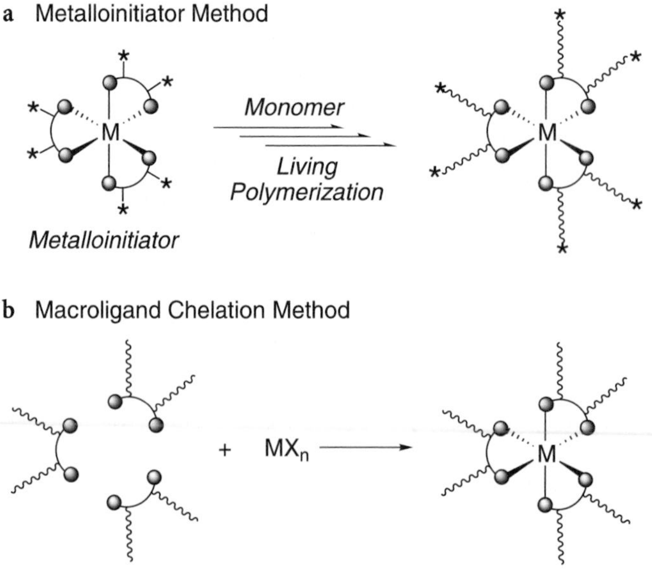

a   Metalloinitiator Method

Metalloinitiator

b   Macroligand Chelation Method

Scheme 14.2. General synthetic methods for metal-centered star polymers with a small and well-defined number of arms

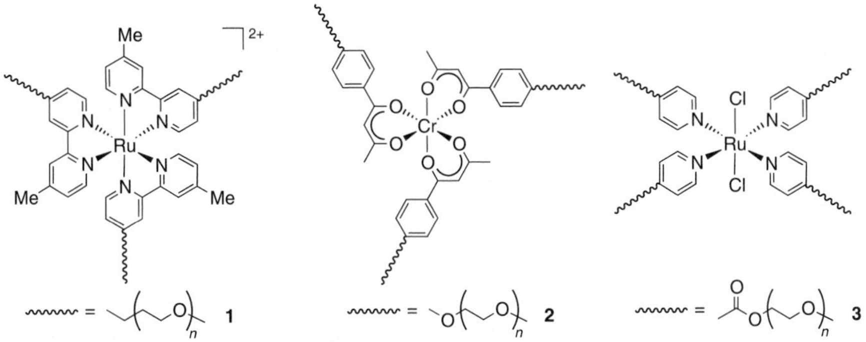

Fig. 14.2. Metal-centered star polymers with poly(ethylene glycol) arms

Fraser's group extensively prepared a series of metal-centered star polymers by combining living polymerization techniques with coordination chemistry of bipyridyl groups onto metal ions [10, 14–22]. These star polymers are varied in terms of not only central metals (Ru and Fe) but also arm numbers (2, 3, 4, and 6) and arm polymers [hydrophilic and hydrophobic polyoxazolines, polystyrene, poly(methyl methacrylate), poly(ε-caprolactone), poly(lactide), block copolymers between them, their heteroarm polymers, etc.]. There are two general methods for the synthesis: the metalloinitiator (divergent synthesis) and macroligand chelation (convergent synthesis) methods. In the former method, ligation of the bipyridyl ligands onto the metals was done prior to the formation of the arm polymers, while the ligation was conducted after the formation of the arm polymers in the latter. The living radical polymerization and living ring-opening polymerization have permitted the synthesis because they are compatible with the bipyridyl groups and/or the metals.

Fraser et al. first reported the iron-centered star polyoxazoline with six arms (**4**) by living ring-opening polymerization of 2-ethyl-2-oxazoline initiated with the metalloinitiator of Fe(II) coordinated with three ligands of (4,4′-bromomethyl)-2,2′-bipyridine (Fig. 14.3) [14]. The star polymer solutions and films showed violet, which was lost by addition of aqueous K_2CO_3 solution or by heating to $\sim 210\,°C$, respectively, due to the labile nature of the ligand for iron. The violet color returns on addition of $(NH_4)_2Fe(SO_4)_2$ to the solution or by cooling the film. Similar iron-centered star polymers with oxazoline-based block copolymer arms consisting of glassy-crystalline or hydrophilic-hydrophobic combinations were prepared using 2-ethyl-, 2-phenyl-, and 2-undecyl-2-oxazolines as monomers (**5–8**) [16]. The block copolymers showed red-violet and T_g and T_m values that correlate well with those observed for the respective homopolymers. Biocompatible polyesters such as polylactide [poly(LA)] and polycaprolactone [poly(CL)] (**9–12**) with an iron core were also obtained by the hydroxy-functionalized initiator instead of the bromomethyl group and exhibited similar thermochromic reversible bleaching [21].

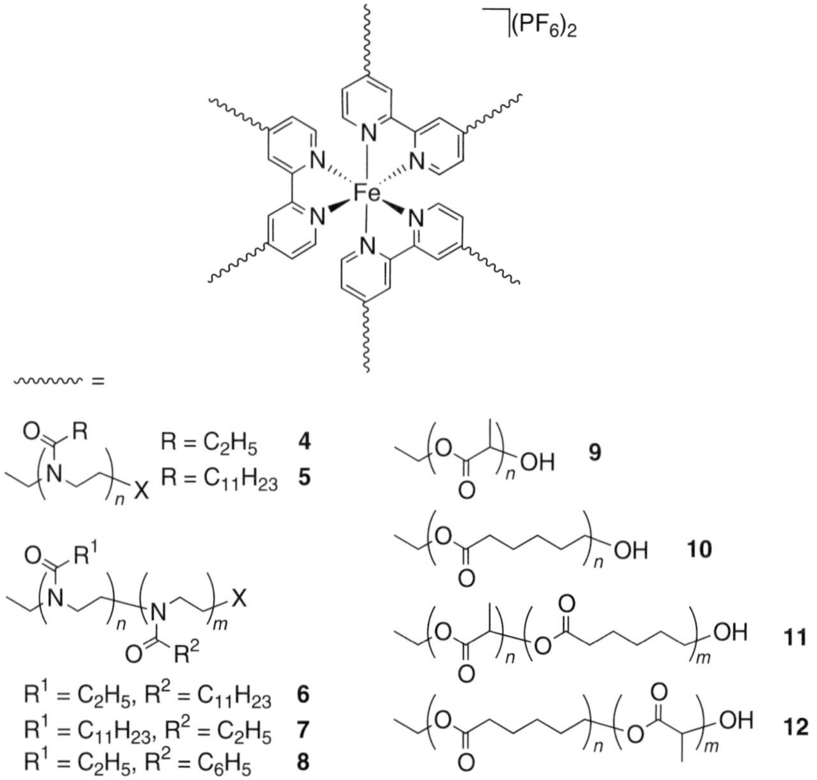

Fig. 14.3. Iron-centered star polymers with 6 arms

A series of ruthenium-centered star polymers (13–28) have been synthesized not only by the metalloinitiator method but also by the macroligand chelation method (Fig. 14.4). In the former method, the researchers prepared the multifunctional ruthenium-centered initiators and then employed them for the copper- and nickel-catalyzed living radical polymerization of styrene, methyl methacrylate (MMA), and *tert*-butyl methacrylate (tBMA) or for the ring-opening polymerization of oxazoline and LA [15, 17, 19, 20, 22]. These orange polymers are luminescent and thus potentially suitable for sensor applications. Biocompatible poly(LA) and poly(acrylic acid), derived from the hydrolysis of poly(tBMA), and their amphiphilic block copolymers were also used as the arm polymers in conjunction with the luminescent ruthenium tris(bipyridine) center to be applicable to drug delivery and imaging functions. The macroligand chelation method was also employed for ruthenium-centered star polymers with various arms including polystyrene, poly(MMA), poly(CL), poly(LA), and poly(CL)-b-poly(LA) [10, 18, 21]. The method has further permitted the synthesis of heteroarmed star polymers (20, 21) by using solvent polarity to control chain conformation and reactivity at the metal core [18].

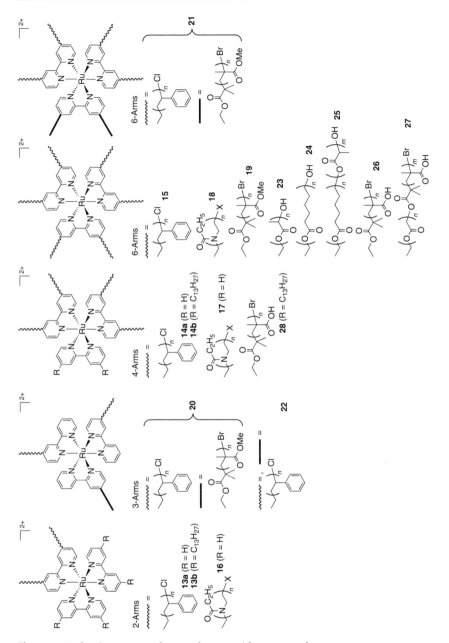

Fig. 14.4. Ruthenium-centered star polymers with 2, 3, 4, and 6 arms

An europium-centered triarmed star poly(LA) (**29**) was prepared by the macroligand chelation method between dibenzoylmethane-terminated poly(LA)s and Eu(III)Cl$_3$ (Fig. 14.5) [23]. The three-armed star polymer was

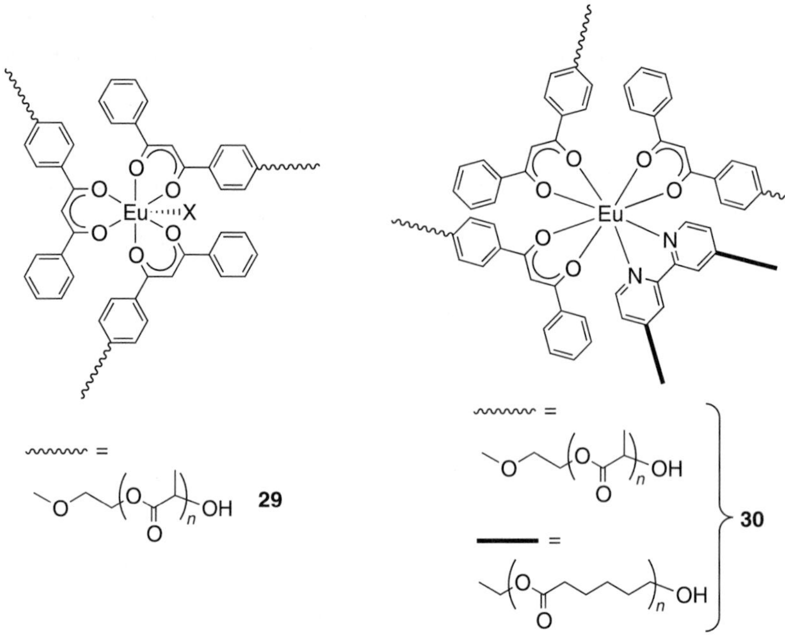

Fig. 14.5. Europium-centered star polymers

further converted into a five-armed one (**30**) on addition of poly(CL)s possessing the bypyridyl group in the middle of the chain. The resulting heteroarm star polymers can be regarded as a kind of block copolymer [poly(LA) and poly(CL)] with a luminescent europium complex at the block junction. The film of the block copolymers underwent microphase separation due to the incompatibility of poly(LA) and poly(CL) to result in the localization of the metal at the microdomain boundary.

14.2.2
Metal-Containing Star Polymers with a Large and Statistically Distributed Number of Arms

The third star polymer synthetic method (c) in Scheme 14.1, based on the linking reaction of living polymer chains with divinyl compounds, can give relatively large-sized star polymers with a large number of arm chains in comparison with those obtained with the former two methods, although the number of arms is statistically distributed. This method has thus been employed for putting a large number of metal ions or often metallic nanoparticles into star polymers rather than one metal atom into strictly well-defined sites.

The first example was reported of a star-block copolymers prepared from the linking reaction of poly(styrene)-*block*-poly(2-vinylpyridine) with ethylene glycol dimethacrylate in living anionic polymerization [24]. The inner poly(2-vinylpyridine) segment can become a ligand for metal ions and a scaffold for the preparation and stabilization of metallic nanoparticles, while the outer polystyrene can help solubilize the metals in organic solvents and prevent aggregation of the nanoparticles. Mixing the solution of the star polymers and $HAuCl_4 \bullet H_2O$ induced coordination of gold ions to the vinylpyridine-units to give a yellowish colored solution (**31a**) (Fig. 14.6). Addition of a reducing agent, N_2H_4, yielded a deep purple-red colored solution, attributed to the formation of small gold nanoparticles (**31b**) (4–7 mm diameter). The gold nanoparticles were stable for a long time without aggregation.

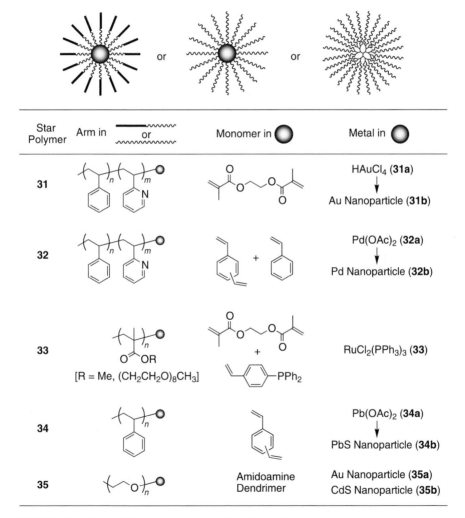

Fig. 14.6. Metal-containing star polymers with microgel core

A similar star block copolymer was synthesized by living radical block copolymerization of styrene and 2-vinylpyridine followed by linking of the block copolymers with divinylbenzene in the presence of styrene [25]. Fréchet and Hawker introduced $Pd(OAc)_2$ into the inner poly(2-vinylpyridine) segment and reduced the metal ions by ethanol at elevated temperature to form palladium nanoparticles (**32b**) with diameters of 2–3 nm. The nanoparticles stabilized by the star polymers possesed a sterically isolated internal nanoenviroment in which a variety of reactions could take place. For example, the palladium-catalyzed hydrogenation and Heck reaction were studied with the Pd-containing star polymers, which showed moderate activity (Eqs. 14.1 and 14.2). The star polymer permitted reaction in nonpolar solvents due to a nonpolar polystyrene corona and facilitated catalyst removal and recycling by simple precipitation of the star polymers.

$$\text{(14.1)}$$

$$\text{(14.2)}$$

Eqs. 14.1, 14.2

Another star polymer containing ruthenium in its core was prepared by $RuCl_2(PPh_3)_3$-catalyzed living radical polymerization [26]. The strategy was based on the direct encapsulation of the ruthenium catalyst during the linking reaction of the living poly(MMA) chains with ethylene glycol dimethacrylate in the presence of a phosphine-substituted styrenic monomer $[CH_2=CHPhPPh_2]$. The phosphine monomer was copolymerized with a divinyl compound upon linking to be incorporated into the core where the ruthenium complexes were captured via coordination to the phosphine monomer units. The star polymers (**33**) had diameters of 10–20 nm and 20–30 arm chains on average, while the diameter of the core was around 2–3 nm. The ruthenium-containing star polymers catalyzed the oxidation reaction of alcohols into ketones and vice versa with moderate activities (Eqs. 14.3 and 14.4). The effects of star polymer sizes and the nature of the arm chains, hydrophobic or hydrophilic, on the redox reactions were also investigated [27].

A carboxylate-functionalized star polymer with polystyrene arms was synthesized by living radical polymerization and subsequently used as a scaffold for the formation of PbS nanoparticles, which may be applied to the optoelectronic materials [28]. Pb ion was loaded by refluxing the star polymers in the presence of $Pb(OAc)_2$, and PbS nanoparticle (**34b**) with an average diameter of 3.0 nm was obtained by exposure of the star polymers to a H_2S atmosphere. Nanoparticle stabilization was also carried out by a dendrimer-star polymer, prepared by grafting poly(ethylene glycol) arms

$$\text{(14.3)}$$

$$\text{(14.4)}$$

Eqs. 14.3, 14.4

onto a poly(amidoamine) dendrimer, which gave Au and CdS nanoparticles (**35**) 1 to 6 nm in diameter [29].

14.3
Metal-Containing Hyperbranched Polymers

A hyperbranched polymer can be distinguished from a star polymer by the presence of branching in the arm chains and is considered the intermediate between a star polymer and a dendrimer in terms of degree of branching. The structure of hyperbranched polymers is less perfect than that of dendrimers and their size and number of arms and branching points are statistically distributed similar to star polymers. They can be easily prepared by single-step reactions, which may be beneficial for applications, in contrast to tedious multistep synthesis of the dendrimers.

The first example of metal-containing hyperbranched polymers was re-ported by Frey et al. [30], who has developed a convenient pathway to well-defined hyperbranched poly(glycerol)s by the anionic polymerization of gly-cidol under slow monomer addition conditions [31]. They used poly(glycerol) and further esterified about 60% of the OH groups with palmitoyl chloride ($C_{15}H_{31}COCl$) to prepare the amphiphilic hyperbranched polymers, which was then utilized as a scaffold for Pd(II) ion. The captured palladium ions were reduced by hydrogen to afford a palladium cluster (**36b**) 2 to 5 nm in diameter, which was dependent on the size of the amphiphilic hyperbranched polymer. The polymer-stabilized Pd colloids can be employed as the catalyst for the hydrogenation of cyclohexene with moderate activity, and these encapsulated catalysts can be separated easily by dialysis without significant loss of activity.

Another encapsulation of metal catalysts was done by noncovalent inter-action of a sulfonated pincer platinum(II) complex with an amphiphilic hy-perbranched polymer (**37**) [32]. The Pt-containing hyperbranched polymer catalyzed double Michael addition reactions between ethyl cyanoacetate and methyl vinyl ketone (Eq. 14.5). A similar hyperbranched polymer (**38**) with covalently immobilized pincer platinum(II) catalysts was also prepared and used for the same reaction [33]. Although the catalytic activity was gradu-ally decreased in a specific order – nonimmobilized > covalently immobi-lized > noncovalently-immobilized catalysts – the immobilized catalysts were recovered in near quantitative yield by dialysis. An optically active hyper-

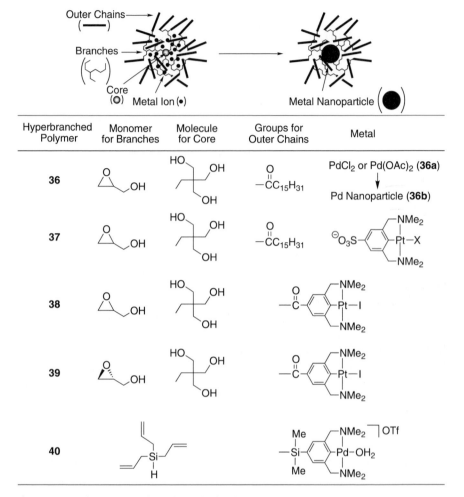

Fig. 14.7. Metal-containing hyperbranched polymers

branched poly(glycerol) was synthesized by anionic polymerization of a chiral monomer, (−)- or (+)-glycidol, and was employed as chiral microenvironments and as scaffolds for noncovalent and covalent immobilization of the platinum(II) complexes [34]. The immobilized catalysts (**39**) showed moderate activity in Michael additions between methyl vinyl ketone and ethyl α-cyanopropionate (Eq. 14.6), although no enantiomeric excess was observed. The researchers also synthesized a silane-based hyperbranched polymer with covalently immobilized platinium(II) catalyst (**40**) and used it for aldol condensation (Eq. 14.7) between benzaldehyde and methyl isocyanoacetate [35]. The catalytic activity was comparable to that of the dendrimer-immobilized platinum catalyst.

Eqs. 14.5–14.7

There are also other reports on the metal-containing hyperbranched polymers by other researchers. A metal-containing hyperbranched polyamidoamine with a structure and molecular weight similar to the fourth-generation polyamidoamine dendrimer was employed for a scaffold for gold nanoparticles [36]. The size of the nanoparticles was larger than those derived from the fourth-generation dendrimer (4 vs. 2 nm) probably due to the open structure of the former than the latter. An aramid-based hyperbranched polymer was also used as a polymeric support for palladium(0) nanoparticles and further employed as a hydrogenation catalyst [37]. Another metal-containing hyperbranched polymer was synthesized from a monomer containing two chlorotricarbonyl rhenium(I) bypyridine moieties and a stilbazole ligand via the coordination reaction between the rhenium and the ligand in one single step [38]. This polymer may be used as an optoelectronic device. A dendriraft polystyrene, a kind of hyperbranched polymer, was also hybridized with ferrocenes to give a hyperbranched polymer with ferrocenyl units at the periphery [39]. The electrochemical properties studied by cyclic voltammetry showed that all ferrocenyl groups in dendrigrafts can participate to the redox cycles while only a small fraction of ferrocene groups is active in linear polystyrene.

14.4
Concluding Remarks

Various metal-containing star and hyperbranched polymers have been synthesized and employed as functional polymers based on the redox functions of the metals in their specific nanospaces. Although their functions are still limited and less prominent than those of the metal-containing dendrimers due to their short history and less perfect three-dimensional structures, it is expected that they will prove to be more accessible functional materials because of their less cumbersome synthetic procedures. Further developments in metal-containing star and hyperbranched polymers can be anticipated in terms of practical applications.

14.5
References

1. Wöhrle D, Pomogailo AD (2003) (eds) Metal complexes and metals in macromolecules. Wiley-VCH, Weinheim
2. Abd-El-Aziz AS, Carraher CE Jr, Pittman CU Jr, Sheats JE, Zeldin M (2003) Macromolecules containing metal and metal-like elements, vol 1. Wiley-Interscience, Hoboken, NJ
3. Abd-El-Aziz AS (2002) Macromol Rapid Commun 23:995
4. Crooks RM, Zhao M, Sun L, Chechik V, Yeung LK (2001) Acc Chem Res 34:181
5. Inoue K (2000) Prog Polym Sci 25:453
6. Astruc D, Chardac F (2001) Chem Rev 101:2991
7. van Heerbeek R, Kamer PCJ, van Leeuwen PWNM, Reek JNH (2002) Chem Rev 102:3717
8. Dijkstra HP, van Klink GPM, van Koten G (2002) Acc Chem Res 35:798
9. Astruc D, Heuzé K, Getard S, Méry D, Nlate S, Plault L (2005) Adv Synth Catal 347:329
10. Wu X, Fraser CL (2000) Macromolecules 33:4053
11. Chujo Y, Naka A, Krämer M, Sada K, Saegusa T (1995) J Macromol Sci Pure Appl Chem A32:1213
12. Naka K, Konishi G, Kotera K, Chujo Y (1998) Polym Bull 41:263
13. Naka K, Kobayashi A, Chujo Y (1997) Macromol Rapid Commun 18:1025
14. Lamba JJS, Fraser CL (1997) J Am Chem Soc 119:1801
15. Collins JE, Fraser CL (1998) Macromolecules 31:6715
16. McAlvin JE, Fraser CL (1999) Macromolecules 32:1341
17. McAlvin JE, Fraser CL (1999) Macromolecules 32:6925
18. Fraser CL, Simith AP, Wu X (2000) J Am Chem Soc 122:9026
19. Johnson RM, Corbin PS, Ng C, Fraser CL (2000) Macromolecules 33:7404
20. Wu X, Collins JE, McAlvin JE, Cutts RW, Fraser CL (2001) Macromolecules 34:2812
21. Corbin PS, Webb MP, McAlvin JE, Fraser CL (2001) Biomacromolecules 2:223
22. Johnson RM, Fraser CL (2004) Biomacromolecules 5:580
23. Bender JL, Corbin PS, Fraser CL, Metcalf DH, Richardson FS, Thomas EL, Urbas AM (2002) J Am Chem Soc 124:8526
24. Youk JH, Park M-K, Locklin J, Advincula R, Yang J, Mays J (2002) Langmuir 18:2455
25. Bosman AW, Vestberg R, Heumann A, Fréchet JMJ, Hawker CJ (2003) J Am Chem Soc 125:715
26. Terashima T, Kamigaito M, Baek K-Y, Ando T, Sawamoto M (2003) J Am Chem Soc 125:5288
27. Terashima T, Ando T, Kamigaito M, Sawamoto M (2003) Polym Prepr Jpn 52:1530
28. Du J, Chen Y (2004) Macromolecules 37:3588
29. Hedden RC, Bauer BJ, Smith AJ, Gröhn F, Amis E (2002) Polymer 43:5473
30. Mecking S, Thomann R, Frey H, Sunder A (2000) Macromolecules 33:3958
31. Sunder A, Hanselmann R, Frey H, Mülhaupt R (1999) Macromolecules 32:4240
32. Slagt MQ, Stiriba S-E, Klein Gebbink RJM, Kautz H, Frey H, van Koten G (2002) Macromolecules 35:5734
33. Stiriba S-E, Slagt MQ, Kautz H, Klein Gebbink RJM, Thoman R, Frey H, van Koten G (2004) Chem Eur J 10:1267
34. Slagt MQ, Stiriba S-E, Kautz H, Klein Gebbink RJM, Frey H, van Koten G (2004) Organometallics 23:1525
35. Schlenk C, Kleij AW, Frey H, van Koten G (2000) Angew Chem Int Ed 39:3445
36. Pérignon N, Mingotaud A-F, Marty J-D, Rico-Lattes I, Mingotaud C (2004) Chem Mater 16:4856

37. Tabuani D, Monticelli O, Chincarini A, Bianchini C, Vizza F, Moneti S, Russo S (2003) Macromolecules 36:4294
38. Tse CW, Cheng KW, Chan WK, Djurisic AB (2004) Macromol Rapid Commun 25:1335
39. Bernard J, Schappacher M, Ammannati E, Kuhn A, Deffiuex A (2002) Macromolecules 35:8994

Electronic Properties of Helical Peptide Derivatives at a Single Molecular Level

Shunsaku Kimura · Kazuya Kitagawa · Kazuyuki Yanagisawa · Tomoyuki Morita

Department of Material Chemistry, Graduate School of Engineering, Kyoto University, Kyoto-Daigaku-Katsura, Nishikyo-ku, Kyoto 615-8510, Japan

Summary Electron transfer in helical peptide systems which were self-assembled on gold is described in this chapter. Electron transfer mechanism for the long distance through helical peptides is explained by the electron hopping between amide bonds. When naphthyl groups were linearly arranged along the helix scaffold, the electron hopping between naphthyl groups via anion radical was also observed. The aromatic linker between gold and the peptide accelerated electron transfer in the molecular system, and an intervening redox species in the helix also promoted electron transfer. The effects of helix dipole moment on electron transfer are also discussed.

15.1
Molecular Electronics

In the field of microelectronics, the downsizing of electronic components and their dense integration are making constant progress year by year to attain faster operation, higher performance, lower price, lower power consumption, etc. In the 1960s, Gordon Moore, one of the founders of Intel, predicted that the number of transistors in an integrated circuit would double every 18–24 months [1, 2]. This prediction has become known as "Moore's Law," and in fact, development in the semiconductor industry has followed this law over the past 40 years, owing to many scientific and technological innovations [3]. Now the feature size of the chip has decreased to below 100 nm. Miniaturization of silicon-based semiconductor devices, which is essential for supporting ongoing modern information technology, is, however, going to face intrinsic problems such as the technical limitations of microfabrication using photolithography and the basic principle of device operation. Conventional photolithographic methods, on which microelectronics is based, are limited in resolution due to optical diffraction. It is possible to achieve finer patterning using advanced lithographic methods such as X-ray lithography and extreme UV (EUV) lithography [4–6]. However, the high cost of these methods is economically prohibitive. Another operational problem arises from the quantum effects due to nanometer-sized devices. For example, in the case of field-effect transistors (FETs), the gate electrode is insulated by an oxide layer from the current channel. This oxide layer should also continue to decrease in thickness if FETs continue to shrink. According to Moore's Law, FETs should be equipped with 1 nm thick insulating layer by 2012. At that time, quantum tunneling of a current across the oxide layer cannot be ruled out, meaning that it will no longer insulate [7]. There is another problem related to interconnecting wires to electrodes at such dimensions. It is therefore necessary to develop a new approach to device fabrication to replace the current method of "top-down" lithography.

The ultimate size of microelectronic circuit components will reach the molecular or atomic scale. These single or groups of molecules should be equipped with some useful function to become molecular-based devices. Further, these molecular components should be integrated into circuits, probably by means of a "bottom-up" method, because of the technical limitation of the "top-down" method as described above. The study of electronic and assembling properties of molecules is thus key for the emerging field of molecular electronics.

Richard Feynman, the prominent physicist and visionary, was the first to propose a concept of molecular electronics in his famous speech in 1959.[1] The first example of a single molecular electronic device was proposed by Aviram and Ratner in 1974 [8]. They suggested that individual molecules consisting of a donor-bridge-acceptor (D-B-A) moiety would behave as molecular rectifiers under an electrical bias between two electrodes. Since their proposal, various molecules and nanostructured materials have been intensively examined as electrical wires, diodes, switching components, etc.

Molecular electronics, in a broad sense, can be defined as technology utilizing single molecules, small groups of molecules, carbon nanotubes, or nanoscale metallic or semiconductor wires to perform electronic functions [9–14]. Molecular electronic devices, which are believed to replace current silicon-based semiconductor devices, should include wire, switch, memory, and rectifier at the nanometer scale. The molecules should assume a well-defined structure to elicit the electronic functions. Further, the molecules should be self-organized to form specific structures through specific intermolecular interaction such as hydrogen bonding and van der Waals interaction [15, 16]. Molecular recognition also plays an important role in the formation of specific interconnections between component molecules. There are several excellent reviews on the current state of knowledge of molecular electronics [9–14, 16–21].

15.2
Electron Transfer Through Molecules

Electron transfer through organic molecules should be the most basic process for molecular electronics, and it has already been extensively explored

---

[1] Richard Feynman gave the speech entitled "There's Plenty of Room at the Bottom" on 29 December 1959 at the annual meeting of the American Physical Society at the California Institute of Technology. The transcript of this speech has been made available at http://www.zyvex.com/nanotech/feynman.html.

to rationalize the electron transfer processes, for example, involved in photosynthesis and oxidative phosphorylation. Now electron transfer through a single molecule can be measured directly by scanning probe methods, and the electron transfer mechanism through single or small groups of molecules inserted between two electrodes has become a highly controversial topic.

Studies of electron transfer from a donor (D) to an acceptor (A) through a molecular bridge (B) (D-B-A) system in solution have provided a rate of electron transfer between D to A [22,23]. A large number of experimental data have been accumulated to relate electron transfer rates with molecular parameters. Generally, the rate of electron transfer (k_{ET}) decreases exponentially with the distance between D and A, as shown by (15.1):

$$k_{ET} = k_0 \exp(-\beta d_{DA}), \tag{15.1}$$

where k_0, β, and d_{DA} are the preexponential factor, the structure-dependent attenuation factor, and the distance between D and A, respectively. The attenuation factor β is considered to represent the molecular ability for electron transfer as medium.

A pioneering study of electron tunneling through molecular films between two electrodes is Mann and Kuhn's investigation [23]. They fabricated tunnel junctions from multilayered Langmuir–Blodgett films of fatty acids and showed that the junction conductance decreased exponentially with the number of layers.

Scanning probe microscopy (SPM), including scanning tunneling microscopy and conducting probe atomic force microscopy, allows us to both study imaging of an organic single molecule or a group of molecules on the surface and evaluate their current-voltage characteristics at the same time. The organic compounds that are subjected to these studies range widely from synthetic compounds of alkanethiol and oligo(phenyleneethynylene)s to naturally occurring molecules of DNA, etc. In the present article, we focus our attention on helical peptides, because nature uses helical peptides in the proteins involved in electron transfer processes.

15.3
Electronic Properties of Helical Peptides

Helical peptides have two characteristic electronic properties: a large dipole moment and a low attenuation factor. The dipole moment of helical peptides increases with molecular length because of the additivity of the dipole moment arising from the peptide units along the molecule. The dipole moment of α-helical hexadecapeptide amounts to more than 50 debye with a molecular chain length of 2.4 nm [24]. On the other hand, the attenuation factor of α-helical peptides is reported to be 0.66 Å^{-1} [25], which is much smaller than that of alkyl chains. Notably, the electron transfer path, which was clarified along the helical peptide, involves hydrogen bonds.

The large dipole moment of α-helical peptides should generate a large electric field in the molecule. To evaluate the strength of the electric field, the self-assembled monolayers (SAMs) of α-helical peptides were prepared on gold with vertical alignment and the surface potentials measured (Fig. 15.1) [26]. When α-helical hexadecapeptides were immobilized on gold via the N terminal, a negative surface potential of -120 mV was observed. The extension of the molecular chain length from a 16-residue peptide to a 24-residue peptide decreased the surface potential to -200 mV. Since the negative polar of the dipole moment is exposed to the surface in these SAMs, the experimental data of the helical peptides are qualitatively agreeable. This potential is applied on the helical peptide SAM 4-nm thick, and the electric field across the SAM is thus around 500 000 V/cm.

The small attenuation factor of helical peptides is favorable to a long-distance electron transfer. In order to confirm the electron-mediating ability of the α-helical peptide, the electron transfer rates along α-helical long peptides carrying a redox group at the molecular terminal were analyzed. Two kinds of helical peptides, one carrying a ferrocene unit at the N-terminal end and a disulfide group at the C-terminal end and the other carrying the same at opposite terminal ends, were synthesized (Fig. 15.2) [27]. The peptides formed well-packed SAMs on gold. The electron transfer rates through the mono-layer were examined by chronoamperometry. The oxidative current from the ferrocene unit to gold was evaluated properly from the current–time curves,

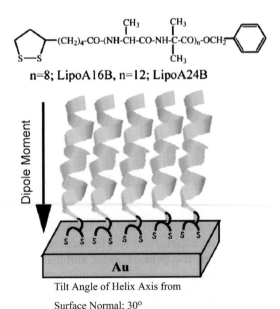

$$\text{(CH}_2)_4\text{-CO-(NH-CH-CO-NH-C-CO)}_n\text{-OCH}_2$$

with CH₃ and CH₃ substituents shown, S—S.

n=8; LipoA16B, n=12; LipoA24B

Dipole Moment

Au

Tilt Angle of Helix Axis from

Surface Normal; 30°

Fig. 15.1. α-Helical 16-residue and 24-residue peptides self-assembled on gold via the N terminal. The helices take a vertical orientation, and the negative polar ends are exposed to the surface

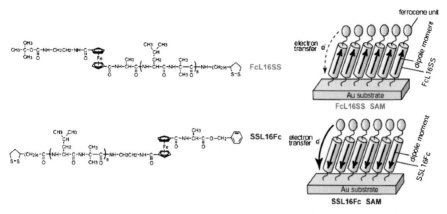

Fig. 15.2. Hexadecapeptides having a redox group at molecular terminal. The helical peptides are self-assembled on gold to evaluate electron transfer rates and the effect of the dipole moment on the rates

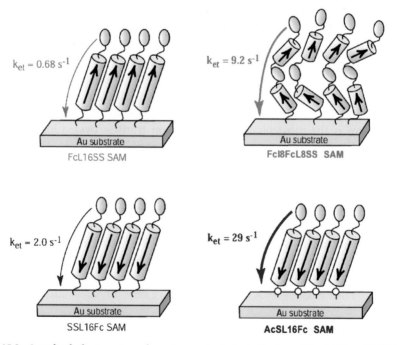

Fig. 15.3. Standard electron transfer rate constants in helical peptide SAMs of FcL16SS, Fc18FcL8SS, SSL16Fc, and AcSL16Fc. The effects of the dipole moment, the electron-mediating redox group in the molecular chain, and the linker moiety on the electron transfer rates are examined

which were measured by changing the applied overpotentials. Standard rate constants, obtained by extrapolation of the data to zero overpotential, were $0.68 \, s^{-1}$ for the FcL16ss SAM and $2.0 \, s^{-1}$ for the SSL16Fc SAM (Fig. 15.3). Considering that the electron transfer occurred over a long distance (more than 4 nm), the α-helical peptides possess a good ability to mediate electrons through molecules.

15.4
Electron Transfer Mechanism over a Long Distance

Two distinct mechanisms of two extreme cases are involved in long-distance electron transfer reactions in biological systems. One is an elastic superexchange mechanism, where a direct molecule-mediated tunneling takes place as an elastic tunnel process. In this mechanism, the electron transfer rate decreases exponentially with the distance between a pair of redox entities. The other is an inelastic hopping mechanism, where electron tunneling is accompanied by energy dephasing at the boundary sites of the molecular terminals as well as in the internal molecular units. In this mechanism, the molecular conductance obeys the ohmic behavior with the molecular length.

In Sect. 15.3, the electron transfer rates of ferrocene-attached peptides showed no significant dependence on the overpotential, suggesting that the inelastic hopping mechanism should dominate over the elastic tunneling mechanism for electron transfer along the peptides. Further, theoretical calculations based on the elastic tunneling mechanism, where an attenuation factor of 0.66 for the helical peptide is used, predict significantly lower electron transfer rates than the experimentally obtained values. Taken together, the inelastic hopping mechanism is the most plausible for electron transfer over a long distance through helical peptides under a zero overpotential.

The electron transfer rate for the SSL16Fc SAM was three times greater than that for the FcL16SS SAM (Fig. 15.3). The oxidative current from the ferrocene moiety to gold in the SSL16Fc SAM flows in the same direction of the dipole moment of the helical peptide. The dipole moment should cause the difference in the electron transfer rates between the two kinds of peptide SAMs. However, this difference was much smaller than what we had expected. This discrepancy suggests there is another rate-determining step in electron transfer, as discussed in the following section.

15.5
Effect of Linkers on Electron Transfer

Electron transfer at metal–molecule–metal junctions should be composed of three steps: electron transfer at two interfaces, metal–molecule and molecule–metal, and electron transfer through the molecule. To evaluate the electron transfer at the metal–molecule interface, the linker moiety of the alkyl chain attached at the peptide end was replaced with a thiophenyl group (Fig. 15.4). The phenyl group was expected to accelerate the electron transfer at the interface due to the rich π electrons.

AcSL16Fc was self-assembled on gold and subjected to chronoamperometry. The standard rate constant from the data extrapolated to zero overpotential was 29 s^{-1}, which is 15 times faster than that of the SSL16Fc SAM (Fig. 15.3). The introduction of the thiophenyl group was therefore successful in improving the electron transfer.

The dipole moment of the α-helical peptide should generate a potential gradient along the molecular axis, which may influence electron transfer through the molecule. However, the effect of the dipole moment on the overall electron transfer in the metal–molecule–metal junction does not manifest directly in the standard rate constants probably because of the slow process at the metal–molecule interface.

What, then, is the rate-determining step in the electron transfer reaction through the metal–molecule–metal junction? To get information on the rate-determining step, a redox group was introduced into the middle of the α-helical peptide (Fig. 15.4). The redox group was expected to mediate the electron transfer reaction through the molecule to accelerate the transfer rate. Fc18FcL8SS was self-assembled on gold, and the standard rate constant was determined by chronoamperometry. A rate of 9.2 s^{-1} was obtained, which was 14 times faster than that of FcL16SS (Fig. 15.3). The chemical modification in the middle of the molecule is therefore as effective as that in the linker moiety at the molec-

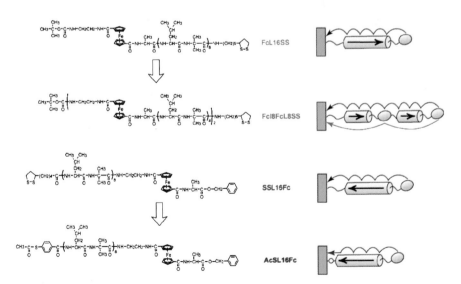

Fig. 15.4. Molecular structures of FcL16SS, Fc18FcL8SS, SSL16Fc, and AcSL16Fc

ular end. Taken together, the electron transfer process at the metal–molecule junction happens to be as significant as that through the molecule in these peptide systems.

15.6
Helical-Peptide Scaffold for Electron Hopping

Under a zero overpotential, an electron transfer reaction occurs in a helical peptide according to the inelastic hopping mechanism. Probably, electrons jump via amide bonds that are connected convalently and each other by hydrogen bonds. Using amide bonds as the hopping sites, however, may not be so favorable for efficient electron transfer in terms of the large HOMO–LUMO gap and the high potential energy of the LUMO level. A novel helical peptide was designed where the helical peptide adopts a 3_{10} helical structure and residues carrying an aromatic group at the side chain appear at every third residue in the primary sequence. This molecular design makes it possible to align the aromatic groups straight along the helical molecule.

A nonapeptide composed of three repeats of naphthylalanine-Aib-Aib (SSN3Fc) was synthesized (Fig. 15.5). The high content of Aib residue in the sequence is known to induce a 3_{10} helical structure. Indeed, the analogous peptide in solution was shown to assume a 3_{10} helical conformation by NMR measurements. Energy minimization, achieved by taking the 3_{10} helical conformation as an initial state, was carried out to obtain information on the space arrangement of the naphthyl groups along the 3_{10} helix. The three naphthyl groups were spaced in a linear array along the helical axis with a face-to-face orientation (Fig. 15.6a) [28]. The center-to-center distance between the naphthyl groups was around 6 Å on average. This distance is longer than the critical distance for excimer formation. The space arrangement of the naphthyl groups was thus suitable for electron hopping without an energy-dissipating trap site. Strictly speaking, the space arrangement of the three naphthyl groups rotated slightly in an anticlockwise direction (Fig. 15.6b). This arrangement was agreeable with the result of CD measurement of the analogous peptide in solution, which showed an exciton coupling at around 223 nm due to the rotating arrangement of the three naphthyl groups in an anticlockwise direction.

SSN3Fc was self-assembled on gold with the help of BA3SS, which was effective in the vertical alignment of the helices, and the standard rate constant was determined by chronoamperometry. A rate of $952 \ s^{-1}$ was obtained, which was nearly 500 times faster than that of SSL16Fc. The major reason for this enhancement is the shorter molecular length of SSN3FC compared with SSL16Fc. To evaluate the effect of naphthyl groups aligned along SSN3Fc upon electron transfer, SSA3Fc (Fig. 15.5), in which naphthylalanine residues of SSN3Fc were replaced with Ala residues, was studied. SSA3Fc was self-assembled on gold with the help of BA3SS, and the standard rate constant was determined by chronoamperometry. A rate of $741 \ s^{-1}$ was obtained, which was slightly slower

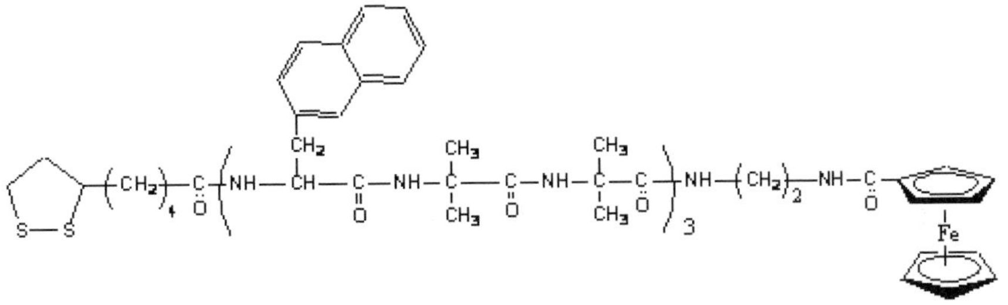

SSN3Fc

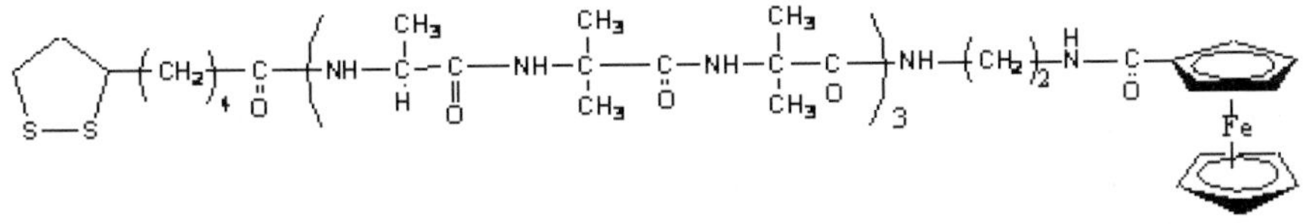

SSA3Fc

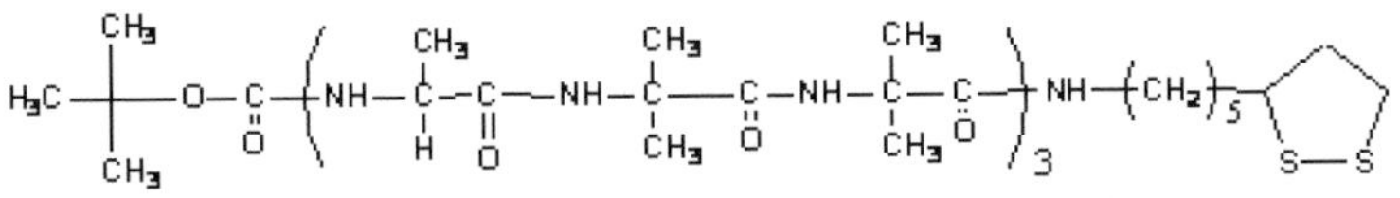

BA3SS

Fig. 15.5. Molecular structures of 3_{10} helical peptides carrying a redox group at molecular terminal and reference compound

than that of SSN3Fc. The effect of naphthyl groups is therefore insignificant in the present system.

In the SSN3Fc SAM, the dependence of the electron transfer rate on overpotential was small. Further, the standard rate constant of SSN3Fc was calculated to be $0.19\,\mathrm{s}^{-1}$ using attenuation factors of 1.2 for methylene groups and 0.66 for the helical-peptide moiety. This calculation under the assumption of elastic electron tunneling cannot explain the fast rate constant of the SSN3Fc SAM. Therefore, the inelastic hopping mechanism is also applicable to the SSN3Fc SAM.

There are several plausible reasons for the inefficiency of the naphthyl alignment for electron hopping. However, the major reason should be the spacing between the naphthyl groups, which is in contrast to successive

a

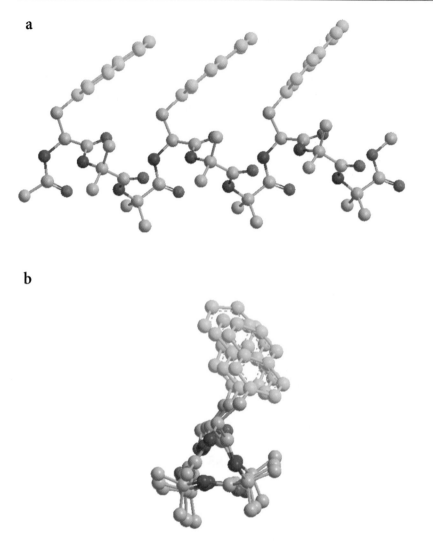

b

Fig. 15.6. *Side view* (**a**) and *top view* (**b**) of the 3_{10} helical peptide, Ac-(Nal-Aib-Aib)$_3$-OMe (Nal and OMe represent 2-naphthylalanine and methyl ester, respectively)

amide bonds through helix by covalent bonds and hydrogen bonds. The electron hopping via naphthyl groups should compete with that via amide bonds in the SSN3Fc SAM, but the latter exceeds the former due to structural advantages. Then, if the electron transfer via amide bonds are negligible compared with that via naphthyl groups, the effect of the naphthyl alignment on electron transfer could be evaluated. This thinking leads us to check the photocurrent generation of the 3_{10} helical peptide carrying naphthyl groups.

15.7
Photocurrent Generation with Helical Peptides Carrying Naphthyl Groups

SSN3B (Fig. 15.7), in which the ferrocene unit of SSN3Fc was replaced with benzyl ester, was self-assembled on gold, and photocurrent generation was investigated by photoexcitation of the naphthyl groups in an aqueous solution containing triethanolamine as an electron donor [28]. Significant anodic photocurrent generation was successfully observed on the SSN3B SAM in response to photoirradiation. In contrast, the control SAM composed of SSA3B (Fig. 15.7), in which three naphthylalanine residues of SSN3B were replaced with Ala residues, did not show such a photoirradiation response. The SSN3B SAM is thus a suitable molecular system for evaluating the effect of naphthyl groups involved in the electron transfer path.

Two reference compounds, SSNA2B and SSA2NB (Fig. 15.7), in which one naphthyl group was introduced in the molecule at the near and the distant end to the disulfide group of the helices, respectively, were studied. Noticeably, the SSNA2B SAM showed almost no photocurrent signal, but a weak photocurrent generation was observed in the SSA2NB SAM. The other possible reference monolayer, where one naphthyl group was located at the middle site, was not examined. But the photocurrent can be roughly estimated to be between those of SSNA2B and SSA2NB SAMs. The photocurrent of the SSN3B SAM is then at least twice as large as the sum of the photocurrents of these three reference SAMs. It is therefore concluded that electron hopping among thee linearly spaced naphthyl groups effectively promotes photocurrent generation in the SSN3B SAM.

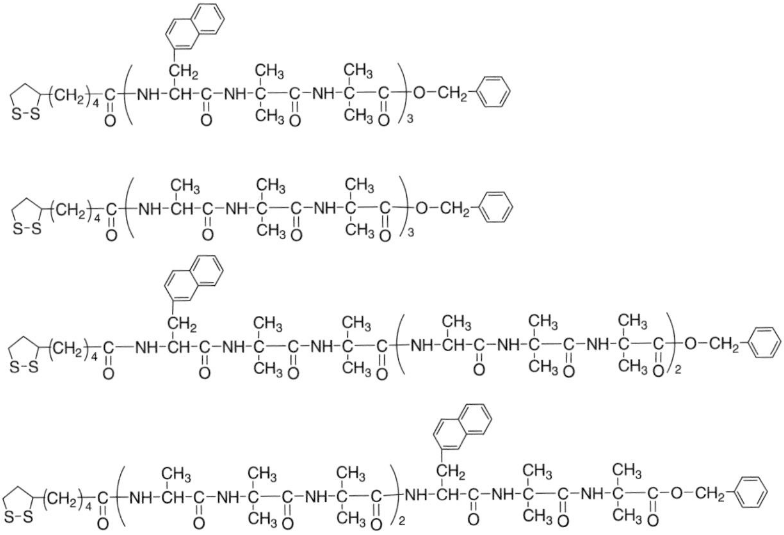

Fig. 15.7. Molecular structures of 3_{10} helical peptides for photocurrent generation

A representative case of photocurrent generation with the SSN3B SAM is as follows (Fig. 15.8). When the naphthyl group at the site nearest to the gold is photoexcited, electron transfer from the excited naphthyl group to the gold occurs to generate a radical cation. Then, the radical cation hops away from the gold via the naphthyl groups and finally is quenched by electron donation from triethanolamine in an aqueous solution. In addition, energy migration

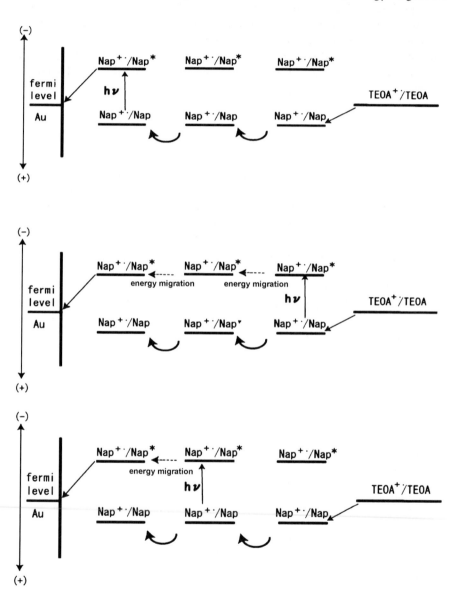

Fig. 15.8. Energy diagram for anodic photocurrent generation by SSN3B SAM

among the naphthyl groups may promote photocurrent generation because the distance between the neighboring naphthyl groups is shorter than the critical distance between naphthyl groups for energy migration of the Förster type. Thus, the photocurrent generation experiment revealed that the linearly spaced naphthyl groups along the helical axis effectively increase the photocurrent by photosensitization and electron hopping between the naphthyl groups.

A precise analysis of the kinetics reveled that the photoexcited naphtyl groups at the site nearest to the gold and in the middle were quenched immediately by gold. On the other hand, the photoexcited naphtyl group near the water phase could receive an electron from triethanolamine to generate an anion radical. The electron of the anion radical should hop between the naphtyl groups to reach to gold.

15.8
Conclusion

Helical peptides are shown to be unique electronic molecular elements because of their large dipole moment and their electron-mediating ability. Further, helical peptides can be utilized as a scaffold to arrange functional groups regularly in space. Helical peptides are thus promising basic compounds for the emerging field of molecular electronics. For example, two kinds of helical peptides, both carrying their own photosensitizer at the terminal but with opposite directions of dipole moment, were coassembled on gold. With the mixed helical peptide SAM, the photocurrent direction was determined by the dipole moment direction at a certain range of the applied potentials. Thus helical peptides with photosensitizers act as molecular photodiodes, and the photocurrent direction is switchable by choosing a wavelength to excite one of the two photosensitizers [29].

15.9
References

1. a) Moore GE (1965) Electronics 38,114–117; b) Moore GE (1998) Proc IEEE 86,82–85
2. Moore G (1997) Electrochem Soc Interf 6,18–23
3. Data from Intel, available at: http://www.intel.com/research/silicon/ mooreslaw.htm
4. Moreau WM (1988) Semiconductor Lithography: Principles, Practices, and Materials, Plenum Press: New York
5. Rai-Choudhury P (ed) Handbook of Microlithography, Micromachining, and Microfabrication SPIE Optical Engineering Press: Bellingham WA, 1997, Vol 1
6. Xia Y, Whitesides GM (1998) Angew Chem Int Ed 37,551–575
7. Muller DA, Sorsch T, Moccio S, Baumann FH, Evans-Lutterodt K, Timp G (1999) Nature 399,758–761
8. Aviram A, Ratner M (1974) Chem Phys Lett 29,277–283
9. Carroll RL, Gorman CB (2002) Angew Chem Int Ed 41,4378–4400
10. Metzger RM (2003) Chem Rev 103,3803–3834

11. Joachim C, Gimzewski JK, Aviram A (2000) Nature 408,541–548
12. Aviram A, Ratner M (eds) Molecular Electronics: Science and Technology. New York Academy of Sciences, New York, 1998, Vol 852
13. Aviram A, Ratner M (eds) Molecular Electronics: Science and Technology. New York Academy of Sciences, New York, 1998, Vol 852
14. Reimers JR, Picconatto CA, Ellenbogen JC, Shashidhar R (eds) Molecular Electronics III. New York Academy of Sciences, New York, 2003, Vol 1006
15. a) Lehn J-M (1995) Supramolecular Chemistry: Concepts and Perspective, VCH, Weinheim; b) Lehn J-M (1988) Angew Chem Int Ed 27,89–112
16. Heath JR, Ratner MA (2003) Phys Today 43–49
17. Adams DM, Brus L, Chidsey CED, Creager S, Creutz C, Kagan CR, Kamat PV, Lieberman M, Lindsay S, Marcus RA, Metzger RM, Michel-Beyerle ME, Miller JR, Newton MD, Rolison DR, Sankey O, Schanze KS, Yardley J, Zhu X (2003) J Phys Chem B 107,6668–6697
18. Nitzan A, Ratner MA (2003) Science 300,1384–1389
19. Salomon A, Cahen D, Lindsay S, Tomfohr J, Engelkes VB, Frisbie CD (2003) Adv Mater 15,1881–1890
20. James DK, Tour JM (2004) Chem Mater 16,4423–4435
21. McCreery RL (2004) Chem Mater 16,4477–4496
22. Adams DM, Brus L, Chidsey CED, Creager S, Creutz C, Kagan CR, Kamat PV, Lieberman M, Lindsay S, Marcus RA, Metzger RM, Michel-Beyerle ME, Miller JR, Newton MD, Rolison DR, Sankey O, Schanze KS, Yardley J, Zhu X (2003) J Phys Chem B 107,6668–6697
23. Mann B, Kuhn H (1971) J Appl Phys 42,4398–4405
24. Hol WGJ (1985) Prog Biophys Mol Biol 45,149–195
25. Sisido M, Hoshino S, Kusano H, Kuragaki M, Makino M, Sasaki H, Smith TA, Ghiggino KP (2001) J Phys Chem B 105,10407–10415
26. Miura Y, Kimura S, Kobayahsi S, Iwamoto M, Imanishi Y, Umemura J (1989) Chem Phys Lett 315,1–5
27. Morita T, Kimura S (2003) J Am Chem Soc 125,8732–8733
28. Yanagisawa K, Morita T, Kimura S (2004) J Am Chem Soc 126,12780–12781
29. Yasutomi S, Morita T, Imanishi Y, Kimura S (2004) Science 304,1944–1947

Construction of Redox-Induced Systems Using Antigen-Combining Sites of Antibodies and Functionalization of Antibody Supramolecules

Hiroyasu Yamaguchi · Akira Harada

Department of Macromolecular Science, Graduate School of Science, Osaka University, Toyonaka, Osaka 560-0043, Japan

Summary We have prepared monoclonal antibodies against water-soluble porphyrins or viologen derivatives. The specific binding of monoclonal antibodies to porphyrins was found to control photoinduced electron transfer from porphyrin to electron acceptor molecules. One of the antibodies against a cationic porphyrin bound Fe-porphyrin strongly, and the complex of Fe-porphyrin with the antibody showed peroxidase activity. We expanded our study to include the construction of antibody supramolecules using their specific molecular recognition to antigens. We constructed dendritic antibody supramolecules. An amplification method of the detection signals for the target molecule in the biosensors based on the surface plasmon resonance (SPR) was devised using the signal enhancement in the supramolecular assembly of the antibody with multivalent antigens.

Abbreviations

AFM	Atomic force microscopy
ABTS	2,2′-Azinobis(3-ethylbenzthiazolin-6-sulfonicacid)diammonium salt
BSA	Bovine serum albumin
cis-2MPyP	[5,10-Bis(4-carboxyphenyl)-15,20-bis-(4-methylpyridyl)]porphine methylester
CD	Circular dichroism
CDI	Carbonyldiimidazole
DNA	Deoxyribonucleic acid
ELISA	Enzyme-linked immunosorbent assay
Fe-TCPP	*meso*-Tetrakis(4-carboxyphenyl)porphyrin iron(III) chloride
Fe-TMPyP	[5,10,15,20-Tris-(4-methylpyridyl)]porphine iron(III) chloride
HRP	Horseradish peroxidase
IgG	Immunoglobulin G
IgM	Immunoglobulin M
KLH	Keyhole limpet hemocyanin
3MPy1C	(5-(4-Carboxyphenyl)-10,15,20-tris-(4-methylpyridyl)-porphine iodide
MV^{2+}	Methyl viologen
M_w	Molecular weight
Np^-	1-Naphthalenesulfonate
ns	Nanosecond

P-TPP   Phosphorus(V)-tetraphenylporphyrin
P-TPP(EG) (5,10,15,20-Tetraphenylporphinato)bis(2-hydroxyethoxy)phos-
           phorus chloride
QCM   Quartz crystal microbalance
SOD   Superoxide dismutase
SPR   Surface plasmon resonance
TCPP   *meso*-Tetrakis(4-carboxyphenyl)porphyrin
TMPyP   *meso*-Tetrakis-(4-*N*-methylpyridyl)porphyrin
Zn-TCPP   *meso*-Tetrakis(4-carboxyphenyl)porphyrin zinc complex

16.1
Introduction

The immune system's ability to generate antibodies against virtually any molecule of interest has resulted in the widespread use of antibodies not only as diagnostic agents but also as catalysts in chemical laboratories and industry. They are also used as probes for isolating and unraveling the structures of complex biological molecules. Monoclonal antibodies have become increasingly important and show great potential as new chemical materials. Pauling recognized the similarity between enzymes and antibodies nearly 60 years ago [1]. While it has not been considered that antibodies catalyze reactions, Schultz [2] and Lerner [3] found about 20 years ago that antibodies might possess intrinsic catalytic activity through the unique sizes and shapes of their binding pockets. We have focused our attention on the special behavior of antibodies, especially monoclonal antibodies, because they can recognize a larger and more complex compound with higher specificity than can enzymes. Recently, with the advent of cell technology [4], it has become possible to prepare individual immunoglobulins that are called "monoclonal antibodies" in large amounts and in homogeneous form.

In this paper, we have prepared monoclonal antibodies for porphyrins or viologens to construct specific catalysts or supramolecular materials for biosensing systems [5–10]. It is well known that porphyrins play an important role as functional groups in a wide variety of biological systems. In nature, there are a number of functional molecules that have porphyrin derivatives as cofactors in their active sites or their reaction centers, for example, oxygen carriers such as hemoglobin and myoglobin, redoxidase such as catalase and peroxidase, electron transfer carriers, and cytochromes [11, 12]. Special functions of metalloporphyrins in those systems appear in an environment where metalloporphyrin moieties are included in proteins [13]. In order to realize the functions of porphyrins artificially, attempts have been made to modify hemoprotein complexes. A large number of artificial hemoprotein model systems are being developed in an attempt to provide insights for structure–activity relationships, understand the minimal requirements for function, and construct tailor-made molecules [14]. De novo designed hemoproteins have provided keen insight into the fundamental governors of protein secondary and tertiary structural specificity [15–18]. The use of proteins with greater biological homology has facilitated the incorporation of hemes. A rearrangement of the amino acid residues by the site-directed mutagenesis or a chemical modification by the introduction of a functional group at the active residue on the protein surface is also one of the engineering challenges in the functionalization of hemoprotein [19, 20]. However, it seems to be difficult to construct an artificial hemoprotein with a novel function by these methods because of the disfavored conformational changes of polypeptides or the nonselective reaction of chemicals on the surface of the protein. Recently, reconstruction of hemoproteins has been carried out by the introduction of a functionalized metalloporphyrin into naturally occurring hemoproteins to control chemical reactivity or photochemical properties [21–24]. However, the use of naturally occurring proteins might be limited due to the decrease of affinity of the proteins for artificial porphyrins. One of the most convenient methods of incorporating artificial porphyrins into protein matrices is thought to be the preparation of monoclonal antibodies for porphyrins [25–34]. Strategies that allow incorporation of cofactors into antibody-combining sites should expand the scope of antibody catalysis. In chlorophyll, a number of porphyrin molecules and electron acceptors are incorporated into a protein domain, which plays an important role in the regulation of electron flow for the charge separation in the photochemical system. There are many studies concerned with electron transfer from porphyrins to covalently linked electron acceptors, but there remain unsolved problems with respect to electron transfer. In order to solve the problems, it is important to study the noncovalent electron donor–acceptor system because the chromophores of the in vivo photosynthetic reaction center are not linked covalently through spacer groups but are simply held in space by the protein environment. Monoclonal antibodies, which can bind both porphyrins and electron acceptor molecules, are most appropriate for the study of noncovalent electron donor–acceptor systems. Viologens are well-known functional molecules that serve as electron acceptors, although they are harmful [35, 36]. Antiviologen antibodies [37–41] may be expected to be useful not only as highly sensitive reagents of viologens but also as functional materials to control electron transfer from electron donors to viologens.

Figure 16.1 shows the structures of porphyrin and viologen molecules used in this study as haptens. Monoclonal antibodies against tetracarboxyphenylporphyrin, phosphorus(V) porphyrin with two axial ligands, water-soluble cationic porphyrin, and a viologen derivative have been prepared. The electron transfer reaction from porphyrin molecules to electron acceptors can be controlled by the binding of the antibodies to porphyrin molecules [5,6]. Peroxidase activity of Fe porphyrin-antibody complexes has been investigated [7]. The construction of supramolecular structures between antibodies and por-

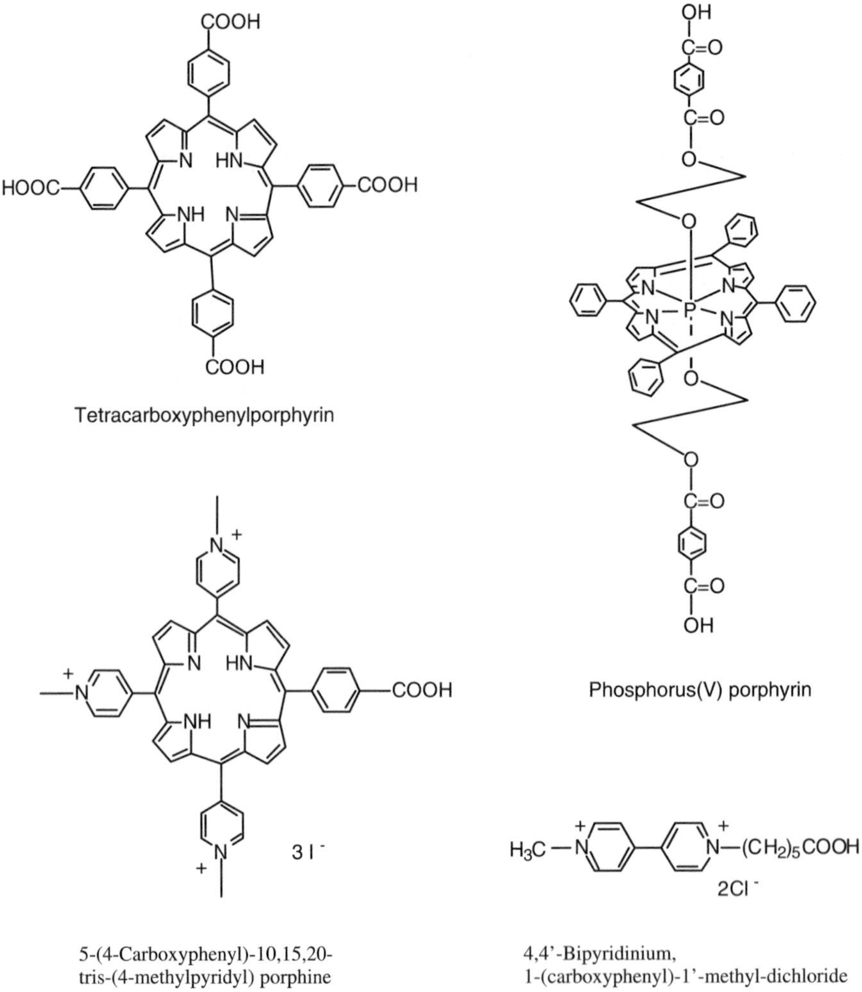

Tetracarboxyphenylporphyrin

Phosphorus(V) porphyrin

5-(4-Carboxyphenyl)-10,15,20-
tris-(4-methylpyridyl) porphine

4,4'-Bipyridinium,
1-(carboxyphenyl)-1'-methyl-dichloride

Fig. 16.1. Structures of porphyrins and viologens used as haptens

phyrins or viologens has been studied, and their supramolecular formations were applied to devise a highly sensitive biosensing system [8–10].

16.2
Photoinduced Electron Transfer from Porphyrins to Electron Acceptor Molecules

Monoclonal antibodies (03-1 and 13-1) for *meso*-tetrakis(4-carboxyphenyl) porphyrin (TCPP) were prepared [28] and their binding affinities to porphyrins were studied [33]. Electron transfer from TCPP zinc complex (Zn-TCPP) to methyl viologen (MV^{2+}) in the presence of antibody 03-1 was compared with that in the presence of another antibody 13-1.

16.2.1
Monoclonal Antibodies for *meso*-Tetrakis(4-carboxyphenyl)porphyrin (TCPP)

Antibodies 03-1 and 13-1 bound Zn-TCPP strongly with a dissociation constant of 10^{-7} M. The absorption spectra of TCPP in the presence of antibody 03-1 showed that the Soret band shifted to a longer wavelength by about 10 nm (Fig. 16.2a). However, in the TCPP-antibody 13-1 system, no changes were observed in the absorption spectra (Fig. 16.2b). Similar results were obtained in Zn-TCPP-antibody complexes. Although Zn-TCPP in the presence of various concentrations of the antibody showed an isosbestic point at 427 nm until the molar ratio reached one-to-one, the spectra showed further increase in the absorption at higher concentrations of the antibody. These results indicate the existence of a higher-order association. Figure 16.3 shows emission spectra of TCPP at a fixed concentration (2.5×10^{-7} M) in the presence of various amounts of the antibodies. In the TCPP-antibody 03-1 system, the spectra showed an increase in the fluorescence bands. Such changes were not observed in the TCPP-antibody 13-1 system. Figure 16.4a,b shows circular dichroism (CD) spectra for TCPP and Zn-TCPP, respectively, in the presence of antibody 03-1. Both spectra showed very strong induced Cotton effects. In the TCPP-antibody 03-1 system, the positive Cotton effects were observed at low concentrations of the antibody until the molar ratio reached one-to-one. Then the positive Cotton effects decreased with an increase in the antibody concentration. When the molar ratio exceeded one-to-one, the positive induced CD

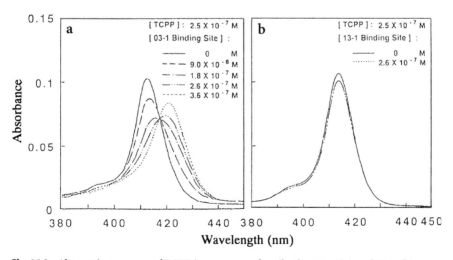

Fig. 16.2. Absorption spectra of TCPP in presence of antibodies 03-1 (**a**) and 13-1 (**b**)

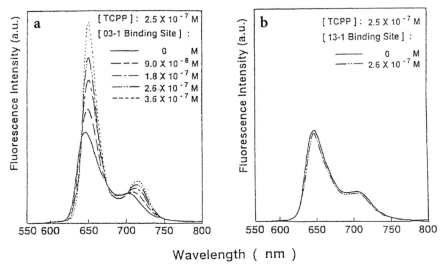

Fig. 16.3. Fluorescence spectra of TCPP in presence of antibodies 03-1 (**a**) and 13-1 (**b**)

decreased, and finally negative induced CDs appeared. These results clearly indicate the existence of at least two kinds of species in the complexes according to the concentration of the antibody. On the other hand, the spectrum of Zn-TCPP in the presence of antibody 03-1 showed sharp splitting at the Soret bands where negative and positive bands were obtained in the longer and shorter wavelength sides, respectively. These bands can be assigned to an exciton coupling of the band. This result can be interpreted as indicating that two Zn-TCPP molecules are close together in the combining site of the antibody. It is suggested that a one-to-one complex is first formed between antibody and porphyrin. Then, another antibody binds TCPP (or Zn-TCPP) to form a two-to-one complex, which is composed of two antibodies and one porphyrin.

The lifetimes of the excited states of TCPP are elongated on addition of antibody 03-1 (Table 16.1). The lifetime of the singlet excited state of TCPP was the longest in the presence of antibody 03-1 at a two-to-one molar ratio (two antibody combining sites and one porphyrin). A similar elongation of lifetime was observed in the triplet excited states of TCPP bound to antibody 03-1. Although most of the catalytic antibodies have been limited to ground-state reactions, this antibody 03-1 raised against TCPP was found to stabilize excited states of porphyrins.

Figure 16.5 shows spectral changes upon titration of Zn-TCPP with MV^{2+} in the absence and presence of antibodies (03-1 and 13-1). When MV^{2+} was added to an aqueous solution of Zn-TCPP, the Soret band of Zn-TCPP decreased and shifted toward a wavelength that was about 5 nm longer, indicating that Zn-TCPP had interactions with MV^{2+} in the ground state. This is probably due

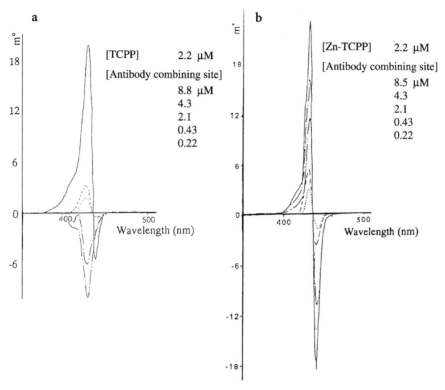

Fig. 16.4. CD spectra of TCPP (**a**) and Zn-TCPP (**b**) in presence of antibody 03-1

to the electrostatic interactions between anionic TCPP and cationic methyl viologen molecules. In contrast, when MV^{2+} was added to an aqueous solution of Zn-TCPP in the presence of twofold equivalent of antibody 03-1 to Zn-TCPP, the Soret band did not shift at all and showed no hypochromicity. This result indicates that there were no interactions between MV^{2+} and Zn-TCPP incorporated into the combining site of antibody 03-1 in the ground state. This may be due to the fact that the antibody prevents MV^{2+} from contacting with Zn-TCPP.

Table 16.1. Lifetime of excited singlet (τ_S) and triplet (τ_T) states of TCPP in absence and presence of antibodies

	No antibodies	Antibody 03-1	Antibody 13-1
τ_S (ns)	8.97	10.2 (1:1) [a]	8.97 (1:1) [a]
		12.1 (1:2) [a]	8.97 (1:2) [a]
		12.1 (1:3) [a]	8.97 (1:3) [a]
τ_T (ms)	0.34	2.7	0.26

[a] [TCPP]:[Antibody-combining site]

Fig. 16.5. Spectral changes upon titration of Zn-TCPP with MV^{2+} in absence (**a**) and presence (**b**) of antibodies 03-1 and 13-1 (**c**)

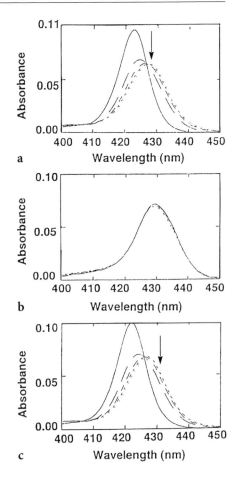

Upon excitation at the Soret band of Zn-TCPP (λ = 427 nm), Zn-TCPP emitted fluorescence, peaking at 610 and 660 nm. The fluorescence of Zn-TCPP was quenched by MV^{2+}. The emission spectra of Zn-TCPP in the presence of twofold equivalent of antibody 03-1 to Zn-TCPP showed that addition of MV^{2+} resulted in a decrease in the intensity of emission, although the ground-state interactions were negligible. Figure 16.6 shows Stern–Volmer plots for quenching of the emission from the Soret band of Zn-TCPP by the addition of MV^{2+} in the absence and presence of antibodies. The plots for the Zn-TCPP-MV^{2+} system showed efficient fluorescence quenching. In contrast, the fluorescence quenching was saturated at $[MV^{2+}] > 0.5$ mM in the presence of antibody 03-1, where the ratio I_0/I was almost constant at 2.0. Considering the result in Fig. 16.6, this saturation behavior indicates the electron transfer from Zn-TCPP in the antibody to MV^{2+} outside the antibody. The complex formation of antibody 13-1 with Zn-TCPP was effective in the fluorescence quenching of Zn-TCPP.

Fig. 16.6. Stern–Volmer plots for quenching of emission from Soret band of Zn-TCPP by addition of MV^{2+} in absence (a) and presence (b) of antibody 03-1 or 13-1 (c)

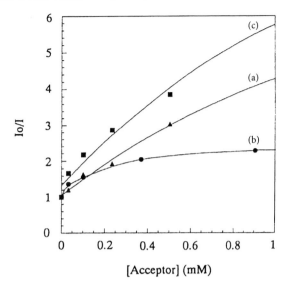

Although two monoclonal antibodies, 03-1 and 13-1, have the same affinity to Zn-TCPP, the photochemical behavior of Zn-TCPP on addition of MV^{2+} in the presence of antibody 13-1 was similar to that of Zn-TCPP in the absence of antibodies. On the other hand, in the case of antibody 03-1, the electron transfer from Zn-TCPP to MV^{2+} was controlled by the complex formation between antibody 03-1 and Zn-TCPP with a two-to-one ratio. When the Soret band of Zn-TCPP in the combining site of antibody 03-1 for a 1:2 complex was continuously irradiated in the presence of triethanolamine, the solution turned blue, distinguishable from the color of Zn-TCPP-MV^{2+} system without antibodies, indicating that the methyl viologen radical cation was formed [42,43]. The fluorescence quenching of Zn-TCPP in the combining site of antibody 03-1 by MV^{2+} is suggested to be due to the result of a long-range photoinduced electron transfer through the antibody-combining site.

The fluorescence decay of Zn-TCPP in the absence of the antibodies was expressed by a monoexponential curve with a lifetime of 1.8 ns, whereas addition of twofold equivalent of antibody 03-1 to Zn-TCPP brought about an increase in the lifetime to 2.5 ns. Addition of MV^{2+} caused the shortening of the lifetime to $\tau_S = 0.5$ ns. The short-lived component of the decay curve is assignable to the complex of Zn-TCPP with MV^{2+}. In the presence of antibody 03-1 (2 molar excess of Zn-TCPP), a biphasic decay curve was observed: $\tau_L = 2.5$ ns (72%) and $\tau_S = 0.5$ ns (28%) at $[MV^{2+}] = 5.0 \times 10^{-3}$ M. The longer- and shorter-lived components are assignable to Zn-TCPP incorporated into an antibody and the complex with MV^{2+}, respectively. Taking into account the negligible contribution from donor–acceptor complexation at the ground state together with the saturation behavior in the fluorescence quenching in the presence of antibody 03-1, the rate constant (k_{et}) for the electron transfer from Zn-TCPP in the antibody-combining site to MV^{2+} was estimated, using the equation $k_{et} = (\tau_{[MV^{2+}]})^{-1} - (\tau_{[none]})^{-1}$, to be $1.6 \times 10^9 \, s^{-1}$, where $\tau_{[MV]}$ and $\tau_{[none]}$ are the fluorescence lifetimes of Zn-TCPP for a 1:2 Zn-TCPP-antibody complex in the presence and absence of MV^{2+}, respectively.

These results indicate that the photoinduced electron transfer from Zn-TCPP to MV^{2+} can be controlled by the incorporation of porphyrins into the antigen-combining site of the antibodies. Figure 16.7 shows the proposed structure of the complex of Zn-TCPP with antibody 03-1 or antibody 13-1 and the possible positions of methyl viologen molecules approaching the complexes.

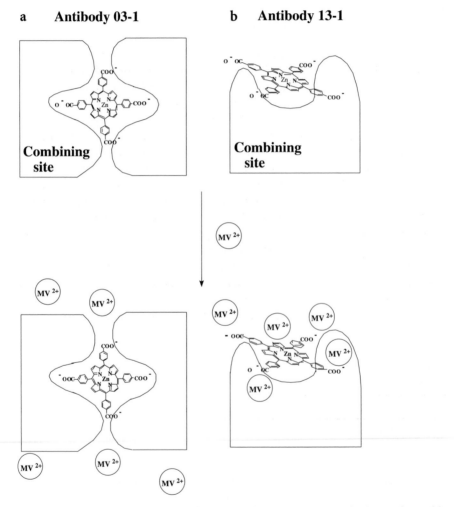

Fig. 16.7. Schematic representation of complex of Zn-TCPP with antibodies and possible positions of methyl viologen molecules approaching the complexes. **a** Antibody 03-1. **b** Antibody 13-1

16.2.2
Photoinduced Electron Transfer from a Porphyrin
to an Electron Acceptor in an Antibody-Combining Site

It is more important to design an antibody-combining site that can accommodate not only metalloporphyrins but also organic substrates [44] or electron acceptors. We succeeded in preparing antibodies capable of binding both a porphyrin and an electron acceptor molecule. Monoclonal antibodies were raised against a porphyrin that was linked to an electron acceptor by a spacer. A monoclonal antibody was then used to bind the porphyrin and the acceptor without a formal covalent linkage.

Phosphorus(V)-tetraphenylporphyrin (P-TPP) in Fig. 16.1 was used as a hapten. Phosphorus porphyrins are able to form stable axial bonds from the central phosphorus atom. Water-stable, axial dialkoxy hypervalent phosphorus(V) porphyrins were synthesized by efficient substitution of axial chloride ligands in dichlorophosphorus(V) tetraphenylporphyrin. Five monoclonal antibodies were obtained by immunization of KLH-conjugated P-TPP to Balb/c mice. One of these antibodies, 74D7A, bound antigen strongly (dissociation constant $K_d = 2.2 \times 10^{-7}$ M) but not TCPP, as shown by an enzyme-linked immunosorbent assay (ELISA) as well as by the absorption and emission spectra. Table 16.2 shows the dissociation constants of the complexes between antibody 74D7A and benzoic acid or phthalic acid derivatives. Although the dissociation constant of the antibody with benzoic acid was 1.1×10^{-2} M, that of the antibody with terephthalic acid was 1.2×10^{-5} M. The affinity of the antibody for terephthalic acid was about 1000-fold greater than that for benzoic acid. The antibody showed an excellent specificity to part of the axial ligands.

Terephthalic acid and its ester derivatives act as an electron acceptor for porphyrins. Fixation of terephthalic acid to the axis of porphyrin induces the changes in photochemical properties of phosphorus(V) porphyrin. For example, the lifetime of the singlet excited state of the antigen (the conjugate

Table 16.2. Dissociation constants of complexes between antibody 74D7A and benzoic acid or phthalic acid derivatives

Substrates		K_d(M)
HOOC—◯		1.1×10^{-2}
HOOC—◯—COOH	o-	$> 10^{-1}$
	m-	3.0×10^{-2}
	p-	1.2×10^{-5}
HOOC—◯—COOCH$_3$		4.5×10^{-5}

Table 16.3. Degree of fluorescence quenching of porphyrins on addition of terephthalic acid or isophthalic acid in the presence of antibody 74D7A

Porphyrin	Substrate	$(1 - I/I_0) \times 100$ (%)[a]
P-TPP(EG)	HOOC-◯-COOH	20
P-TPP(EG)	HOOC-◯-COOH	2
TCPP	HOOC-◯-COOH	5

[a]I_0: Fluorescence intensity of P-TPP(EG) or TCPP in the presence of antibody 74D7A (without substrates). I: Fluorescence intensity of the porphyrin in the presence of the antibody and substrates

P-TPP(EG)   TCPP

of carrier protein with P-TPP was shorter than that of porphyrin P-TPP(EG) (conjugate: 0.4 ns; P-TPP(EG): 4.4 ns in phosphate borate buffer-CH$_3$CN). The fluorescence quenching and the shortened excited-state lifetime of the conjugate were considered to be due to an intramolecular electron transfer between the porphyrin and the terephthaloyl moiety. When the complex between P-TPP(EG) and antibody 74D7A is formed, it is expected that the antibody may provide a space where acceptor molecules may enter. Table 16.3 shows the degree of fluorescence quenching of porphyrins on addition of terephthalic acid or isophthalic acid in the presence of the antibody. A 20% fluorescence quenching of P-TPP(EG) in the presence of antibody 74D7A was observed on addition of terephthalic acid. This degree of quenching on addition of terephthalic acid is tenfold larger than that of added isophthalic acid. Under the same conditions [P-TPP(EG) (8.0×10^{-7} M)] and terephthalic acid (1.6×10^{-6} M)], the quenching behavior was not observed in the absence of the antibody. The titration experiments of fluorescence quenching of P-TPP(EG) with terephthalic acid showed that terephthalic acid–antibody binding has a 1:1 molecular ratio.

The fluorescence decay of P-TPP(EG) in the presence of an antibody was expressed by a monoexponential curve with a lifetime of 4.5 ns. Addition of terephthalic acid shortened the lifetime to 0.3 ns. On the other hand, the life-

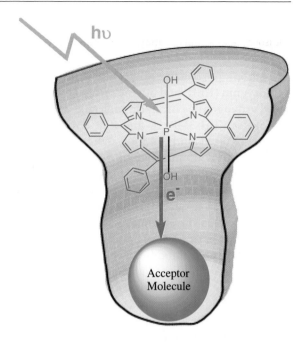

Fig. 16.8. Schematic representation of photoinduced electron transfer from porphyrin to electron acceptor molecule in antigen-combining site of antibodies

time of P-TPP(EG) with terephthalic acid was 4.5 ns in the absence of antibody 74D7A. These results show that the electron transfer from porphyrin to an electron-accepting molecule occurred in the antibody-combining site. Figure 16.8 shows a schematic representation of a donor–acceptor system in the antibody-combining site.

16.3
Peroxidase Activity of Fe-Porphyrin-Antibody Complexes

It is well known that in nature, anionic porphyrin derivatives (dicarboxylic compounds) are used as a cofactor in various functional proteins. However, there are no cationic porphyrins in a naturally occurring system. Cationic porphyrins such as *meso*-tetrakis-(4-*N*-methylpyridyl)porphyrin (TMPyP) and its related analogs are known to bind as well as cleave DNA [45]. They are known to exhibit a marked anticancer activity with relatively low toxicity and a highly superoxide dismutase (SOD) activity. It is suggested that they may increase the concentration of H_2O_2 in cancer cells [46].

We have prepared monoclonal antibodies for a cationic porphyrin (5-(4-carboxyphenyl)-10,15,20-tris-(4-methylpyridyl)-porphine iodide (3Mpy1C) and investigated the peroxidaselike activity of the complex between the antibody and the corresponding iron porphyrin. The catalytic activity of the antibody–porphyrin complex on the oxidation of pyrogallol was compared with that of horseradish peroxidase (HRP).

16.3.1
Preparation of Monoclonal Antibodies Against Cationic Porphyrins

The cationic porphyrin 3MPy1C was synthesized and coupled to keyhole limpet hemocyanin (KLH) and bovine serum albumin (BSA) via activation of the carboxyl group in the porphyrin molecule using carbonyldiimidazole (CDI). The conjugates KLH-3MPy1C and BSA-3MPy1C were purified by size exclusion chromatography and used for an antigen for the immunization to mice and enzyme-linked immunosorbent assays (ELISA), respectively. Balb/c mice were immunized with KLH-3MPy1C in saline emulsified 1:1 in Freund's complete adjuvant four times at 2-week intervals. Three days after the final injection, the spleen was taken from the mouse and the spleen cells were fused with the SP 2/0 mouse myeloma cells. The hybridomas secreting antibodies for the cationic porphyrin, 3MPy1C, were detected by ELISA. The hybridomas secreting anti-3MPy1C antibodies were cloned two times by limiting dilution. Three monoclonal antibodies (34A1F, 12E11G, and 83B5D) specific to 3MPy1C were obtained, and their subclasses were found to be immunoglobulin M (IgM), IgG_1, and IgG_1, respectively. One of these monoclonal antibodies, 12E11G, was found to have the highest affinity for 3MPy1C and was used for further experiments. The binding of monoclonal antibody 12E11G to metalloporphyrins was investigated by ELISA. Monoclonal antibody 12E11G bound to the hapten (3MPy1C) with a dissociation constant of 1.7×10^{-8} M. The Soret band of 3MPy1C shifted to a longer wavelength (5 nm) and the hypochromism was observed by the addition of antibody 12E11G, indicating that porphyrin is placed in the low-polar environment. Figure 16.9 shows the CD spectrum of tetramethylpyridylporphyrin (TMPyP) bound to the antibody. Although porphyrin TMPyP itself showed no CD peak, the complex showed induced Cotton effects on TMPyP in the region of the Soret band, suggesting that the porphyrin molecule was incorporated into the chiral environment of the combining site of the antibody. The split CD on TMPyP in the presence of the antibody was ascribable to an exciton coupling of the band of porphyrins. This result can be interpreted as indicating that two porphyrin molecules bound to the antibodies were close together. It was suggested that the antibody bound one porphyrin molecule and one-to-one complexes of the antibody with the porphyrin associated together to form a two-to-two complex. Figure 16.10 shows Klotz plots for the binding of antibody 12E11G to cationic porphyrins bearing the different numbers of charges in a molecule. The affinity of the antibody to dicationic porphyrin (cis-2MPyP) was lower by about one order of magnitude over that to tri- and tetracationic porphyrins. It was suggested that more than a half part of the porphyrin molecule might be incorporated into the antibody-combining site. Although the antibody was elicited against free-base porphyrin, the antibody bound not only to TMPyP but also to Fe-TMPyP. The dissociation constants of antibody 12E11G with Fe-TMPyP were found to be 2.6×10^{-7} M. The binding mode of antibody 12E11G to cationic porphyrins

Fig. 16.9. CD (**a**) and absorption (**b**) spectra of TMPyP in the presence of antibody 12E11G

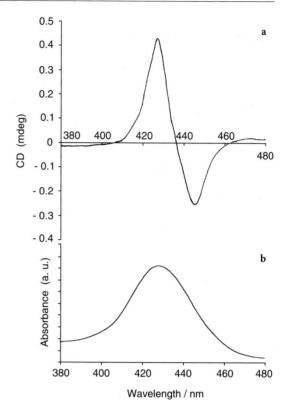

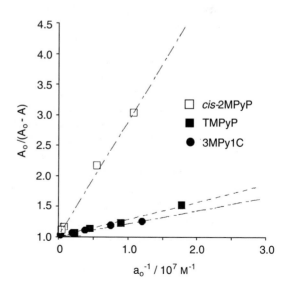

Fig. 16.10. Klotz plots for binding of antibody 12E11G to cationic porphyrin derivatives bearing different numbers of charges in a molecule

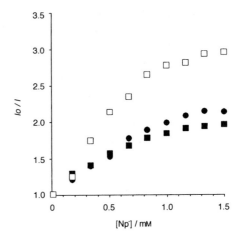

Fig. 16.11. Stern–Volmer plots for quenching of emission from Soret band of TMPyP by Np⁻ in the absence and presence of antibody 12E11G

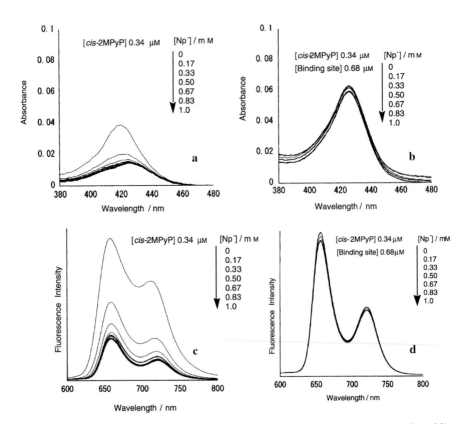

Fig. 16.12. Absorption (**a, b**) and fluorescence (**c, d**) spectral changes of *cis*-2MPyP by addition of Np⁻ in the absence (**a, c**) and presence (**b, d**) of the antibody

was examined by the study of electron transfer from a series of cationic porphyrins to electron acceptors 1-naphthalenesulfonate (Np⁻) in the presence of the antibody. Figure 16.11 shows Stern–Volmer plots for quenching of the emission from the Soret band of TMPyP by Np⁻ in the absence and presence of antibody 12E11G. It was indicated that TMPyP was incorporated into the antibody with stoichiometric 1:1. Figure 16.12a shows the spectral changes upon titration of *cis*-2MPyP with Np⁻ in the absence and presence of antibody 12E11G. When Np⁻ was added to an aqueous solution of *cis*-2MPyP, the Soret band of *cis*-2MPyP decreased and shifted toward a longer wavelength, indicating that *cis*-2MPyP had interactions with Np⁻ in the ground. This is probably due to the electrostatic interactions between cationic porphyrin and anionic Np⁻. In contrast, when Np⁻ was added to an aqueous solution of *cis*-2MPyP in the presence of an equimolar amount of antibody 12E11G, the Soret band did not shift at all and showed no hypochromicity (Fig. 16.12b). Figure 16.12c,d shows fluorescence spectral changes of *cis*-2MPyP by the addition of Np⁻ in the absence and presence of the antibody, respectively. Without

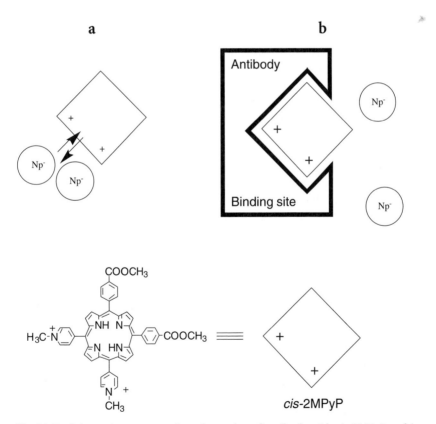

Fig. 16.13. Schematic representation of complex of antibody with *cis*-2MPyP and interactions between *cis*-2MPyP and anionic electron acceptor (Np⁻) molecules

antibody, the fluorescence intensity of *cis*-2MPyP decreased as well as TMPyP. However, no fluorescence quenching was observed in the presence of the antibody, indicating that there were no interactions between *cis*-2MPyP and Np^-. No electron transfer from *cis*-2MPyP to Np^- in the presence of antibody 12E11G suggested that at least two cationic moieties in *cis*-2MPyP were covered with the binding pocket of the antibody. Figure 16.13 shows the schematic representation of the complex of *cis*-2MPyP with the antibody. Taking into account dissociation constants of the complexes between the antibody and a series of cationic porphyrins with the results of the photoinduced electron transfer from cationic porphyrins to anionic electron acceptors, more than a half part of porphyrin molecule was incorporated by the antibody-combining site.

16.3.2
Peroxidase Activity of Antibody-Fe-TMPyP Complex

The catalytic effects of the complex of Fe-TMPyP with antibody 12E11G on the oxidation of substrates, which are oxidized by horseradish peroxidase (HRP), were investigated. The complex of antibody 12E11G with the cationic iron porphyrin (0.5 µM) was dissolved in tris-acetate buffer (90 mM, pH 8.0) and incubated for 2 d at room temperature. Hydrogen peroxide (50 mM) was then added, followed by a substrate. The changes of absorbance at λ_{max} for the oxidation product of each substrate were monitored. Results showed that the Fe-TMPyP-antibody 12E11G complex had catalytic effects on catechol, guaiacol, and pyrogallol. In particular, the complex markedly catalyzed the oxidation of pyrogallol. Peroxidation catalyzed by the antibody-Fe-TMPyP complex was faster than oxidation in the presence of Fe-TMPyP alone. Further addition of the substrate caused a further catalytic reaction in the presence of the antibody-Fe-TMPyP complex, indicating that the catalyst was still active. On the second addition of pyrogallol and hydrogen peroxide to the solution of the complex of the antibody with Fe-TMPyP, the catalytic oxidation reaction was observed with 80% activity, compared with that on the first addition of these substrates. On the other hand, the catalytic activity of Fe-TMPyP alone disappeared. It was suggested that the porphyrin catalyst in the absence of the antibody would be destroyed by an excess amount of hydrogen peroxide.

The complex of the antibody with Fe-TMPyP accelerated the oxidation of smaller substrates such as catechol, guaiacol, and pyrogallol; however, it had no effect on the oxidation of the substrates with a large molecular size such as ABTS or *o*-dianisidine. The substrate specificity of the antibody-Fe-TMPyP complex on the catalytic oxidation might be due to the limitation of space around the active site by the binding of the antibody to the porphyrin molecule. The complex of Fe-TMPyP with antibody 83B5D, which was also obtained by the immunization of the 3MPy1C-conjugate, had no catalytic effect on the oxidation of pyrogallol and the other substrates. Antibody 12E11G was suggested to have catalytic residues in its antigen-combining site and antibody 83B5D was thought to have no binding space for the substrates or no catalytic residues.

Figure 16.14a shows a Lineweaver–Burk plot on the oxidation of pyrogallol by the Fe-TMPyP-antibody complex. The concentration of hydrogen peroxide was fixed and set as the same as that in the previous study [25]. From the plot, the Michaelis constant K_m and the catalytic constant k_{cat} values were calculated. The K_m value in the presence of Fe-TMPyP-antibody complex was 8.6 mM. The k_{cat} was 680 min^{-1}, which was ca. eight times as high as that in the absence of the antibody (83 min^{-1}). We compared these kinetic parameters with those catalyzed by the complex between Fe-tetracarboxyphenylporphyrin (Fe-TCPP) and anti-TCPP antibody 03-1 [34]. The K_m and k_{cat} values of the Fe-TCPP-antibody 03-1 complex on the oxidation of pyrogallol were 4.0 mM and 50 min^{-1}, respectively. The k_{cat} of the Fe-TMPyP-antibody 12E11G complex was 13-fold higher than that of the Fe-TCPP-antibody 03-1 complex. Although the Fe-TMPyP-antibody 12E11G complex was found to have lower affinity for pyrogallol than the Fe-TCPP-antibody 03-1 complex by comparison of the K_m, the catalytic activity of the cationic porphyrin-antibody system ($k_{cat}/K_m = 7.9 \times 10^4$ M^{-1} min^{-1}) was higher than that of the anionic porphyrin-antibody system (1.2×10^4). The increase of the catalytic activity of the Fe-TMPyP-antibody 12E11G complex was ascribable mainly to the

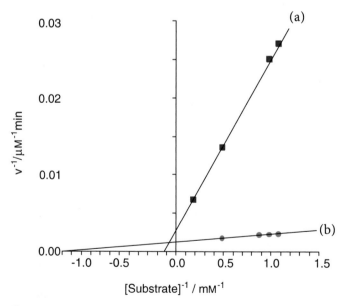

Fig. 16.14. Lineweaver–Burk plots for oxidation of pyrogallol in the presence of complex of antibody 12E11G with Fe-TMPyP (a) and HRP (b)

high reactivity of the cationic porphyrin itself and stabilization of Fe porphyrin by the binding of the antibody. Figure 16.14b shows a Lineweaver–Burk plot in the oxidation of pyrogallol in the presence of HRP with 50 mM H_2O_2. From the plot, the K_m and k_{cat} values for HRP were estimated to be 0.8 mM and 1750 min^{-1}, respectively. The differences in k_{cat} values of the Fe-TMPyP-antibody complex and HRP are within a factor of three. Although the k_{cat}/K_m value was 2.2×10^6 M^{-1} min^{-1} for HRP, being 28 times higher than that of the Fe-TMPyP-antibody complex, the Fe-TMPyP-antibody complex was highly reactive (Table 16.4). The catalytic activity of HRP decreased at higher concentrations of H_2O_2; however, that of the antibody-porphyrin complex was retained. Naturally occurring catalyst, HRP, catalyzes the oxidations of various substrates, not only pyrogallol but also hydroquinone, catechol, resorcinol, guaiacol, ABTS, and o-dianisidine. The reactions promoted by HRP are nonspecific. In contrast, the catalytic oxidation by the Fe-TMPyP-antibody complex was selective for small molecular substrates such as catechol, guaiacol, and pyrogallol.

Table 16.4. Kinetic parameters for oxidation of pyrogallol by Fe porphyrins in the absence and presence of antibodies for porphyrins or HRP

Catalysts	K_m [mM]	k_{cat} [min^{-1}]	k_{cat}/K_m [$M^{-1} min^{-1}$]
Fe-TMPyP		83	
Fe-TMPyP-Ab.12E11G	8.6	680	7.9×10^4
Fe-TCPP[a]		8.7	
Fe-TCPP-Ab.03-1[b]	4.0	50	1.2×10^4
HRP	0.81	1750	2.2×10^6

[a] Tetrakis-*meso*-(4-carboxyphenyl)porphyrin iron complex
[b] Monoclonal antibody elicited against TCPP

16.4
Dendritic Antibody Supramolecules

IgG is a basic type of antibody, consisting of two heavy peptide chains with a molecular weight of about 50 000 and two light chains with a molecular weight of about 25 000. There are two identical binding sites at the top of Fab fragments of IgG that are bound by flexible hinges with a single constant stem (Fc). IgG takes a Y or T shape. IgG is generated in a final stage of immunization, so it is matured and highly selective. IgM has a pentameric structure of IgG and ten antigen binding sites in a single molecule. The presence of ten antigen binding sites enables IgM to bind tightly to antigens containing multiple identical epitopes. However, IgM is generated in an initial stage of immunization, so it is unmatured and less specific for the antigen than IgG. In order to design an antibody system with a high specificity and a high affinity, a combination of the functions of both IgG and IgM seems to be important. We designed and prepared dendritic antibody supramolecules, in which IgM is placed in a core and many IgGs are bound around the IgM.

A monoclonal antibody (IgM) for cationic porphyrin was prepared using 3MPy1C as a hapten. IgG specific for anionic porphyrin, TCPP, was prepared. The cationic porphyrin 3MPy1C was attached to the IgG via activation of carboxylic acid in 3MPy1C using the condensation agent carbonyldiimidazole. The IgG-cationic porphyrin conjugate was purified by column chromatography using Sephadex G-150 to remove the porphyrins that did not react to the antibody. The characteristic binding ability and specificity of IgG were found to remain during the chemical modification of IgG with 3MPy1C. Scheme 16.1 shows the route for the construction of the dendritic supramolecules. When IgM for 3MPy1C is treated with IgG covalently bound cationic porphyrin, IgM binds the cationic porphyrin attached on the IgG to give a dendritic antibody supramolecule "antibody dendrimer" (**G1** in Scheme 16.1).

The binding property of the antibody dendrimer (**G1**) with a cationic or anionic porphyrin was measured by ELISA. Figure 16.15a shows the binding properties of IgG, IgM, and **G1** with the cationic porphyrin, TMPyP. Although IgG did not bind the cationic porphyrin and IgM bound the cationic porphyrin, the dendrimer did not bind TMPyP. These results show that the cationic porphyrin attached to IgG occupies the binding sites of IgM in the dendrimer; thus there are no free binding sites against TMPyP on IgM. Figure 16.15b shows the binding of IgG, IgM, and **G1** to TCPP. The IgM used in this study can bind both anionic and cationic porphyrins, due to the low specificity of IgM against porphyrins. Both IgM and IgG bound TCPP, while **G1** bound TCPP more efficiently than IgM or IgG. The increase in affinity of **G1** for the anionic porphyrin indicates that many IgG molecules attach to the surface of the IgM molecule.

The biosensor technique based on surface plasmon resonance (SPR) shows that the antibody dendrimer has an advantage in its amplification of detection

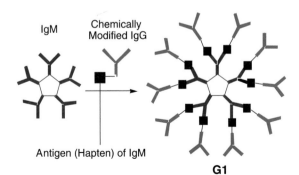

Scheme 16.1. Preparation of antibody dendrimers using IgM and chemically modified IgGs. An ideal structure of the dendrimer is shown as **G1**

Fig. 16.15. Binding affinities of IgG, IgM, and dendritic supramolecule (**G1**) with TMPyP (**a**) and those with TCPP (**b**) estimated by ELISA

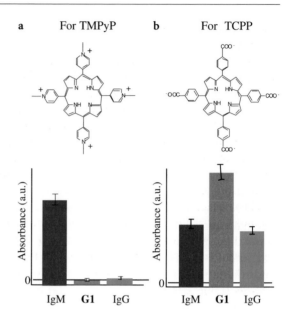

signals for antigens. A solution of **G1** was added to the sensor chip on which TCPP was precoated by the coupling with hexamethylenediamine as a spacer. The sensorgram for the binding of the antibody dendrimer to TCPP was compared with that of IgG to TCPP as shown in Fig. 16.16. The signal intensity increased by the injection of the antibody dendrimer was sufficiently larger than that of simple addition of IgG. Taking into account the change of the binding property of the antibody dendrimer for porphyrins with the increase in the amount of bound antibody to the anionic porphyrin on the SPR biosensor,

Fig. 16.16. Sensorgrams for binding of G1 (**a**) or IgG (**b**) to corresponding antigen immobilized onto surface of sensor chip of SPR biosensor. 60 s after injection of antibody solutions, flow cell was filled with buffer

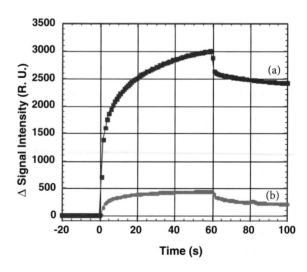

the antibody dendrimer had many IgG molecules successively bound to IgM molecule.

The structural observation of the antibody dendrimer was carried out by using atomic force microscopy (AFM) [47]. The sample surface was observed under the suitable conditions that any damage caused by scanning the cantilever is minimized and that any nonspecific assembly among antibodies does not occur [48]. Figure 16.17 shows AFM images of the dendrimer and starting IgM. The image of the dendrimer was twice as large as that of starting IgM. Some branches (IgGs) can be seen outside of the IgM core. Such an assembled structure was not observed in a chemically modified IgG solution or an IgM solution alone.

The characteristic features of the antibody dendrimer are that they (1) are composed of proteins, (2) are large with a molecular weight of about 2 million, (3) are composed of noncovalent bonds, and (4) bind antigens strongly with high specificity. The antibody dendrimer will be used as functionalized materials for sensitive detection of many kinds of chemicals, for diagnosis and for drug-delivery systems.

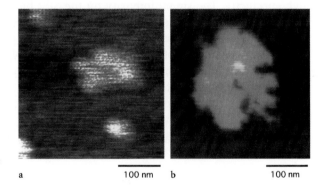

Fig. 16.17. AFM images of IgM (**a**) and dendritic antibody supramolecule (**b**) on surface of graphite plate

a   100 nm   b   100 nm

16.5
Linear Antibody Supramolecules: Application for Novel Biosensing Method

An optical technique based on SPR [49] or a microgravimetric quartz-crystal-microbalance (QCM) [50] technique is useful for measuring and characterizing macromolecular interactions in the increasingly expanding area of biosensor technology. In particular, SPR has great potential for macromolecular interaction analysis in terms of sensitivity and signal translation. The use of biosensors based on SPR has made it possible to determine kinetic parameters in real time and without any labeling of biomacromolecules for detection. However, the SPR response reflects a change in mass concentration at the detector surface as molecules bind or dissociate; the specific sensing of substrates with low molecular weight is difficult. In such a case, functional molecules with a high molecular weight such as antibodies have a great potential for amplification of the response signals expressing a molecular recognition event. In this study, methyl viologen was selected as one of the target molecules to be detected. We prepared supramolecular antibody complexes and applied these complex formations for a highly sensitive detection method. The complex formation between antibodies (IgG) and divalent antigens (viologen dimer) was investigated. The amplification of detection signals for methyl viologen by using the antibody supramolecules on the biosensor techniques is discussed.

16.5.1
Antiviologen Antibodies

Monoclonal antibody 10D5 (IgG$_1$) has been elicited for the viologen derivative 4,4'-bipyridinium 1-(carboxyphenyl)-1'-methyl-dichloride and was found to bind this hapten with the dissociation constant K_d = 2.0 × 10^{-7} M. The dissociation constant between antibody 10D5 and methyl viologen was also 2.0 × 10^{-7} M. Antibody 10D5 recognized the bipyridinium moiety with high specificity. The signal intensity on the SPR biosensor increased on the addition of an aqueous solution of antibody 10D5 to the sensor chip on which the viologen dimer-antibody complex was precoated. Figure 16.18 shows the sensorgram of the repeated injection of the aqueous viologen dimer and antibody solutions. The signal intensity was enhanced by the binding of antibody to the viologen dimer-antibody complex. The viologen dimer molecule is considered to act as a connector between antibodies. The same behavior was observed by ELISA measurements. The antibody was mixed to the aqueous solution of viologen dimer and the solution was deposited onto the ELISA plate that was coated with the hapten-

Fig. 16.18. Sensorgram of repeated injection of solutions of viologen dimer (**a,c,e**) and antibody (**b,d,f**) into flow cell of SPR biosensor. The antibody for viologen was immobilized onto the surface of the sensorchip. The surface of the sensor chip was subsequently washed with buffer after the injection of aqueous solutions of viologen dimer and the antibody

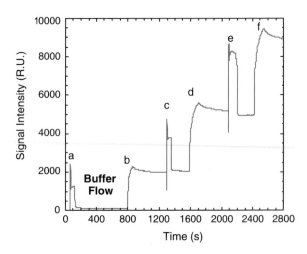

BSA. The ELISA signal in the mixture of the antibody and viologen dimer was higher than that in the absence of viologen dimer. It is suggested that the phenomena observed in the SPR and ELISA measurements are ascribable to the higher-order complex formation between antibodies and viologen dimer.

16.5.2
Applications for Highly Sensitive Detection Method of Methyl Viologen by Supramolecular Complex Formation Between Antibodies and Divalent Antigens

We found that the supramolecular formation of antibodies and viologen dimer effected an SPR signal enhancement of the biosensor. The additional binding of the antibody to the viologen dimer-antibody complex gave a remarkable increase in signal intensities. On the other hand, the addition of methyl viologen (viologen monomer) instead of viologen dimer was expected to block the antigen-binding sites and to inhibit the additional antibody binding. A small amount of methyl viologen can be detected as a decrease in signal enhancement due to the inhibition of the complex formation between viologen dimer and antibody by methyl viologen, compared with the signal intensity of complete supramolecular formation between the antibody and viologen dimer. Scheme 16.2 shows the strategy for the amplification of detection signals for methyl viologen based on the signal enhancement by the supramolecular formation between the antibody and viologen dimer using the biosensor technique. This system includes a two-step procedure as follows. (i) The aqueous solution of antiviologen antibody with methyl viologen is injected into the sensor chip, whose surface is modified with the antibody-viologen dimer complex, and then (ii) an antibody-viologen dimer (1:2) complex is

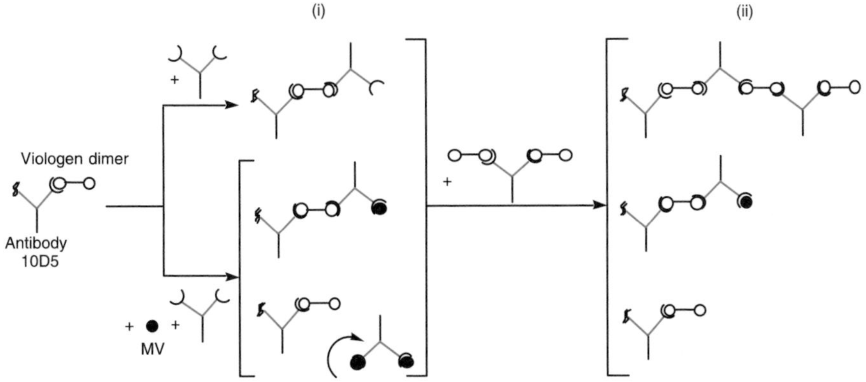

Scheme 16.2. Schematic representation of the highly sensitive method for the detection of methyl viologen based on the SPR biosensor technique

added to the previous state. The changes in the signal intensities in the presence of methyl viologen are compared with those in the absence of methyl viologen.

The amount of antibody immobilized to the sensor chip decreased as the concentration of methyl viologen increased. To enlarge the difference in the signal intensities in the presence of a small amount of methyl viologen, a solution of the antibody-viologen dimer complex was added to the previous state (i). Figure 16.19 shows the differences in the response signal intensities (I_0–I) between the complete supramolecular system in the absence of methyl viologen (I_0) and that in the presence of various concentrations of methyl viologen (I) at each step. The differences in signal intensities due to the binding of the additional antibody (0.9 µM) in the presence of methyl viologen ranging the concentration from 0.2 to 1.1 µM was slight in step (i). However, further addition of viologen dimer and antibody solutions in step (ii) caused a clear difference in the response signal intensities in the same concentration range of methyl viologen. The total changes in the signal intensities were found to have a linear relationship against logarithm of the concentration of methyl viologen as shown in Fig. 16.19. The sensitivity in this system was 140-fold larger than that in the simple addition of methyl viologen to the antibody immobilized to the surface of the sensor chip. It is clear that this system can be utilized for the quantitative detection of methyl viologen. Amplification of methyl viologen sensing processes was realized by the inhibition of complex formation between the antibody and viologen dimer-antibody complex and signal enhancement due to the supramolecular formation of the antibody and viologen dimer. The amount of methyl viologen

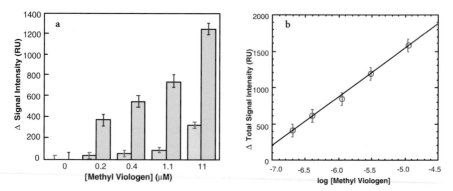

Fig. 16.19. The differences in the SPR response signal intensities between the complete supramolecular formation system in the absence of methyl viologen and that in the presence of various concentrations of methyl viologen. The *left colomn* at each concentration of methyl viologen in (**a**) shows the change in the signal intensity in the step (i) in Scheme 16.2, and the *right* one is that in the step (ii). (**b**) The relationship between total changes in the signal intensities of SPR response and the concentration of methyl viologen

(M_W = 257) was expressed as the amount of the antibody (M_W = 150 000) that could not form the supramolecules between the viologen dimer and the antibody.

16.6
Conclusions

Photoinduced electron transfer from Zn-TCPP to an acceptor molecule (methyl viologen) has been found to be controlled by the complex formation of monoclonal antibodies with Zn-TCPP. Although there are no ground-state interactions between Zn-TCPP and MV^{2+} for a 2:1 complex of antibody 03-1 and Zn-TCPP, the fluorescence of Zn-TCPP is quenched by the addition of MV^{2+}. The Stern–Volmer plots and emission lifetime studies presented show that there is a long-range electron transfer through antibody 03-1. One of the antibodies elicited by an antigen containing both phosphorus(V) porphyrin and terephthalic acid positioned axially shows high selectivity for both porphyrins and axial parts. Moreover, the antibody was found to bind both the porphyrin and terephthalic acid simultaneously and that the specific insertion of an electron acceptor into the antibody combining site makes it possible to facilitate the electron transfer from the porphyrin to the acceptor molecule.

We have obtained monoclonal antibodies for a cationic porphyrin. One of the antibodies, 12E11G, bound the cationic porphyrin selectively and formed stable complexes with the porphyrins and metalloporphyrins. The complex of antibody 12E11G with Fe-TMPyP was found to have catalytic activity for the oxidation of catechol, guaiacol, and pyrogallol. The metalloporphyrin-antibody complex was stable enough to show catalytic activity in the presence of an excess amount of H_2O_2. The catalytic activity of the cationic porphyrin-antibody complex for the oxidation of pyrogallol was higher than that of the anionic porphyrin-antibody complex and active as a catalyst even under the conditions where porphyrin alone or HRP should lose its catalytic activity. The high durability against H_2O_2 and the ability to generate H_2O_2 of antibodies were considered to be the cause of high catalytic activity on the oxidation.

New antibody dendrimers were designed and prepared by the combination of IgG and IgM, that is, using IgM as a core and IgG as branches. Many binding sites of IgG were arranged radially on the surface of one object by noncovalent bonds, and the resulting artificial antibodies bound antigens more selectively than IgM and more strongly than IgG.

An amplification method of the detection signals for a target molecule has been devised using the signal enhancement in the supramolecular assembly of antiviologen antibodies and divalent antigens. Target substrate added to the flow cell of SPR can be detected quantitatively by monitoring the total amount of the antibody bound to the surface of the sensor chip. The sensitivity in this system was found to be two orders of magnitude larger than that in the simple addition of target substrate to the antibody immobilized to the surface of the sensor chip. This method can be potentially applied for many compounds to be detected with a high sensitivity and specificity using corresponding antibodies and dimers of the target molecule (divalent antigen).

16.7
References

1. Pauling L (1948) Am Sci 36:51
2. Pollack SJ, Jacobs JW, Schultz PG (1986) Science 234:1570
3. Tramontano A, Janda KD, Lerner RA (1986) Science 234:1566
4. Kohler G, Milstein C (1975) Nature 256:495
5. Harada A, Yamaguchi H, Okamoto K, Fukushima H, Shiotsuki K, Kamachi M (1999) Photochem Photobiol 70:298
6. Yamaguchi H, Kamachi M, Harada A (2000) Angew Chem Int Ed 39:3829
7. Yamaguchi H, Tsubouchi K, Kawaguchi K, Horita E, Harada A (2004) Chem Eur J 10:6179
8. Harada A, Yamaguchi H, Tsubouchi K, Horita E (2003) Chem Lett 32:18
9. Yamaguchi H, Harada A (2002) Biomacromolecules 3:1163
10. Yamaguchi H, Harada A (2003) Top Curr Chem 228:237
11. Dolphin D (1978) The Porphyrins. Academic, New York
12. Kadish KM, Smith KM, Guilard R (1999) The Porphyrin Handbook. Academic, San Diego
13. Reedy CJ, Gibney BR (2004) Chem Rev 104:617
14. Lombardi A, Nastri F, Pavone V (2001) Chem Rev 101:3165
15. Mofft DA, Hecht MH (2001) Chem Rev 101:3191
16. Rau HK, Haehnel W (1998) J Am Chem Soc 120:468
17. Takahashi M, Ueno A, Mihara H (2000) Chem Eur J 6:3196
18. Mofft DA, Certain LK, Smith AJ, Kassel AJ, Beckwith KA, Hecht MH (2000) J Am Chem Soc 122:7612
19. Lu Y, Berry SM, Plister TD (2001) Chem Rev 101:3047
20. Watanabe Y (2002) Curr Opin Chem Biol 6:208
21. Hayashi T, Takimura T, Ogoshi H (1995) J Am Chem Soc 117:11606
22. Hamachi I, Tanaka S, Tsukiji S, Shinkai S, Shimizu M, Nagamune T (1997) J Chem Soc Chem Commun 1735
23. Hu YZ, Tsukiji S, Shinkai S, Oishi S, Hamachi I (2000) J Am Chem Soc 122:241
24. Hayashi T, Hisaeda Y (2002) Acc Chem Res 35:35
25. Pollack SJ, Nakayama GR, Schultz PG (1988) Science 242:1038
26. Pollack SJ, Schultz PG (1989) J Am Chem Soc 111:1929
27. Schwabacher AW, Weinhouse MI, Auditor M-TM, Lerner RA (1989) J Am Chem Soc 111:2344
28. Harada A, Okamoto K, Kamachi M, Honda T, Miwatani T (1990) Chem Lett 1990:917
29. Cochran AG, Schultz PG (1990) J Am Chem Soc 112:9414
30. Cochran AG, Schultz PG (1990) Science 249:781
31. Harada A, Okamoto K, Kamachi M (1991) Chem Lett 1991:953
32. Keinan E, Benory E, Sinha SC, Shinha-Bagchi A, Eren D, Eshhar Z, Green BS (1992) Inorg Chem 31:5433

33. Harada A, Shiotsuki K, Fukushima H, Yamaguchi H, Kamachi M (1995) Inorg Chem 34:1070
34. Harada A, Fukushima H, Shiotsuki K, Yamaguchi H, Oka F, Kamachi M (1997) Inorg Chem 36:6099
35. Liou HH, Tsai MC, Chen CJ, Jeng JS, Chang YC, Chen SY, Chen RC (1997) Neurology 48:1583
36. Brooks AI, Chadwick CA, Gelbard HA, Cory-Slechta DA, Federoff HJ (1999) Brain Res 823:1
37. Niewola Z, Hayward C, Symington BA, Robson RT (1985) Clinica Chimica Acta 148:149
38. Hogg PJ, Johnston SC, Bowles MR, Pond SM, Winzor DJ (1987) Mol Immun 24:797
39. Bowles MR, Pond SM (1990) Mol Immun 27:847
40. Bowles MR, Hall DR, Pond SM, Winzor DJ (1997) Anal Biochem 244:133
41. Yamaguchi H, Harada A (2002) Chem Lett 2002:382
42. Watanabe T, Honda K (1982) J Phys Chem 86:2617
43. Sadamoto R, Tomioka N, Aida T (1996) J Am Chem Soc 118:3978
44. Nimri S, Keinan E (1999) J Am Chem Soc 121:8978
45. Ward B, Skorobogaty A, Dabrowiak JC (1986) Biochemistry 25:6875
46. Pasternack RF, Halliwell B (1979) J Am Chem Soc 101:1026
47. Yamaguchi H, Harada A (2003) In: Ueyama N, Harada A (eds) Macromolecular Nano-Structured Materials. Springer, Berlin Heidelberg New York
48. Harada A, Yamaguchi H, Kamachi M (1997) Chem Lett 1997:1141
49. Rich RL, Myszka DG (2003) J Mol Recog 16:351
50. Niikura K, Nagata K, Okahata Y (1996) Chem Lett 1996:863

Subject Index